D0577791

# *Trackings*

# Trackings

Composers Speak with

RICHARD DUFALLO

New York Oxford
OXFORD UNIVERSITY PRESS
1989

Oxford University Press

Oxford New York Toronto
Dehli Bombay Calcutta Madras Karachi
Petaling Jaya Singapore Hong Kong Tokyo
Nairobi Dar es Salaam Cape Town
Melbourne Auckland

and associated companies in
Berlin Ibadan



Published by Oxford University Press, Inc.,
200 Madison Avenue, New York, New York 10016

Oxford is a registered trademark of Oxford University Press



Library of Congress Cataloging-in-Publication Data
Dufallo, Richard.
Trackings : composers speak with Richard Dufallo /
Richard Dufallo.
p. cm. Includes index. ISBN 0-19-505816-X
1. Composers—Interviews.
2. Music—20th century—History and criticism.
I. Title ML390.D815 1989
780'.92'2—dc20 89-9372 CIP MN

9 8 7 6 5 4 3 2 1

Printed in the United States of America
on acid-free paper

*To Pamela Mia Paul*

# Acknowledgments

The composers in the book, for their music, for taking their valuable time to speak with me, and for their cooperation in assessing and editing their dialogue.

Pamela Mia Paul, my wife, for her tenacity in transcribing these talks from cassette tape, translating and editing where necessary, for editorial advice, and for her encouragement to write this book.

Jeffrey Bishop, who first perceived the preliminary manuscript as a potential book and for his invaluable insights in regard to its form and content.

Pia Gilbert, for reading the manuscript at its various stages of development, offering significant editorial comments, translations, and reflections along the way.

Kurt Oppens, for officially translating from German the Cerha (prepared by Gertraud Cerha), Kagel, and portions of the Reimann conversations.

# Contents

# Introduction

*Trackings* is a collection of "tracks" or "marks" left in passing by the twenty-six composers in this book. I have performed the music of all these composers and have shared a special professional intimacy with each of them. I didn't want to present them in alphabetical order, nor did I want to construct an arbitrary hierarchy. I perceive no such thing. To present them in the order that I actually interviewed them seemed too accidental. Therefore, I decided to present each composer in the approximate sequence of having met them, sometimes adjusted slightly by the weight of that first meeting versus a subsequent meeting.

While I placed no restrictions on what these composers wished to say, in a number of instances I tried to extract what they felt their musical genetic code was. Naturally, the Stravinsky/Schoenberg axis figured prominently, as did the influence of Nadia Boulanger and Olivier Messiaen. Interesting exceptions came from Eastern bloc composers who, through their musico-political isolation, had to develop along slightly different lines.

In addition to this line of questioning, I wanted those composers who were involved in the Darmstadt International Summer School for New Music (Internationale Ferienkurse für Neue Musik), which was established in the early 1950s, to speak about their experiences and/or their reactions as seen from today. Something that began to emerge in the talks was the sense of crisis that many felt had occurred in their compositional lives and that was occurring in music in general. This crisis, no matter of what origins, invariably was expressed as the crisis of tonality. To embrace tonality or some form of new harmonic control seemed to be on so many minds. This of course is a reaction to the dominance of non-tonal thinking of the past thirty years.

In the period of post-World War II, the ordering of the twelve pitches (attributed to Arnold Schoenberg) became an obsession with many composers. Beyond that, the ordering of everything connected with those pitches, i.e., rhythm, dynamics, timbre, etc., also became an integral part of this organizing obsession, which came to be known as "total organization." Not unlike "totalitarian" socialism.

On the other hand, for some composers, "non-organization" in the form of "chance" was one counterpart. Not unlike anarchy. Not unlike trying to harness chaos. Side by side, but not the same, was the concept of the freedom of "choice," which in most cases was at least pre-ordered. Not unlike democracy.

For those composers espousing "choice," it meant giving the conductor or performer the possibility of choosing from given material in order to guide the direction or consequence of a piece. For those composers espousing "chance," it meant handing over the result to an outside, extra-musical element, much like throwing dice or following the principles of the Chinese *I-Ching.*

These two areas—"total organization" and "chance" versus "choice"—perhaps define the extremes that occurred in the 1950s and '60s in attitudes toward composition. It was not, however, as cut and dried as this. These extremes only reflected the internal polarities, dualities, and opposites that were being observed within a larger context.

In the case of "total organization," almost any polarities, extremes, or opposites could be conceived as potential sources for mediation: a mediation that had only to be constructed on a scale of observable increments which then was manipulated according to the composer's sense of creativity.

In other words, seemingly endless lists of criteria that helped define an area of compositional exploration were required as raw material in order to construct the work of art. A multiplicity of approaches existed, but all were very determinate, fixed; and the outcome was totally controlled.

Within the area of "choice" and "chance" a wide variety of approaches also existed. There appeared to be no one way to solve these problems, since it would represent neither choice nor chance to be dogmatic. Improvisation, indeterminacy, aleatoric procedures were the words most often used to describe these approaches.

For example, the path and duration of a piece were left "open" due to the pre-ordained variables of specific musical material the composer gave to the performer or conductor. Hence the term "open form." Open form could be accomplished using a mobile principle which allowed the recurrence (or not) of various sections of given material, the setting in motion and reacting to this material, depending on the guidance and choices of the conductor.

It could also be accomplished by simply providing options in the direction or progression of given material within a piece. In the option or "maze-like" situation, alternate paths were provided which led to an unforeseeable outcome. In a "mobile-like" situation, while also providing alternatives and an unforeseeable outcome, the degree of unpredictability was increased along the way.

Another method used was to alternate between conventionally written music that used bar lines and meters, and areas ("boxes") or sections of indeterminate or approximate lengths. Each of these methods provided an elasticity of formal shape which intrigued many composers.

At the extreme end of indeterminacy were methods that would simply give a set of instructions to the performer(s) without any written notes. Or if notes were present, their random appearance on the page indicated that the composer did not wish to control the outcome at all. In this case of pure chance, the "accident" or "coincidence" level was extremely high. However, it was precisely the "accident" that was looked upon by some as the freest "expression."

The look of notation on the page became important. Proportions that the eye perceived on the page precluded the idea of absolute rhythm, pitch, or tempo. This proportional notation became highly graphic and indicated a system of relationships between things: which is to say, a basic principle common to all written music from any period of history. Some examples of this proportional notation took on an almost "painterly" look. Visual perception, translated into sound, gave the performer a certain latitude of interpretive freedom.

In addition, a new and extensive array of graphic symbols was introduced that provided short cuts to the articulation and timbre of vocal and instrumental execution. These symbols went far beyond those found in the conventional articulation practiced in the 19th century: from near-speech to a note sung; from air blown through an instrument to a note sounded; from bowing the plates of a vibraphone with a bass bow to a note struck, etc.

Alongside all of these were those composers who reflected a rather neoclassic attitude left over from early "Stravinskian" and "Hindemithian" compositional practices which in part incorporated conventional attitudes toward melody, harmony, counterpoint, and style. This was an attitude that avoided serialism and the "chance"/"choice" department completely. It also produced a lyricism somewhere between a tune, pretty or not, and declamation, heavy or light.

*Abstract* was a word that was often linked to the creative efforts of many composers, painters, sculptors, and writers in the '50s and '60s. This set up yet another apparent polarity: *the real* and *the abstract,* which is difficult to support. Where reality leaves off (if it does) and abstraction begins (if it does) is a vague territory that may be looked upon as one and the same.

For the reader, viewer, or listener to enter that uncharted territory, it is necessary to experience a shock, a disorientation, and to let go of familiar concepts, for example, of narrative, object, and form. While no one period of history defines or answers all of our questions, the post-World War II period at least defined or re-defined many of them.

The questions of transformation, irreducibility (purity), chronology, rhetoric, order, dis-order, all provided the spiritual base that made things happen. The mark of this spirituality was in the intensity of *invention* and *discovery,* and not in the false god *entertainment.*

*Trackings*

*A track . . . where one*
*stands there and then*

# Narrative I

I never thought that a kid who grew up in a small refinery town called Whiting, Indiana, just outside of Chicago, would be, later in his life, in the middle of promoting the cause of contemporary music and digesting the complexities of newly written scores. That growing-up period had little to do with such lofty idealism. It had to do with living in the bleak, gray atmosphere of a small town that was, like many towns in America, a result of the Industrial Revolution. Acrid, smelly air that was at times unbearable and brought unprovoked tears to one's eyes. . . . A polyglot of humanity struggling to survive, and almost entirely dependent on the success of steel mills, chemical factories, oil refineries, soap factories, etc. . . . Living in monotonous or hastily built dwellings, enshrouded in the insensitive environment created by an untalented group of industrial planners.

What was there to look at that could turn a young man's mind to fancy or to beauty or to something beyond this crude reality? At the very least, one should have been able to rely on nature for relief. But the ever-present image of pipes, valves, storage tanks, railroad tracks, machinery bellowing and clanking non-stop, all precluded the possibility of an external sense of beauty or aesthetic. The people, turning inward, produced the only glimmer of a landscape that would sporadically offer some solace.

My people were of sub-Carpathian Ruthenian origin and were typical of the many immigrants who came to the United States from Central Europe, bringing with them religion, customs, and attitudes of the Old World. I was born in the New World, and the culturally barren wasteland called the Calumet Area was not a source of inspiration to me. The only outlet was the big city: Chicago.

My early unconventional musical background included a variety of performing experiences and initiatives, all of which had nothing to do with serious contemporary music. These professional experiences began at age twelve when, pretending to be age sixteen and therefore a member in good standing of the American Federation of Musicians, I played my first nightclub engagement in notorious Calumet City at a place called the Riptide, a former strip joint. From that moment on and after being reinstated in the Musician's Union at age fourteen or fifteen, I encountered an enormous variety of musical ensembles that ranged from a fife and jug band, to a high school band and orchestra, to dance bands. All of this led to performing for civic clubs, fraternal organizations, veterans' hospitals, proms, country clubs, weddings, church functions, Mexican dances, black dances, Rumanian dances, Jewish dances, and Sunday afternoon concerts in the parks, etc.

In studying my principal instrument, the clarinet, I was easily able to play all of the saxophones and early on acquired a reputation as a baritone saxophonist in the "big bands," which were fashionable after the end of World War II. I enjoyed "rolling" into the bass notes of those popular dance band arrangements and acting as the harmonic bottom of the band. Upon graduation from high school in 1950, I had heard my first opera, *Cosi fan tutte,* and after a summer of working in a steel mill, I entered the American Conservatory of Music in Chicago.

The next three years expanded my musical horizons in a very traditional and primarily Germanic way, using the Adolf Weidig text of musical theory, listening to Friedrich Stock's Chicago Symphony, and becoming principal clarinetist of the Chicago Civic Orchestra (a training orchestra sponsored by the Chicago Symphony). Also Germanic was my training in the Lindemann "school" of clarinet playing.

The musical atmosphere in Chicago was conservative. Stravinsky was regarded as a renegade, Wagner was looked upon with suspicion, and Strauss seemed to represent advanced thinking. Rachmaninoff was considered mundane, Debussy a mystery. Upon hearing the music of Poulenc for the first time on an "esoteric" (red vinyl) record label, I had the impression that he represented the avant-garde for most of my associates. Schoenberg wasn't even discussed, Ives was considered a joke!

Courses in composition with Leo Sowerby, and theory and conducting with Irwin Fischer (a pupil of Nadia Boulanger, Bernhard Paumgartner, and Pierre Monteux) completed a rather formal and fundamental musical education that was reflected in the first second-hand scores of Beethoven, Brahms, Mozart, and Mendelssohn that I purchased. My nomadic wanderings through various areas of music and music making slowly began to focus.

The Chicago Symphony at that time was under the direction of Raphael Kubelik, who had just taken over from Artur Rodzinski; and the musical press was dominated by the vitriolic critic Claudia Cassidy of the *Chicago Tribune.* I attended concerts and rehearsals of the orchestra diligently. "New music" for me was hearing Bartók's Concerto for Orchestra and *Bluebeard's Castle,* which I found extremely stimulating.

Playing in the Chicago Civic Orchestra under George Schick, the associate conductor of the Chicago Symphony, broadened my knowledge of the standard orchestral repertory. With the exception of the music of Milhaud, there was no great adventure into the 20th century. The culmination of this orchestral training was my first major performance of the Mozart Clarinet Concerto with Schick. I had most recently studied the concerto with Clark Brody, principal clarinetist of the Chicago Symphony and a former pupil of Daniel Bonade. My conducting experience was topped off by a performance at the graduation ceremonies of the American Conservatory in Orchestra Hall with the Conservatory Orchestra of which I was the assistant conductor, and consequently I was awarded my first baccalaureate degree.

It wasn't until after a nearly two-year stint in the U.S. Navy as a commissioned officer that, as I now realize, I began to be slowly plugged into a musical genetic code that in America was largely the immediate product of two people: Nadia Boulanger and Serge Koussevitzky. Upon enrolling at the University of California in Los Angeles in 1955 (on the GI Bill), my eyes, ears, and mind began to open to the world of new music; and it was there that I met Lukas Foss.

*James Tuttle*

# *Lukas Foss*

BORN AUGUST 15, 1922, BERLIN

**May 20, 1986** **New York**

I haven't seen Lukas often enough in recent years to have the kind of lengthy musical conversations we used to have in the Los Angeles and Buffalo days. So, in order to conjure a bit of that past, I brought with me some memorabilia of those early California times, of which he probably has very little since his house in Beverly Hills burned to the ground in 1961. This small collection included his first statement on our work in the field of improvisation: "A Beginning, Report from a Workshop for Composers and Performers in Search of a New Mode of Making Music Together."[1] And his last statement on this subject: "Improvisation vs. Composition."[2]

It is a Tuesday afternoon in the Fosses' Fifth Avenue apartment. The windows are open and the sounds from the street and the sky serve as a background to our talk.

RD I'm speaking to a man who has been called a few things; let's start with composer, conductor-pianist, programmer, festival director, teacher, lecturer, innovator, writer, traditionalist, and avant-gardist. I've known you for thirty years; and I want to talk to you, Lukas, about four or five of those areas, beginning first with a small history. I don't think I ever asked you what it was like, as a child, to grow up in Germany.

LF Well, my parents were very protective. They made sure that my brother and I would not be embarrassed by the anti-Semitism that was prevalent in '32. I went to the French Gymnasium and there were some Nazi teachers I got to know at age eleven; but that was the first few months, and then we left for Paris. Of course, I witnessed some of the tragedy. My father went back and forth between Germany and Paris; and every time he left for Germany my mother was suicidal and anxious and afraid he would never come back, but he did. Then he decided we were still too close to trouble, and in '37 we left for America. By that time of course he had lost all of his money. He lost it in a business venture here. There wasn't much left anyway, so then he had to look for work and

1. *The Juilliard Review,* Volume V, No. 2, Spring 1958.
2. *Musical America,* May 1962.

we had to look for work. As a result, I didn't go to school in America. I didn't get a high school diploma; but much, much later I became the youngest full professor at UCLA.

**RD** Do you have recollections of America, arriving that first time?

**LF** Oh, it was fascinating, of course. It was our new home, it was wonderful; people were generous and helpful. The first two people to help me were Adolf Busch and Rudolf Serkin. They got in touch with Barber and Menotti, and they insisted that I go to Curtis. While I was a student at Curtis, I would go back and forth between New York and Philadelphia. I lived in New York and I made a living accompanying dancers; and by teaching, giving some piano lessons, I could just eke out an existence. I remember staying in a little one-room apartment at the age of sixteen; and the couch was literally under the piano, because there was no room for anything more than that. Those days were very exciting. My parents had moved to Haverford, Pennsylvania, where, after some years, my father became professor of philosophy. My mother was a painter. My brother enlisted in the American Army . . . I was 4F because of an asthma condition, which vanished subsequently, luckily for me! And so, that was my life! Those were the early days.

**RD** Was that when you studied with Hindemith?

**LF** I studied with Hindemith, whose music taught me a lot, when I was eighteen, at Yale; but first at Tanglewood, where I became a student of Hindemith and Koussevitzky. Later on, Koussevitzky offered me the position of pianist of the Boston Symphony. By that time I was twenty, I believe. I had just graduated from the Curtis Institute in all three fields—composition, conducting, and piano. I had already studied with Fritz Reiner, conducting; Isabel Vengerova, piano; and Rosario Scalero, composition—but I hasten to add that I didn't learn anything from Scalero.

**RD** Then life stabilized a bit when you became the pianist of the BSO?

**LF** Yes, then I had a real income. That was my first real job, and I had time to compose. Koussevitzky said the piano parts are far between when you are with an orchestra as a pianist. I had a little loft near Symphony Hall; and sometimes I would have weeks off because there was no piano part to play, and yet the salary was there. So that was extremely helpful! He also played my music. That was where I met Stravinsky . . . I became fond of the man and learned a lot from his music . . . Oh, those days were also very exciting.

**RD** Did you meet Bernstein in those days?

**LF** No, I met Lenny in conducting class with Fritz Reiner. He was the "experienced" student, because he was a Harvard graduate, and I was the youngest one. He was like an older brother to me . . . very, very nice and of course, we still are friends. Reiner was a powerful teacher. He made it possible for me to stand in front of an orchestra for the first time and not make a fool of myself. He also let me guest conduct the Pittsburgh Orchestra. He got me started.

**RD** And then, shortly after that, you began to crank out the works. Would you call *The Prairie* among the first?

**LF** No, the first things I published, when I was fifteen, were written on the New York subway. They were piano pieces called "4 two-voiced inventions." Now they are called Four Two-Part Inventions, now that I know English better. They were published by Schirmer. Schirmer also published something called *Grotesque Dance,* for piano. Then I was only fifteen and still a student, in a sense. But *The Prairie* was the first piece that got recognition. I was nineteen when I started, twenty when it was performed by Robert Shaw and the Collegiate Chorale in New York's Town Hall. That was probably the first time that Robert Shaw conducted an orchestra. It was for chorus and orchestra and soloists, and it was the first big performance I had. It drew the attention of the intelligentsia to my work; and ever since then, I've had the mixed blessing of . . . well, no longer working in anonymity. *The Prairie* did that for me.

**RD** Looking at your more recent work, *Solo* for piano solo and *Solo Observed* for piano and chamber musicians, Lukas, can you talk about any change in style as you've progressed?

**LF** Yes. Style to me always means personality, so I would say that my personality evolved; but it's the technical changes that are most apparent to someone who hears my work, because I was always adding new techniques to my musical vocabulary. I was of course completely satisfied to remain a tonal composer, but then a door opened through this experiment that I began in the spring of '57 at UCLA, where we met, you and I. I wanted to free my students from the tyranny of the printed note, and started an improvisation program born out of envy, really, of jazz improvisation, because they were the great improvisors and we never got to improvise. Little did I realize that this project, which had nothing to do with the music of jazz, would change me as a composer far more, probably, than my students! And so, turned me into a student again! I became quite a different composer as I applied what I had learned in improvisation to my non-improvised works. My style changed radically, and I suddenly found myself in that "no-man's-land" called the "avant garde." This wasn't my intention, it just happened. The doors opened.

**RD** You made proclamations about the new performer-composer or the new composer-performer, or at least about the two getting closer together. Do you think that's realized today?

**LF** No, but it certainly was true at that time when we worked together. For example, I would watch you walk past one of the timpani and a strange resonance happened as your bell of the clarinet came close to that instrument. Suddenly, it was like it was miked. There were these weird overtones, and I worked that quickly into the first improvisations, and then I worked it into *Echoi,* calling it "the Dufallo." In other words, it also became like an orchestration workshop for me. We worked closely together, and then I thought that this must be the "new idea," a small combo, a small group of virtuosi who really understand each other; and maybe together with a composer or several composers, they would travel together as we traveled together through the world and gave improvisation concerts . . . and I felt that indeed it would be the most wonderful relationship, the composer-performer relationship. And it was! Shortly thereafter other groups became known who did the same thing. Berio had a group, but he doesn't now. Boulez had a group, the Domaine Musical, but now he has IRCAM, in Paris. . . . It was a time also, when none of us wrote big symphonies. . . . The reason was that there wasn't enough rehearsal time to do experimental music with orchestras, I thought. You couldn't use the orchestra for experimenting, there was no time. Whereas in our living room, in our studio, we could. We had all the time we wanted. The same is true in computer and electronic studios. There is time there for experimentation. And so it probably is true that electronic music and the new chamber music which we devised, and which I sometimes call "combo" music, for lack of a better word, was the ideal hunting ground for this new chamber music.

RD But did it reflect Stravinsky's interest in the small unit, too?

LF Yes, Stravinsky started it and Schoenberg . . . Schoenberg probably beat Stravinsky to it in *Pierrot Lunaire* by one year, and then came *The Story of a Soldier.* They started this thing of putting virtuosos together . . . of course, chamber music is as old as can be, but there's a vast difference between a string quartet, where there's that huge literature for the amateur, for the connoisseur, and these isolated pieces . . . that only 20th-century virtuosos can handle and are willing to get together.

RD In this time, we had formed the Improvisation Chamber Ensemble, and as you say, other composers had been forming their groups.

LF We were the first to do an unsigned music. In other words, the first to discard that vanity of the composer's signature. It didn't completely catch on, because the signature has an insidious way of finding its way into music, and music became again more tied to the composer. In other words, my dream, that there would be a kind of performers' music that everybody would go for, not necessarily replacing composers' music, but as a valid other way of making music, is no longer intact today.

RD But there was an offshoot, I think you'll agree, in the sense that composers made it part of their language. In other words, freedom for a performer was being born there. So that eventually, we have "aleatoric" practice.

LF That's right, and it's important to point out that we preceded that "jargon" word "aleatoric" with our experiments. At that point, the "aleatoric" was not yet "in," although John Cage had already begun his indeterminate music, and he credits me with having opened up improvisation to the cause and made it interesting again for the composer, because he hadn't really thought of it as a possibility for indeterminate music. Strangely enough, I always was a little bit dubious of chance music, and still am. I never felt that I really wanted to relegate my composition to chance; but rather, I wanted to play it safe and make choices available, simply because it wouldn't matter whether it's a 16th note or an 8th note. It wouldn't matter whether the register is this one or another. In other words, I felt that it was wrong to pin down something that I didn't mean. It was morally incorrect, and it would mislead the player into practicing the wrong thing. I wanted to make sure that they knew what was important to me. For instance, if I just want the percussionist to hit everything in sight, then I think it is better to do what Berio does, which is to write "tutti" and have him hit every percussion instrument, than to, in traditional notation, suddenly pin down the exact way in which these instruments should be hit. I felt it was more exact to leave that freedom if that wasn't where the importance of the result was lying. So it was really a matter of notation and exactness to reveal my intention as a composer to the performer, so he would practice the essentials and not the unessentials.

RD At the same time there was the notion of graphic notation in the air, and we used some of it.

LF Yes, we developed graphic notation, we developed a lot of things. More and more, though, I began to feel, in my own work as a composer, that I didn't want "freedoms" that would actually influence the result of the composition.

RD I can understand that. You were almost finished with *Time Cycle* when we were at our heaviest work . . . I saw you finish it and later, I also conducted it. It was a rather pivotal point, don't you think? Were you looking backward and forward at the same time?

LF Yes, I think it's a transitional piece, but in a positive way.

RD I also think it's fair to say that *Echoi* was served a little by our experiments, that is, of sound technique. I don't know that notation was necessarily solved. I think that was your own, but you did allow a certain kind of "box-like" or predetermined freedom as a result.

LF Oh, very much so. That's right, especially at that time I did. And in some subsequent works, too, even more so.

RD Is that kind of freedom something that you still would dare to give a performer anymore?

LF Well, my wildest piece in that direction is *Map*. It's rather a book of many pages, but no notes ... all "instructions" ... what the musicians have to do ... the game rules. Because it's a game; and you win via following the game rules, and the music comes to pass. That is probably the most extremely chance-like and mixed-media type of piece that I've ever done.... *Map* was reviewed by one critic in the *New York Times* as if it were a sports event ... so-and-so is now ahead, now this person is, now the clarinetist is ahead ... I had a lot of fun with that. But ever since then, I would say I've probably retreated to more traditional, what is called mainstream, type of expression in music, for better or for worse.

**RD** In that period, while we were improvising, while you were finishing *Time Cycle,* going on with *Echoi* and *Elytres* ... were you faced with the question of serialism at all?

**LF** In a very interesting way, yes, I was. Because once I had found my way into this no-man's-land of atonality, the question became how to make choices. Schoenberg of course pointed out a very valuable way of making a choice by producing a substitute for tonality; and that was the twelve-tone row. And that was applied later on by others to other parameters like rhythm, dynamics, register, and so forth.

**RD** The totalitarian aspect ...

**LF** Right! And what happens now is the composer can establish a row in all these parameters, then apply it, and then find out what he has wrought, which then gives him a huge surprise, because what he has wrought he could never have imagined in terms of sounds. He suddenly gets the surprise of now, this is it. Now he has the choice of whether he is going to work with it, or say this is IT and just leave it. The typical serial composer just left it. I have always felt, *now* my work begins. Now I'm going to compose with these groups of notes, rhythms; and these whole jungles that I'd obtained, I am going to clear for habitation, and make a piece of music by choice. Well, that was my choice, to handle the whole scene like that. But the interesting part is, a composer like John Cage who threw dice at that time, also wound up with surprises and was going to have to either work with that or say, this is *it.* In other words, the two opposite camps, chance and serial composition, really were just opposite in their methods; but very much the same in their way of trying to deal with the overwhelming amount of choice.

**RD** When you wrote *Orpheus* and those "instruction pieces," was your thinking influenced by Cage?

**LF** I would think that he, being the wonderful human being that he is, had an influence on me the way a prophet had an influence, even if the music actually doesn't ... I mean, he found a way out of harmony. Somebody else would have simply given up, and he simply said, all right, I'll turn it around this way. This kind of wonderful elasticity ...

**RD** Invention ...

LF Inventiveness . . . out of dire needs, so to say, you become an inventor. You simply define problems out of existence. That is one way of becoming an inventor. You know, being Germanic by nature, I always have to suffer and beat my head against the wall, and I remember John laughing and saying, "Lukas is still suffering when he composes. I don't suffer anymore . . . ha ha ha . . ." laughing that wonderful John Cage laugh! I tried to learn that . . . there were lots of things I tried to learn from John. They were usually not from notes, but from his wonderful attitude and his mind.

RD It's very provocative.

LF But you know, everyone was doing it at the same time. Mixed media was "in" and therefore instructions were "in." And I really don't remember any particular page in Cage that would have given birth to any particular page in *Orpheus*. I don't think it worked that way. Rather, you had a project, and you found your own way of notating and your own way of instructing. You had to do it by yourself.

RD Which might be the time to bring up the word "originality." How do you feel about originality?

LF It's something that in my case developed late, I would say, because I think that it develops in spite of yourself. Our love for music is what's there from the start, it's the first thing you develop . . . a love for music; and love for music means love for other people's music. It means love for music written before you; and in your earliest works you are trying to imitate. So for a long time composition was for me writing the music I loved . . . traditional music. But originality creeps in, even into that. And then of course those doors open and you enter that no-man's-land, then it sort of . . . inundates you, maybe! I don't know. Originality is a very interesting word.

RD How does it relate to the word "robbing" or "stealing"? Do composers steal or rob?

LF The most original have the easiest time stealing, because they don't steal—they make it their own, which is a wonderful phrase. Only in the English language do you "make it your own." And if you have a style of your own, you can steal freely.

RD The writer Milan Kundera, in observing the West, speaks of the virtue of relativity and ambiguity existing in our society. Does that apply to the artistic process?

LF It's true that art is a form of metaphor, and that's where the ambiguity would come in right away. It's curious, we were talking about John Cage; because I once went mushroom hunting with him, and we didn't find any mushrooms. But we found a wonderful vegetable which he proceeded to cook. Now, I once asked him at a symposium in Buffalo . . . I believe you were present . . . it was in our great festival of the arts that we did together . . . and in that symposium I asked him, since we played a game together of asking each other the most difficult questions, I said, "In your book, John, *A Year from Monday,* you said composing is easy. The real hard work is revolution." I said, "If that is true, as a revolutionary, can you afford to find something other than what you're looking for? Like the vegetable!"

**RD** That's a very tough question.

**LF** In other words, this is what Kundera probably meant. . . . That in politics you cannot ever afford to find something other than what you're looking for. And in the arts you constantly find something else . . . and you give up looking for what you were looking for. There's that rigidity, that stiffness, that comes from living up to the demands of a totalitarian government.

**RD** Well, in analogy, totalitarian serialism kind of lived itself out, in the eyes of composers a generation younger than you. . . .

**LF** Well, yes, at that time, every young ambitious composer acted and talked as if *we no longer must compose this way, but now we must compose that way,* much the way people came up with new economic theories . . . no longer this, now that the world has changed, we must have that . . . they were full of utopian formulas which ultimately all went bankrupt. I think we know now that there is no longer a "no longer" *this* or *that!* We can even sneak tonality back through the back door, right in, and use everything. We must use everything!

**RD** But side by side with that notion, there was also the notion of irreducibility. Irreducibility was a word used by New York painters a lot. How do you reduce something, for example, just to the paint? I used to hear them say . . .

**LF** That was Webern's way. He was using things to order! Order became a mystical concept for Webern. Not for Schoenberg; but for Webern, it was definitely that. And then he ends up with, what for me are occasionally, very beautiful tombstones.

**RD** Did irreducibility, as a notion in composition, touch your mind? Did you think in those terms, for example, with *Elytres?* You were working with tiny units, I remember.

**LF** Maybe a little bit. I have this endless curiosity. I have to try everything. That's why I had to try improvisation and push it to its ultimate conclusion.

**RD** What about the use of collage? I know that entered your mind.

**LF** Oh, yes, very much. But I immediately tried to transform it, mainly in my piece *Baroque Variations.* And maybe that's why *Baroque Variations* had an influence on the young. Because it was actually overcoming collage in favor of a more artistic thing to do with the materials. I really was in favor of the variation form . . . dream technique.

**RD** That was the piece that faded in and out . . .

**LF** Fading in and out . . .

RD And the silences occurred . . .

LF 16th notes, baroque 16th notes washed ashore rather than collaging in a bit.

RD But can't you conceive of it as a "linear" collage?

LF If you wish to call it collage. To me it was more variation technique. But obviously it had its origin in collage, or it was a combination of collage and variation technique.

RD You have written opera. Do you have the urge to write an opera again?

LF I always want to write another one! I feel time is ripe, and I might start soon. I have the usual libretto problems.

RD How did you look for your sources in the past?

LF In my opera *Griffelkin,* which is my only full-evening opera . . .

RD And I have conducted that, by the way.

LF Right . . . and there I went back to my youth. My mother told me a fairy tale when I was seven, and I loved it, so that she made a little libretto out of that fairy tale and I composed music for it when I was eight. And when I was nine, I decided that what I wrote was childish and I discarded it. Then when my mother died (I was thirty-five), I commissioned Alastair Reid, the English poet, to write me a grown-up, mature libretto. I gave him my mother's libretto and he came up with a fairy-tale libretto, *Griffelkin.* And of course my ideal was Mozart for the opera, and so I was trying to do a kind of American Mozart Opera. And that is what that was. My next one will be very different, indeed.

RD You've handled a lot of poets, and that's something else I'd be interested in having you speak about. Because just from the works I've mentioned, we have Sandburg, Rilke, Auden, Houseman, Kafka, Nietzsche . . .

LF And Wallace Stevens, more recently, in *13 Ways of Looking at a Blackbird,* the most set-to-music American poem ever. And I have ideas as to why, probably; but I've had that in my files for a long time prior to setting it, because I always wanted to set it. I think when a text is mysterious and visual, then it attracts my music.

RD And how do you conceive of your application of composition to a text?

LF It helps me the way a twelve-tone composer is helped by a twelve-tone row. It gives me something to start with.

RD The imagery?

LF The imagery gives me ideas. And actually this is what happens with most of these things. Every composer latches on to something that gives him ideas. With Wagner, it was the Germanic myth that gave him ideas.

RD Do you have any reverence for the actual word or sequence of the poet's words?

LF There is a form of reverence whenever I set words to music. It's a kind of homage. But I found out, after thinking for years that it was just homage, that I was actually raping the poem. In other words, though an homage prompted it, what I'm actually doing when I set a poem to music is a form of rape, a violation; because the poem doesn't really need me and my music, and doesn't want to be labeled and pinned down. So there is something ruthless and willful there and at the same time it's a form of homage.

**RD** Well, that's easier to do with a dead poet. What about a live one?

**LF** The live ones very often want it very badly. Everybody labors under the illusion that music plus words must be better than just words. But I don't think it works that way. Just like music and dance isn't better than just music . . . mostly one parameter will dominate. With dance it's the dance and the music begins to be a servant; and with poetry and music it's the other way around—the poem is becoming the servant and the music dominates.

**RD** When you look for a poet, is there any method in your madness?

**LF** As little as when one looks for a mate. It's just . . . of course you want to find the right mate in life, and so you look. But ultimately it's an accident when it happens, right? And it's the same with a poem.

**RD** What about the influence of Stockhausen? You invited him to Buffalo in 1964 to do the American premiere of his *Momente,* I remember vividly. He appeared with the New York Philharmonic in 1971, and about that time I had conducted his *Carré* in Europe. He spoke about things that opposed the "mechanical" in terms of performers and their attitude, whether they're performing his music or not; whether they're performing old music; in other words, that so-called "spiritual" side of performance. . . . He found that a musician who was not concentrated was like someone performing some sort of streamlined factory work.

**LF** I think this is a kind of performer's mysticism and I'm for it. I don't practice it, but I'm for it. I can see that it means that there is real concentration, and of a spiritual kind. Of course, there is a little bit of showmanship involved. It's like spreading the word beforehand that "This is going to be *spiritual!*" Personally, I like composers that just think in terms of their work; and they can do it in hotel rooms or wherever and don't necessarily wear their mystical hat in such an obvious manner. But some have. . . . Wagner leaned towards a kind of theatrical holiness in his concept of music and became more so as he became older. It's a German thing. It's a German form of mysticism and it's a performer's mysticism.

**RD** Of course, Stockhausen carried it on with the idea that as a composer, he wanted intuition available to him at all times. *All times,* not just once in a while when one had a good idea.

**LF** I'm sure it's a good routine like any meditation or any kind of discipline that you force on the daily structure of your life. It's going to do just that. It's going to give you a constant flow instead of a sporadic now-and-then sort of thing.

RD Could you make an observation about the current use of "minimal elements"?

LF I have used "minimal elements" already as far back as the Percussion Concerto. You remember how it goes on and on with repetition and so forth. I always make it a practice . . . every technique interests me. And so the minimal technique interested me right from the start. In fact, Steve Reich at one point actually took some lessons from me in Ojai. But at that time he had not developed this; I don't mean to say that he got any of this from me; but there is a kind of constant cross-fertilization and I believe in that. So I pick up everything and make it my own. But in *Solo* I went one step further, and I think a very important step; and that is, I asked myself . . . well, life is repetitious. I mean we get up in the morning, we have lunch, have dinner, work, go to bed, get up again in the morning . . . life is repetitious, but it's not only repetitious. It develops. Every day is different and ultimately we die. And so in *Solo,* I did that. I did a piece that starts one way—twelve-tone . . . and ends totally different, namely diatonic, almost "pop"; and constantly develops, although every bar seems to be the same when you go from one to the next. It develops *towards,* constantly. So it tries to take the minimal idea and lift it out of the drug . . .

RD Dream . . .

LF Dream, and meditation sameness, into the actual way repetition happens in life. We get older, we change.

RD Indeed we do. We've talked about your composition; we've talked about other composers. Let's talk about the conductor for a while. You are the conductor of the Milwaukee Symphony and the Brooklyn Philharmonic; and I knew you as the conductor of the Buffalo Orchestra. You also had ideas about conductors being a new type. I remember an article in 1968, "The New Type of Conductor." We had the new type of *performer,* now we have the new *conductor!* Has that creature evolved?

LF In a sense I'm sure Gustav Mahler was the new type of conductor. The composer-conductor is the type that would be close to my heart because it means the kind of conductor who is close to the source. And therefore he can peek behind the laboratory of the composer and put his loving, intelligent searchlight on the music; and therefore conduct it best . . . maybe. On the other hand, the composer-conductor never has enough time because he wants to compose! So therefore, it turns out that some people who only conduct are making just as much contribution as the composer-conductor, if not more. So it's hard to say. I know that I have to give up Milwaukee because I simply cannot give it enough time. So that's the way it goes. I think that we are still stuck with the typical jet-set conductor who is in fashion nowadays; and it's very strange when the media expects that type of conductor to make news, because there isn't going to be any news! It's going to be the same type who "milks" the classics . . . tries to make things effective, to look good, sound good; and depending on his talent and degree of integrity, he will succeed in doing good performances. But—*there is no news!*

**RD** That's right! Well you spoke about gestures, for instance . . .

**LF** I like *functional* gestures, gestures that mean something to the musicians, and not gestures that "Mickey Mouse" the emotional content of the music.

**RD** And this notion of being a lion-tamer or a hypnotist?

**LF** I think we've got to get away from that. . . .

**RD** Are we, though?

**LF** Not really, but I think it gets in the way of real leadership. And by functional, that doesn't mean unemotional. Functional can be very emotional because you are, after all, communicating the music, every aspect of it . . . the structural and also the emotional aspects. Stravinsky's conducting taught me something; he was all "ears"—no choreography.

**RD** You had a very precise view of the symphony orchestra . . . whether it was dying or not. Bernstein had said it was a museum.

**LF** I like museums; I think they are wonderful. We have a need to make love to the past. We also have a need to discover the future, or at least try to live in the present. So museums can be brought up to the present, to have compartments that are devoted to the present. But they will always concentrate more on the past; and so the symphony being, after all, something that goes, roughly from Haydn to Mahler, with some anachronisms like Shostakovich coming later . . . still basically that's the development . . . from Haydn to Mahler. So the symphony will thrive on that literature.

**RD** You mentioned, at that time, the breaking up into smaller units to perform other literature. Have you used that?

**LF** Yes. I do that in Brooklyn in the Meet the Moderns series. We can use smaller units, because obviously with the audience we get for modern music, we cannot have always the whole orchestra; and I think that works well. And yet those pieces for sometimes five, six, seven, eight people need a conductor very often because they are so complicated. And also the conductor is a timesaver. So I think that I am by no means pessimistic. But at the same time . . . there are a lot of things that could be different and that are reactionary today, and are not really opening up things for the creative mind.

**RD** And lastly . . . your view of critics, in a word?

**LF** I think that critics are the tastemakers of today. They are the princes, what used to be the princes in the 18th century, the people who decided who would exist and who would be employed.

**RD** You think they have that much power?

**LF** I think they have more and more of that power, yes. They are the tastemakers. Whether we need tastemakers . . . obviously we do, but should we, is another question. Probably not, but we have them. They are here to stay.

RD How do you feel about them saying things about your work? I mean, I can understand a performer being under the thumb of a critic . . .

LF The composer is not under the thumb of the critic, but if he is launched by an influential critic, it helps immensely. Similarly, if he is consistently labeled by a critic and labeled in a way that is totally misunderstanding, it can postpone and maybe even ruin a life. I don't know to what extent these things are helpful or harmful. I love critics who have made themselves advocates of new music, like Alfred Frankenstein of former years. Those are wonderful people. I think that if they make themselves advocates of a cause, they themselves will endure more than if they make themselves judges of trends.

# Narrative II

Those enlightening and, for me, important years in Los Angeles lasted until 1963. When Lukas Foss, who had just taken the Schoenberg Chair of Composition at UCLA, invited me to join him in a project of ensemble improvisation in the spring of 1957, I knew I was joining a crusade for new music.

The initial instrumentation of the group was flute, clarinet, cello, piano, and percussion. Our work progressed at a fervent pitch. Our first rehearsals were at the Foss residence in Beverly Hills, which was the former aviary of John Barrymore (complete with a stained glass window of Barrymore in an undershirt and a female companion, Delores Costello, overlooking a scenic view). The work was intense, but pragmatic.

The basics that Lukas wanted to get at were melody, counterpoint, and harmony. How could we improvise in the 20th century using these fundamental elements? In our early stages, melody reflected each individual's contemporary musical past. Counterpoint became a Webernian exercise in playing our note or notes within a bar at prescribed rhythmic points or variance. These two elements were held together harmonically by a series of groundtones, which had to be memorized as they underwent modulatory procedures, and to which melody and counterpoint adjusted, according to preset rules.

For a while it was difficult and demanding to find one's notes: place them in the prescribed rhythmic pattern, pay strict attention to the groundtones. For all of this effort, we had no "pieces" for a long time. It was a marvelous exercise in ear training. Then came the problem of form: the problem of making a piece. We rehearsed three and four times a week. Eventually, we decided to call ourselves the Improvisation Chamber Ensemble (ICE): Lukas Foss, piano; Richard Dufallo, clarinet; Charles DeLancey, percussion; Howard Colf, cello.

Lukas's traditional experience as a composer guided us through things such as fugue, variations, etc.; but when we discovered spontaneous dissonance, our formal conceptions expanded. This enabled us to begin building a repertoire of workable "Studies in Improvisation." The avoidance of "pulse" or the "bar-line," became uppermost in our collective thought. We were eager to destroy the "bar-line" and began to concentrate more and more on irregular, disjunct, and sometimes abrupt methods of execution.

Only when the interludes were conceived for the October 1960 *Time Cycle*[1] premiere performances with the New York Philharmonic, with Leonard Bernstein and Adele Addison in Carnegie Hall (at that time slated to be torn down), did we feel we were working in a more abstract musical area.

I won't forget meeting Bernstein for the first time just prior to those performances. He had recently become the Music Director of the New York Philharmonic and was in Los Angeles with his orchestra to perform Berlioz, Beethoven, and Bartók in the Hollywood Bowl on September 2, 1960. It was during that stay that he came to hear us improvise at Lukas's new Beverly Hills home (the ill-fated one that burned down), one sunny California late morning. I still remember him appearing in a blue blazer and white "everything else." Glamorous was the word! He sat down on the floor of the studio with a plate of scrambled eggs, and we improvised for him. He was very generous, encouraging, and he gave us the impression that he would have liked to join in.

Throughout our whole experience, the notions of something half-composed, half-improvised, or something improvised again and again until it became workable and finally "wore out," were the main procedures of our work. The most exciting thing for me as a performer was to be able to "correct"—on the spot in a very quick moment—what I was playing; and to turn something that might have led to a disaster into something musically viable. Usually we didn't know whether we were succeeding unless the others spontaneously picked up on our initiatives. Conversely, to add support to a colleague's presentation was equally thrilling. And when the group managed, with intensity, to sustain this kind of exchange consistently and for a reasonable duration of time, a "performance piece" was born.

We conducted each other with hands or head or nod of instrument. We explored new sonic possibilities of our instruments. We played in a truly virtuosic fashion at times. And always with the simple guide of 3 by 5 cards, on which our symbols and graphics were notated. Lukas and I developed an outline, complete with exercises, for a book of twenty-plus chapters on how to improvise before we abandoned the idea.

Before the project was terminated in 1963, we had performed in two American tours and one European tour of improvised chamber music, collectively written and performed a *Concerto for Improvising Instruments and Orchestra,* performed the improvised interludes to the orchestral version

1. This premiere was recorded by Columbia Records, ML5680/MS6230, 1961.

of *Time Cycle,* as well as performed and recorded the chamber version[2] of that work (without improvisations), recruited and trained other musicians with the help of a Rockefeller Foundation grant, and performed Foss's *Echoi,* a work that benefited from the experiments of the ICE. The group appeared on CBS television as well as the BBC. Orchestral performances, in addition to the New York Philharmonic, included the Philadelphia Orchestra, Boston Symphony, Los Angeles Philharmonic, and the Berlin Philharmonic.

Not only had Los Angeles been home for Arnold Schoenberg, but Hollywood was where Igor Stravinsky was living. The musical atmosphere was alive. On March 11, 1957, Pierre Boulez conducted the American premiere of *Le Marteau sans maître* at that temple of intellectuality, the Monday Evening Concerts[3] in Los Angeles. I heard the performance and met Boulez for the first time. His music dazzled and perplexed me. I had heard nothing like it before and his conducting was a revelation.

Three months later the Los Angeles Music Festival presented a Gala Concert on June 17, celebrating the seventy-fifth birthday of Igor Stravinsky with an all-Stravinsky program. Festival Director Franz Waxman conducted the *Greeting Prelude* for the eightieth birthday of Pierre Monteux. Robert Craft conducted the *Symphonies of Wind Instruments,* the American premiere of *Canticum sacrum,* and the world premiere of *Agon.* Aldous Huxley spoke during the intermission, and to this day I do not remember what he said, because I was so eagerly awaiting my first live glimpse of Igor Stravinsky. At age twenty-four, as a budding young conductor in the throes of developing a method of ensemble improvisation as a clarinetist, I could think of nothing more inspiring than to see Stravinsky conduct his arrangement of Bach's Chorale and Canonic Variations and his *Symphony of Psalms.*

Stravinsky's stage presence was immediately arresting. His "Jiminy Cricket-like" body contained an energy that was charged by an unusually acute mind. The performance was electrifying.

In the fall of 1958, I met Stravinsky in the flesh. Lukas and I went to Hollywood to catch the end of a recording session that Craft and Stravinsky were doing. At the end of the session I was introduced to I.S. by Lukas, who told the *Maître* that I would be presenting an all-Stravinsky wind concert: *Symphonies of Wind Instruments,* Octet, Concerto for Piano and Winds, and the Mass for chorus and double wind quintet. *Maître* said, "Bon," not knowing that I would be conducting the UCLA Concert Band, whose repertoire mainly contained those insipid band pieces that border on theme music for "B" movies!

That said, the next problem was . . . who would drive Stravinsky home? I was elected. Panic was not the word! I rushed to my car, a black and white Plymouth two-door with light blue seats, and picked up the great man. We

2. Epic Records, BC1286; Library of Congress, R64–1128.

3. Under the direction of the late Lawrence Morton from 1954 to 1971.

began driving in the direction of his Hollywood home and I was simply speechless. There was Igor Stravinsky sitting next to me. What do you say? I said nothing. The only thing I remember him saying was, "Vot's dat?," pointing to the sky where a jet airplane had left a white vapor trail. I became terribly animated. This was my chance to speak! To tell I.S. what that white mark in the sky was! Having done this, we arrived at his home and I vividly recall him tipping his hat several times, in gratitude, as I drove off in a state of wild excitement. It was like a scene from a Jacques Tati movie.

The all-Stravinsky wind concert occurred on December 10, 1958. But prior to that, on December 1, I shared the podium with Karlheinz Stockhausen. This was a joint concert: Monday Evening Concerts and New Music Evenings (UCLA) "honoring the young German composer." He was in Los Angeles to conduct his *Kontrapunkte* and the American premiere of *Zeitmasze,* that most difficult score of bewildering rhythmic complexity. I conducted Stravinsky's Cantata (1951–52). I didn't speak to Stockhausen, although we shook hands.

During this time I had acquired my B.A. and M.A. degrees from UCLA and joined the faculty as Lecturer in Music, as well as becoming Associate Conductor, to Lukas, of the UCLA Symphony. Our collaboration on the improvisation project had worked well, and we were to continue our musical relationship until 1967. In the meantime we had become friends; and I was a frequent visitor to the Foss household, through which flowed a steady stream of distinguished members of the arts community.

After observing and meeting Stravinsky, Boulez, and Stockhausen, there continued to be many "firsts" that enhanced and stimulated my musical development. Time spent in conducting and playing was about equally divided. In addition to the standard repertoire, I began conducting the works of Copland, Ives, Ruggles, Webern, Schoenberg, and young composers.

The musical activity at UCLA was enticing. In 1959, Lukas did the West Coast premiere of Stravinsky's *Threni.* Later, I shared the podium with Roy Harris, who was a visiting composer. In 1961, Carl Sandburg made an appearance reading his poem *Prairie* on a program that included Foss's excerpts from *The Prairie* and Copland's *Lincoln Portrait,* with Sandburg narrating. Benjamin Britten's *Turn of the Screw* was given its West Coast premiere. My opera conducting started in 1962 with a production of Stravinsky's *Rossignol* on a double bill with *The Prisoner* of Dallapiccola, who was present. I also did the Foss opera *Griffelkin.*

With Stravinsky actually on the scene in Los Angeles, it was hard to avoid knowing what was happening in the international music world. He spoke and wrote about it abundantly. In the 1950s his composition had changed; and those works written in the decade from the Cantata (1951–52) to *The Flood* (1961–62) contained both "serial" and "twelve-tone" aspects. The primary distinction between these two aspects is that "twelve-tone" music is intrinsically *serial,* but "serial" music need not be *twelve-tone.*

The Ricercar II of the Cantata was regarded as his first serial movement, and the second movement of *Canticum sacrum* (1955) as his first twelve-tone movement. *Threni* (1958) was considered his first completely twelve-tone work. But his interest in serialism in general seemed to outweigh his interest in the twelve-tone system.

He said, "The rules and restrictions of serial writing differ little from the rigidity of the great contrapuntal schools of old." In his *Movements* (1958–59), he felt that he was "becoming not less but more of a serial composer." His fundamental interest seemed to lie in a new ordering and not in a "system."

The use of serial units of four or five notes, for example, in the first of the *Shakespeare Songs* (1953) and *In Memoriam Dylan Thomas* (1954) was indicative of his non-twelve-tone approach, while at the same time he used operational procedures from the twelve-tone system. He said, "The intervals of my series are attracted by tonality. I compose vertically and that is, in one sense at least, to compose tonally."

Regarding his composing process in his very brief *Epitaphium* (1959), he said, "I heard and composed a harmonic-melodic phrase. I certainly did not (and never do) begin with a purely serial idea; and in fact, when I began, I did not know, or care, whether all twelve notes would be used. After I had written about half the first phrase, I saw its serial pattern however, and then perhaps I began to work towards that pattern."

He also said, "Masterworks aside, it seems to me the new music will be serial." This endorsement of serialism in the '50s made an enormous impact on many composers and the world of music in general. Reactions varied from surprise and approval to disappointment and confusion. In that period, Stravinsky also endorsed two younger composers—Pierre Boulez and Karlheinz Stockhausen.

Until 1945, Stravinsky's own "neo-classicism" had been a major influence, along with Hindemith, and Schoenberg's school of dodecaphony. Post-World War II ushered in a kind of "Big Bang." Serialism began its march toward a totalitarian concept, with or without twelve-tone practices; and chance procedures began to differentiate between "pure chance" and matters of "choice."

Stravinsky felt that this new period of exploration and revolution began with the rediscovery of the masterpieces of 1912: *Pierrot lunaire, Jeux, Altenberg Lieder,* and *Le Sacre du printemps;* and the music of Webern in general. "Boulez's cantatas are representative of this new music of the immediate postwar.... In this new period of exploration the only significant work so far is still Boulez's *Le Marteau sans maître* (1954)." He was also attracted to Stockhausen's *Zeitmasze* (1955–56), *Gruppen* (1955–57), and *Carré* (1959–60). Of both composers he said, "The ordinary musician's trouble in judging composers like Boulez and Stockhausen is that he doesn't see their roots. These composers have sprung full-grown."

During this postwar period, other significant developments occurred. The International Summer School for New Music in Darmstadt was founded and organized by Dr. Wolfgang Steinecke. Domaine Musical was established by Boulez in Paris in 1954. The Fromm Music Foundation[4] established the Fromm Fellowship Players at Tanglewood in 1956, held the Princeton Seminar in Advanced Musical Studies during the summers of 1959 and 1960, and began publishing the journal *Perspectives of New Music,* edited by Arthur Berger and Benjamin Boretz, in the fall of 1962. The periodical *Die Reihe,* edited by Herbert Eimert and Karlheinz Stockhausen, had already appeared in 1955. Both publications were extremely informative; but at times one felt the need of a "mind-sweeper" in order to navigate through the sea of information. I thought, "Is this a crusade I'm on? Or is it simply a path that a musician should naturally follow?"

The fervor and revolution of the postwar period weren't felt by all composers. American composer Ned Rorem, who lived in France from 1949 to 1957, is a particular case in point. I met him in the early '60s in New York.

4. Paul Fromm, director, died July 4, 1987.

*Charles Abbott*

# *Ned Rorem*

BORN OCTOBER 23, 1923, RICHMOND, INDIANA

**March 22, 1986** **New York**

As I approached Rorem's West Side apartment, I thought of his reputation as a writer of diaries, as an essayist, and as a critic. He has always shown a brilliant verbal tongue. When I rang his bell, I thought I might be awakening a sleeping cobra!

RD Ned, I want to ask you some questions in about four areas, first of which has something to do with when you returned from Paris. I believe that was 1957. I remember knowing about you and knowing about your music in the very early '60s because I was playing the clarinet at the time with Lukas Foss, and we were using your Greenwich Village apartment to rehearse in for something we were doing.

NR Lukas sublet my one-room apartment at 247 West 13th Street during the year and a half that I was in Buffalo, from 1959 to 1961. Some weird call girl had sublet it during the first semester, and then when the University of Buffalo invited me back, Lukas took it, unless it was the other way around . . . I don't mean that Lukas was a call girl . . . I mean that Lukas was either the first or the second sublettee . . . and he wrote much of *Time Cycle* in that room. I feel honored, because it has become my favorite piece of his. He had already done a record of improvisations at that time, it seems to me, unrelated to *Time Cycle.*[1] I don't know if I knew you, but I certainly knew your name, and I think we must have met each other soon afterward. I knew your clarinet playing from that record. Indeed, my almost first article of professional prose was a review (quite favorable) of that record for, I think, *High Fidelity.* So that was twenty-five years ago—Jesus! . . . In those improvisations the one thing that always amused me was that no matter which solo player was doing a riff, it all came out sounding like Foss. Performers aren't composers. Thus, in their anxiety at being forced to improvise, they clutch at the strongest straw around, in this case Lukas—and the personality of his music.

1. *Studies in Improvisation,* RCA Victor, LM/LSC 2558, 1961.

RD I brought up the idea of your apartment and these first moments for another reason.... After having spent from 1949 to 1957 in France, what were you facing in Music, USA, when you returned?

NR I went to Europe in the first place because I was attracted toward French culture. I hadn't intended to stay when I went for the summer of 1949, but I remained eight years. I lived a French life, as opposed to an expatriate life in the American colony. Before Europe, it never occurred to me, any more than it did to any American composer, that one was supposed to write a certain kind of music. I've always been pretty much of a lone wolf and still am, never part of the Copland milieu or the Stravinsky milieu or the Varese or Thomson milieu. I was a little of each, but mostly I was my own person. If I didn't get the backing of any of these people, neither did I get stamped as their protégé. But one was allowed during the war years to compose in a pure, diatonic style. It came to me as a big surprise when, almost overnight, once the war clouds had cleared, America, who during the war had come into her own with a native language, a very Coplandesque language, a folkloric language ... and Europe took up from where it had stopped in 1932 when Schoenberg became taboo in Germany. Suddenly Schoenberg was allowed to be played again; and France, of all countries, not Germany at all, was the one that set the tone, largely due to the strong personality of Boulez. Boulez more or less called the shots for the entire world for many years. I have always thought that ... I was going to say I thought it was unhealthy, but nothing is unhealthy. No school is unhealthy in itself; it's only what's done within the school that's unhealthy. Well, perhaps Boulez is the one exception: his school is unhealthy in itself! Anyway, Boulez was prominent in France in the years I was there, though I was never a part of his milieu. I was more a part of the milieu of Les Six. The two milieux had nothing to do with each other. They didn't even hate each other. When I came back home in 1957, I realized that America was already under the yoke of multiple serialism, which was to grow even more rampant by the early 1960s. Still, I had enough of a reputation to get jobs and to get performances.

RD But the atmosphere, at least returning, was conducive to composition for you?

NR It was, because the pieces I wrote got played. I was—knock wood—one of the few at that time writing orchestra music that orchestras were able to play without the number of rehearsals that it took for, say, a piece by Milton Babbitt. In that sense, I profited from the mood of complexity that was in the air.

RD There was an emphasis on smaller instrumentation at that time—

NR Smaller instrumentation was already funded by Paul Fromm with maximum rehearsal time that orchestras lacked. Obviously, musical formats changed from large to small. Song, the art of song, and the song recital stopped almost completely; and it's still stopped. There are no true singers of songs today in America. In this country of 250 million souls there is not one singer—not one—who makes a living primarily as a recitalist. The only people that do are Europeans. Our best singers, the ones that interest you and me, don't give recitals with piano essentially. The art of song got lost along the way.

RD So there are two items there that attracted you, and you took advantage of that attraction with the large symphony orchestra and the songs.

NR They had nothing to do with each other. Symphony orchestras played me because my music was easy. (I like to think it was because my music was also good quality!) Singers sang my songs because there was nothing much else to choose from. They were seduced by grand opera, because that's where fame and fortune lay. Much of contemporary vocal music was with several instruments, like *Marteau sans maître;* and singers didn't always want to sing that . . . or have the occasion. Nobody was writing songs just for voice and piano.

RD And the comparison between social habits at that time, and social habits today. How one dressed, how one behaved, what the manner of society was then and now . . . did that have any influence? Or did it strike your mind as a contrast?

NR The composer has been a pariah for 150 years, since the beginning of the Industrial Revolution and Beethoven, when the composer was no longer a church worker. He became extraneous, and that's when symphony concerts began, and that's when Bohemianism began, with the Romantics dressing funny and being beatniks. If you mention the word "composer" today in polite society, it is assumed you are talking about somebody in the pop world. Even our intellectuals are generally unaware of contemporary classical music, for lack of a better term. It doesn't get recorded, it doesn't get performed on the regular circuit, it doesn't get published by big publishers, and big recording companies aren't the least embarassed to say there isn't any money in it and therefore they can't afford to do it. . . . You can count on the fingers of one hand the American composers of serious music today who are able to support themselves from the just rewards of their labors. The average music critic makes, as a critic, more money than I do as a composer. I support myself teaching and playing the piano and writing prose; but what I get on commissions is probably less than John Rockwell gets at the *Times.* Another composer supports himself in another way, and different composers write different kinds of music: some write for the movies and some write for the stage.

RD What about the concept, brought up in the summer of '85 in Aspen with the critics and the Fromm Foundation, that there is a musical pluralism. You spoke about people writing for theater—there is music that Rockwell describes as near-pop, near-rock, we have minimalism, we still have aspects of serialism, we still have aspects of chance. Do you see us in a state of plurality?

NR There have always been musical pluralities. The only difference now between Europe and America, or let us say France and New York, is that French composers themselves for economic reasons have always been pluralistic, individually pluralistic. A composer in France in the 1950s would write simultaneously for the movies, for the stage; he would write operas, ballets, string quartets; he would write for pop singers; he would write for classical singers. In America, a person who writes operas seldom writes songs; a person who writes quartets doesn't write for the movies; and a person who writes academic music for schools as a rule doesn't write for symphony orchestras. We are specialists in everything, but there are no more general practitioners. However, since the 1950s there has been a relaxation. One school of musicians wouldn't spit on another school of musicians in the 1950s; while today we don't despise each other by nature of the language that we speak. One composer doesn't disapprove morally of another composer's language; he only disapproves of how well that language is uttered.

RD You remember Morton Feldman at one point for instance saying, "The rhetoric is gone." Do you find that good or bad?

NR Perhaps people should boo more at concerts. Ever since *The Rite of Spring,* which was booed and people then realized they were wrong, they are afraid to boo for fear they'll be proved wrong fifty years hence. A couple of weeks ago I went to hear the Boulez *Répons,* and everyone sat around sanctimoniously as though it were Holy Writ or something. A few healthy boos would have been in keeping. . . . Not that the piece was at all that bad. I rather liked it. But I liked it for completely different reasons from what Boulez doubtless thinks is in it. I liked it because it was colorful, amusing, rather vulgar, thoroughly non-intellectual, utterly sensual, completely derivative. The other day Boulez said when asked by a young composer at Juilliard, "What do you think of the so-called New Romanticism in America?"—"You Americans are all so nostalgic!" Nostalgic for what? Nostalgic for Mahler, who is the composer that most of them are imitating? We never had any Mahler in our culture; we're not part of Europe. Boulez's piece meanwhile is vastly nostalgic insofar as it cribs literally from Debussy, the Stravinsky of *Les Noces,* and Messiaen, without their tunes. He takes their filling and ornaments. He takes the gargoyles but not the cathedral. All composers steal.

RD I was going to ask you what you think of this New Romanticism that's being tossed about as a title. I mean, is there such a thing?

NR I remember Aretha Franklin on the Johnny Carson show some years ago. He asked, "What is soul?" She gave him a vague look, meaning don't ask me! It's the same with New Romanticism. These are terms . . . not by composers but by critics after the fact.

**RD** That one was coined by Jacob Druckman.

**NR** Okay, well he coined it insofar as he's a critic. But a composer doesn't get up each morning and say, I shall now write some more New Romantic music, anymore than a singer says, I am now going to sing soul. Soul is nothing more than what blues was a generation earlier. What Boulez doesn't seem to realize, or Druckman, is that trends go by generations. We have contrapuntal generations superseded by harmonic, then contrapuntal, then harmonic. The contrapuntal generations are generally complex, the harmonic generations are generally simple. We're in a harmonic, vertical, simple generation right now, having come out of a horizontal, complicated, polyphonic generation. Music doesn't progress; it continually retrogresses; or rather, music doesn't evolve, it revolves, like a great wheel, and you go from plain to fancy to plain to fancy. The same is true of all the arts.

**RD** It's an unfolding, you mean.

**NR** An unfolding and then a refolding, and then an unfolding and then a refolding. And part of that has to do with economics, and part of that has to do with, we get sick of what we've already had and we've got to let up. And then we've got to tighten up and then we've got to loosen up and then tighten up. It doesn't get better and it doesn't get worse.

**RD** Then in a kind of summary, in light of what's been said, what's your biggest observation?

**NR** The way I fit into the picture. I don't know quite what the Romantic Movement is, except that it never happened in France. The so-called New Romantic composers are imitating the Germans, specifically Strauss and Mahler. I have always been French in my penchants. But you don't hear about any Americans except me being impelled by French musicians. They aren't even despised, they are simply ignored in this whole scene.

**RD** Could we move to the idea of conductors and conducting? Do you enjoy attending rehearsals when you have your music performed?

**NR** I am lucky in that almost every piece I've written has been commissioned, usually by good people. I've heard enough of my music to know pretty much how it's going to sound when it finally gets performed. Let's talk just about orchestra music. Usually I know who the conductor is and the orchestra. But I hate hotels, and traveling. I find, even, that if I just attend the premiere, the performance will be little different than if I'd been at all the rehearsals. Either the conductor gets the point from the score alone and what the composer tells him is superfluous, or the conductor will never get the point so what the composer tells him is also superfluous. Most composers are dead anyway. (I don't completely believe all this.)

**RD** What do you expect in a conductor?

**NR** What do I expect? Probably that he should get the tempos according to the metabolism of my dreams. Every composer has an ideal conductor in his mind when he writes the piece. But reality and ideal are not the same. So sometimes when I've heard a first performance it's sounded . . . not necessarily different, but nevertheless better or worse. That usually has to do with tempo. Now, every conductor errs necessarily by playing a piece either slightly too fast or too slow (or sometimes much too fast or too slow). But there is no *one* way to do a piece; there are as many right ways to do a piece as there are intelligent approaches to a piece of music. There are radical differences in classical music, and each version is kosher.

**RD** Yes, well, speed is a big problem. But surely that's not the only consideration.

**NR** No, but I would say that I'm attracted to conductors who play my music a little faster than it "ought" to be played, rather than a little slower. Lenny, and I worship him, forgive me, takes slow movements too slowly for my taste. I feel them sagging and getting out of control, because Lenny likes to milk things. My music won't bear milking. Ormandy has done five of my pieces, two of them on tour. So in that sense he's been more important to me than Bernstein in terms of dissemination of my music. And yet, I never had a meal with Ormandy, I never called him anything, and he called me Mr. Rorem.

**RD** Do you remember your best performances? Can you single out a few that you felt were your best?

**NR** The first performance is always the one against which all ensuing performances are judged. If the first performance is lame, it's nevertheless the most exciting. The single best performance of an orchestra piece was Stokowski's conducting of *Eagles* with the Boston Symphony (his debut with that group) in 1964. He made the piece tremble and glitter and breathe with an urgent life of its own. Yet the experience was less thrilling—less *necessary* somehow—than the premiere seven years earlier in a far milder presentation by Ormandy.

**RD** It's a strange thing when you get preconditioned . . . do you think that's true?

**NR** Yes. You can get preconditioned by a mediocre singer, or a singer with a certain kind of voice, too.

**RD** That's kind of scary, isn't it?

**NR** But there it is. And that's why I don't like to attend rehearsals. Tomorrow I'm going to Washington to hear *An American Oratorio,* a forty-five-minute piece for chorus and orchestra and soloists, and I won't have heard any rehearsals at all. But it's not a premiere, so I'm sort of blasé. Margaret Hillis is doing it three weeks later in Chicago, with the Chicago Symphony. I won't go to any rehearsals there, either. I do find though that if I am at a rehearsal, I always have something to say. . . . If a piece is going to be wrong, it's going to be wrong in spite of what the composer says, though he can sometimes pull it together at the last minute by saying, "Play it a lot faster."

RD And expression . . . how about that?

NR My music is very transparent—easy to conduct. You'd know the answer to that better than I. But everything I have to say is there on the page. . . . I'm leery of what we call "interpretation" in a performance. Jaime Laredo recently played my new Violin Concerto. It was full of expression, full of little nuances and meaningful pauses. I loved it all, but I was surprised. I'm always surprised when performers seem to care enough to put these things into the music. And yet all these things are 19th-century things.

RD Yes, but they find a way of doing that . . . do you object?

NR I don't object, but I never hear it that way myself. As for singers, any composer will tell you he'll give a singer more leeway then anyone else, for the simple reason that there's a bigger difference between two singers than between two anything else. Blindfolded, you can't really tell whether Conductor A or Conductor B might be male or female, or two oboists, or indeed, whether any instrumentalist is male or female. But the difference between a male and female singer is radical. If they are singing the same song, you'll allow *him* to sing more slowly, because he's singing an octave lower the same tune, which has fewer vibrations per second (or is it more?). I mean there's a difference in kind between human voices that doesn't occur between any other category of music maker.

RD Do you have a group of favorite performers of your music?

NR Well, the performers who have been my friends have probably been my friends because I liked their performing and their unstinting devotion to the music of their time and place. Donald Gramm, Phyllis Curtin, Nell Tangeman, the mezzo-soprano, were big influences on my life. Billie Holiday, whom I knew but who never sang my music, was an influence on the way I wrote vocal music, not the tune itself, but her way with a tune, the way she would dip and swerve. Of conductors, leaving you out of it for the moment, I liked Abravanel quite a bit. I even liked Krips. There are conductors and singers who are famous who have done works of mine, and I've been disappointed; and you're not supposed to say so, because they are famous. For example, Eileen Farrell sang some songs of mine; and her voice is so beautiful that you forgive her in a way. But her point of view was all wrong. It had to do with beauty of voice, rather than sense of words.

RD Sometimes a violinist with a very good instrument does that, too.

NR I have to add to my list of favorites the baritone Will Parker, organist Leonard Raver, pianist Jerry Lowenthal, cellist Sharon Robinson, choral director Gregg Smith. I've composed specifically for each of these faithful geniuses; and the fact of them, by definition, has shaped my style somewhat. I feel possessive about them—they are *mine*, the way Ansermet and Dushkin were Stravinsky's.

RD Do you remember some of the best critics, or some of the worst critics, of your music?

NR I don't approve of any critics, but they seem to be a necessary evil. Music can exist without critics, but critics cannot exist without music. They are parasites by nature. I've never learned anything from critics, never written a piece differently or revised a piece according to their criticism. Bad write-ups make me feel worse than good write-ups make me feel good. (I read good write-ups once and bad write-ups twice.)

RD Was there any outstanding in-depth review that you can remember?

NR No, because it's always wrong. Even when they like it, it's wrong. It has to be. A critic in twenty lines can't say the whole thing.

RD And how about in the worst sense?

NR A critic should never, never be sarcastic. It goes without saying that most new things are terrible, and a critic is duty-bound to say so; but he should say so with regret, not with relish. I think Virgil Thomson is the only critic who ever mattered.

RD And fair-minded in both the positive and the negative sense?

NR The things he could not be fair-minded about he usually did not write up. Like German music, Brahms, that he didn't care much about and couldn't think of anything to say: he would farm them out to others on the staff at the *Herald Tribune*.

RD Well now, let's move on to 1982, when you were in Aspen. Do you have some general remembrance of your stay there as composer-in-residence?

NR I've been a composer-in-residence often, but never for as long a period as at Aspen. It's usually about two days in a university environment, except for Santa Fe Chamber Music Festival, which is a first-rate festival. In 1982 I came directly from Santa Fe to Aspen. I really felt wanted that summer! In Aspen, I guess I was there mainly as your guest. I remember all of the performances. You asked about going to rehearsals, and I don't like to go to rehearsals on some occasions, especially of pieces I've already heard; but when I'm in a working atmosphere like Aspen, which is a school as well as a performance center, I love it. It's not as though the rehearsal were a final thing. In Aspen I also played, and was able to coach, which is lots of fun. I don't know if you were there, but the song program at the end, in a church somewhere . . .

RD No, I think I was gone by then.

NR Adele Addison and Paul Sperry compiled a whole program of my vocal music of various sorts, and without telling me had geared everybody in the audience and everyone on stage, and all the stage hands, page turners and everybody, so that at the very end of the program, they all got up and sang my song called *Early in the Morning* in unison. I was madly touched. Aspen was able to do a lot of impractical pieces. They did a piece called *Some Trees,* which is for three voices and piano. (I was going to say such games don't grow on trees, but they grow on some trees!) And they did other unusual combinations, including three orchestra pieces. Slatkin did *Lions,* you did *Sun,* and that man in San Francisco with a Dutch name . . .

RD de Waart.

NR Incidentally, about composers and conductors, de Waart and I had had some vague intercourse previously, after *Air Music* won the Pulitzer Prize. de Waart played it, and a tape was sent to me. I thought it was fantastically good and wrote him a letter which he never answered. The only time I ever met him was when he did *Remembering Tommy* in Aspen. He was so cold I wanted to go hide in a corner. And yet his performance was very fine. So you don't have to love somebody to love the way they play a piece. It's possible he might hate my music. But whether he hates it or not, I like the way he does it. By the same token, I don't especially like Schubert, and yet I think that I play Schubert . . . knowledgeably and well. Conversely, to love a composer doesn't mean you'll play him "right." Objectivity is crucial. The French, who like only French music, play it all wrong.

RD I would just like to touch a little bit on the pieces *Lions, Sun,* and *Eagles.* Can you just briefly say something . . . *Lions* I believe you said at one point came out of a dream, and *Eagles* had something to do with eagles' mating, and *Sun* of course is associated with words.

NR I am Stravinskian in that I don't think music means anything literary. Which is why composers often take great pains to call a piece *La Mer* or *Francesca da Rimini;* if they don't call it that, nobody else is going to call it that . . . unlike a painter who because he knows there is no such thing as an abstract painting. . . . Every painting is concrete. You always see an identifiable face in the clouds of Jackson Pollock's oils. Which is why painters often take great pains *not* to give a picture a specific title. Just as I don't think there is such a thing as political music, in the sense that music can change our beliefs, moral or political, so I don't think that music can tell a story unless the composer tells you, in words, the story that he means to impart. Therefore, if you explain to someone who has never heard of *La Mer* that it represents three pictures of Paris in the morning, afternoon, and evening, at the slaughterhouse, the Louvre, and Les Halles, they are going to hear those images forevermore—images that have nothing to do with water. People will hear in music what you tell them to hear. That's why, in the two tone poems *Eagles* and *Lions,* I imposed a program before I wrote the pieces. I've often tacked on titles after the fact. *Sun* is different in kind, because *Sun* illustrates, through verbal singing, a pre-existent text. It's a hybrid bastard; it's two arts at the same time. And people will hear what the words that are sung imply. Yet I could have taken the same words and written a radically different kind of music. In fact, I once did that in a cycle, *Poems of Love and the Rain.* Taking the eight poems and setting each of them to music twice, quite contrastingly, I tried to show that there is no one way of doing it, and that two different ways by the same composer can be legitimate, or at least I hoped so.

**RD** And the sound of eagles' mating I think you once referred to?

**NR** Well, it's a fourteen-line poem by Walt Whitman called *The Dalliance of Eagles.* I took each of those fourteen lines, and in a margin noted musical ideas for how to illustrate them—like fluttering of wings, like the shrieking of beaks, the immutable sky, the poet's mood, the gyrating copulation. It's so graphic it's embarrassing (especially at the end, with the brassy postcoital spasm), but only if the listener has read the program.

**RD** I remember one other very vivid thing, and that was your lecture on Stravinsky at 100, which was witty, intelligent, informative, I thought very personal about your views of Stravinsky, and yet with respect. I think you called him a sacred monster. I think you spoke about his nature as being French, as opposed to German; and I think you made the statement that these are the two big areas of definition . . . in music.

**NR** Not only in music, but in the whole universe. Everything is divided between German and French. Oranges are German. Apples are French.

**RD** You listed a number of things that you said to be French is.

**NR** Yes, that's a game that only the French can play. The Germans . . . as I've often said, a German joke is no laughing matter! That phrase "sacred monster" is a French phrase. It simply means "sacred cow." Stravinsky was the last one. There's only one sacred cow left on earth, and that's Martha Graham. It's a 19th-century, Lisztian concept—bigger than life. Our current superstars, like Bruce Springsteen, are not in themselves bigger than life. They are small, but blown up.

**RD** You also took a little bit of cause against how Stravinsky used a poem or set words; not that he didn't write for the voice well, but you did have a view about how somebody shouldn't tamper with a poem, for instance, repeating phrases. Were there any other violations that composers make that upset you in terms of a pure way of setting a poem?

**NR** When it comes to vocal music, I feel proprietary. Whatever my value may be in the long run, I do think I know what I'm doing and why I'm doing it, so far as word setting is concerned. My viewpoint doesn't have to be yours and it doesn't have to be Boulez's. I mention the latter because he is the antithesis of what I am. I have no quarrel with the way he sets words because he's barking up a different tree, and he barks with clarity and purpose. . . . His aim is not to have the words comprehended any more than Gregorian chant is literally understood: he is illustrating the poet, not the poetry. He puts a frame around the poet; whereas I try to *heighten* the poetry. It's already so insolent to take a work of art that is complete in itself, like a poem of Mallarmé or Whitman, and garb it in music, that I've never gotten used to that fact that I'm treading on toes that don't ask to be trodden on. Therefore, if I take a poem and presume to set it to music, the least I can do is not to repeat words the poet has not repeated. Does not a song composer who repeats phrases not repeated in the poem he is setting threaten that poem's integrity? Is he not rewriting the poem—a rewrite which, if read aloud without music, he would reject for its now unbalanced meters, false echoes, too obvious rhythm here, an aural rhyme there? Why not by the same token justify omissions? Stravinsky did just that when he set Shakespeare. He simply forgot to include certain verses. Well, English wasn't his language. Yet he persisted; and with Dylan Thomas and others, he repeated words not repeated by the poets, even as Britten unaccountably repeats words not repeated by Thomas Hardy and others. The resultant verses, sans music, read not like Hardy and Thomas but like Gertrude Stein. Stravinsky was a marvelous composer for the voice. But the marvel had nothing to do with the proper word setting, that is, concern for parsing, declamation, quantities, considerations aimed toward comprehensibility of a sung text. In a text, sound more than sense absorbed him, which is why he was happiest and most convincing in Latin. Since no one knows how that tongue was pronounced, he is free to let accents fall where they may.

RD Also in that lecture I think, you felt that *The Rake's Progress* collaboration with Auden was a tired pastiche.

NR Tired pastiche is excessive, I suppose. By the time 1951 rolled around, the year of *The Rake's Progress,* verbal intelligibility seems to have been his goal. It was his longest theater work by far, his first in English prose, and to my ear his most bloodless. To join the world's best poet with the world's best composer was a coup only in theory; in fact, poets do not the best librettists make; and Auden, despite his easy couplets, his "in" jokes and Dickensian nouns, was too respectful of his collaborator . . . worse, he succumbed to the amateur practice of combining words that are "helpful" to the composer. Stravinsky, for his part, astray in the jargon, despite some unexpected colors and a few touching airs, produced what I then thought was a tired pastiche, with such inadvertently burlesque word-stresses that each character seems to sing with a Russian accent! Today I find the music thrilling, energetic, while the libretto remains embarrassing.

RD Then you also touched on a topic with the question, "Is he original?"

NR The only reason I mention originality with Stravinsky is because he has been so influential himself that nobody ever talks about whom he stole from, and that's why I wanted to put that in. . . . That famous

*Petrouchka* texture, recognizable in a split second, of a pair of clarinets in close-harmony roulades, is actually cribbed from Ravel's *Rhapsodie espagnole,* which predates *Petrouchka* by three years. The entire prologue of *Rossignol* (1911) is almost a note-for-note steal from Debussy's *Nuages* (1895), which in turn comes too close for comfort to Moussorgsky's *Sans soleil.* The tenor solo from *Oedipus Rex* is a rewrite of Verdi's tenor solo *Qui Miriam* in the *Requiem,* while Jocasta's harangue is accompanied seemingly by Bartók's First Quartet, and so forth.

**RD** That leads us to robbing and stealing. You made a very spontaneous remark about how an amateur doesn't know what he steals, as opposed to a pro taking such pain and effort to conceal his stealing that it is a work of art . . . that the process is an art.

**NR** A lot of composers wouldn't agree; or if they did, they wouldn't put it into words. I get a little bored with grandiose statements of composers who say, "I don't like to repeat myself." No artist has more than four or five ideas in his whole life; and he spends that life trying to put those ideas into communicable shape, to hone them and boil them and distill them so that they can be projected from his concept into an audience's concept. Chopin never changed during his entire life, and neither did Ravel. They used the same harmonies, the same sequences, the same kinds of tunes. The only shifts they made were in format. They spoke always a consistent language. So, actually, did Bach and Beethoven, although they verged into peripheral dialects occasionally. Variety is a shallow virtue.

**RD** Well then, composers can have a kind of inherent sound by virtue of their harmonic thinking.

**NR** They can think of different ways in which to put their ideas. They can couch them in a cantata or in a woodwind trio. But to make singers improvise, as Lukas does, on a-e-i-o-u while they are standing on soapboxes is a device, not an idea. Nothing comes from nothing. Nothing is new under the sun. People who have the highest reputations for being original—Debussy, Wagner, Stravinsky—are the ones for whom it is easiest to find less-famous precursors. Spohr is less famous than Wagner, and more original. Debussy took ideas of Rebikov and Satie and made something of them, while Stravinsky took the ideas of anyone who happened to be around, the way Poulenc did. He was simply bigger, grander, and had more talent. But to think up new systems or even new languages is not in itself particularly commendable.

**RD** Well then, in that French area, what about Berlioz?

**NR** He doesn't interest me.

**RD** Is he French or German?

**NR** German, of course. The difference between French and German is the difference between superficiality and profundity. The Germans are deep. The Germans will belabor a notion until it's exhausted. "Da da da

DUM" is reiterated 572 times in the first movement of Beethoven's Fifth. They drive one nail as deep as it will go. They psychoanalyze. Meanwhile, the French are superficial in the highest sense of the word. They skim over the surface to invent Impressionism. Impressionism is the image of a water lily or a peach-cheeked infant glimpsed in one split second before the setting sun has moved ever so slightly through the willow tree.

RD And hence your attraction to the music of Debussy and Ravel?

NR Yes, I'm terribly moved by it. You hear about the French being cold and classical and objective . . . I never knew what people meant by that. For me, hot tears flow from out those icy sounds. They're not cold to me. Superficiality in its true sense is what life is all about. Most of our daily routine is casual, fragmented, perishable, mundane, but the years flow by imperceptibly. It is through such give and take that our life is run. Even with close friends, one doesn't too often sit down and discuss the meaning of the universe. That meaning is reserved for work. For Germans (like Berlioz or Elliott Carter) the work digs deep. For Frenchmen (like me or Virgil Thomson) the work spreads out. The French are not long-winded—but like cheetahs, they cover distance fast.

RD You mentioned Debussy. The notion that he ushered in a good part of the 20th century is fairly well accepted.

NR The claims that people make for Debussy are mostly not correct, namely that he was formally so very inventive. He opened the 20th century in the same way that Strauss closed the 19th (although he was older than Strauss). Strauss was not a modern composer. He made a lot of wild noises, but they're really 19th-century noises made with tonics and dominants and subdominants. The older he got, the more that way he got. Whereas Debussy was more ambiguous harmonically. His modernity is his emphasis on sound, sheer sound. No one hitherto, not even Wagner, was ever so centered on color as opposed to the other variables of music—harmony, counterpoint, rhythm, melody.

RD And formally? *The Afternoon of a Faun,* for instance?

NR *The Afternoon of a Faun* is a tone poem, but so are some of the Tchaikovsky pieces tone poems. I'm thinking for example about the piano Études. Everyone says, "Such innovation!" But of the twelve, only three are innovatively unusual. The other nine are not only four-square, they repeat themselves very literally, and Debussy does that frequently, even in *Jeux.*

RD I was going to bring up *Jeux.*

NR The overall form of that is free. But the melodic content is very square. Debussy will state a premise . . . for example (singing), and then he'll repeat it (singing). He'll repeat his idea, unaltered, twice, clear through a piece like *Jeux.* His germinal ideas, always very short like Beethoven's, are stated at the outset, and they are stated twice, like placing a pair of twins on stage. Then he'll make another statement, and he'll repeat *that* literally. Ravel, on the other hand, wrote long, long tunes, like Puccini. Debussy wrote short motives. *Pélleas* is not an opera of obvious arias. The reason people don't like *Pélleas,* if they don't, is that they are hearing it wrong. They are waiting for the big tunes, not realizing that what they think is recitative is really aria in microcosm. There is no utterance by any character in *Pélleas* that is not, even for its three or four notes, a complete tiny song in itself. Have you ever conducted it?

RD I've conducted the *Pelleas and Melisande* of Schoenberg, but that's a whole other thing.

NR That's a composer I don't understand at all.

RD Well, you've been very vocal about him from time to time, as you are with anything that isn't French.

NR I don't think that Schoenberg is a fake, nor certainly Alban Berg. I'm overwhelmed every time I see—or should I say hear?—*Wozzeck.* Nor Webern, although I don't understand him either. But I don't find the beauty or the charm. I do think Boulez is something of a fake. . . . What he is producing and what he would like to see produced doesn't make any difference. It's an intellectual invention, not the real thing.

RD And Messiaen?

NR Wonderful!

RD What is French there? The Catholicism, or what?

NR The French are economical, as opposed to the Germans. In French orchestration, and you as a conductor will appreciate this, there is nothing that isn't going to sound as it appears on paper, once you've rehearsed it. If the English horn is balanced by three cellos, you are going to hear three cellos and an English horn. In Strauss, everything is doubled and thickened. There is no transparency. Such a lot of tutti writing, even in soft areas, makes it German, overloaded with fat, as opposed to Ravel's scoring wherein the fat has been trimmed off. If you accept that definition with orchestration, then Messiaen, who does use huge agglomerations, madly thick harmonies; nevertheless, everything in the chord can be heard. And he usually doesn't double at the octave as Strauss does. Also, his tunes are French, like modal folk songs.

RD Even the birds?

NR Those birds are ornamentation to something underneath. I didn't hear *St. Francis,* but I do know there's one forty-minute section of nothing but birds twittering, which makes people want to go home and wring their canary's neck! There are only four major French composers of the last hundred years: Debussy, Ravel, Messiaen, and Poulenc. I do not include either Fauré, whom I love, or Boulez, whom I do not. Boulez has been a notable influence, but his influence will be nil in thirty years.

RD You've called France a very unmusical country.

NR France has produced great musicians; and she has produced pride about music. But she lacks a caring musical public. The French like to talk about music more than they like to listen to it. Still, they do name their streets after musicians, which is more than we do. Since Boulez is one of the most intelligent men in our world, since one cannot doubt his integrity (not to mention his sincerity), since at great cost to his government he caused the construction of IRCAM—that gaping aperture leading down to a maze of torture chambers beneath the Beaubourg Center—and since IRCAM is obviously very close to his bone (too close, indeed, for him to focus logically on other structures), it is tempting to suggest that if Boulez can't tell his ass from a hole in the ground, he's absolutely correct. You get what you ask for, and France has asked for Boulez. His *Répons,* in technical terms, was a disappointment to me, because of all that hype about the 4X machine. There was nothing in *Répons,* or in the piece that preceded it for solo clarinet, that indicated that IRCAM had gone beyond the experiments we used to do in our high school gym in 1938, through loudspeakers. I didn't even perceive any bending of tones, any alterations; but there were a lot of Folies-Bergères kind of gimmicks, through hidden microphones, so that the clarinet was heard playing four separate passages at the same time. IRCAM to me would be an expensive embarrassment if it weren't so cute.

RD While we're on the topic of French orchestration, what about the orchestration of Jacob Druckman? Do you see any French in it or in his music? Do you see any French in it?

NR I don't know who Jacob thinks he is.

RD Well, he has a certain affinity toward things French . . .

NR He lived in France. That doesn't mean he has an affinity toward things French. Except for Charles Rosen, none of the American musicians who lived in France, at least in my day, ever played French music at all.

RD Well, what of Jake's music?

NR I'm not against it. It is more complicated than it needs to be to make the point it makes, but it is attractive to the ear. The attraction comes from color rather than from form or content. His *Vox Humana,* for example.

RD That's a recent piece . . .

NR I've looked at it carefully, and I've listened to it thrice. It owes a lot to Orff, believe it or not, with that clangorous orchestra and shouting chorus. It also owes a little bit to, what's his name, Werner Egk as well. In other words, to Germany and Russia, not France. The music sounds basically tuneful; and when you look at the score you say, "Why is he going through all these gyrations for such a simple result?" If I were a conductor, I wouldn't begin to know how to conduct it.

**RD** Well, I've done a number of his pieces, and the tasks are not beyond the means of the performers. I think the way he puts orchestration together makes fascinating rehearsing.

**NR** I would say Jacob is a quite stylish interior decorator rather than an architect.

**RD** You said at one time that you hear tonally. Now if I mention three composers, let's just use that as a touchstone: Crumb, Cage, and Del Tredici.

**NR** I hear everything tonally. I would say that if I hear a piece of Crumb, I will hear a tonic or a subliminal pedal point over which the composer weaves his web. Once the piece has been woven, he'll remove the pedal—the support—so that it sounds modern.

**RD** And then Cage?

**NR** Well Cage is very tonal . . . all those pieces . . .

**RD** The early works?

**NR** The early works, the ones with Lou Harrison, are to my ears like Haydn. But his later music often lacks clear pitch. You've got to have pitch to have a subliminal pedal point. Without pitch—no music, at least for me.

**RD** That's one way, yes.

**NR** But if John simply has an imitation of the wind or something of that sort. . . .

**RD** Well, I remember doing an enormous piece called *Atlas Eclipticalis* in La Rochelle, France, and that was based on pitches in between pitches and all that. You mean that destroys the pedal too?

**NR** Oh, no. I think that I would hear that too, just as I hear Hába's music or Carrillo's, or Harry Partch's thirty-six-tone scale on the special keyboard. I still hear it as a seven-note scale, out of tune. Anyway, John is a master of show biz. His personality is far more powerful than his product. Of course, Bach's passions are show biz too. Let's say John is show biz without the biz.

**RD** And Del Tredici?

**NR** David Del Tredici, his most recent music is tonal as can be. Are you talking about the early pieces?

**RD** No, I think it's obvious he's dealing with a form of tonality and if it's not Strauss it's . . .

**NR** Just as I think that all painting is representative (there is no abstract painting, as I earlier said), so there's no such thing as concrete music. All music is abstract. . . . It could be convincingly argued that all music is tonal, and that insofar as a composer is a composer, he's dealing with the cosmological fact of inherent tonality. Music comes from tonality the way babies come from sperm.

**RD** What do composers say about your music?

**NR** I don't know. Elliott Carter would probably be diplomatic. Maybe not.

He did refer to Messiaen's work as vulgar and meant it negatively. Although to me, vulgarity is a prime ingredient, along with ugliness, of all great music. My God! Look at Beethoven. (A little vulgarity in Elliott might help him a lot.) It's not for me to know what other composers say about my music, but of course I am hurt if they don't like it. Still, music is written mainly for performers and listeners. Composers do want to impress one another, yes. But my career, such as it is, has never been supported by my brothers and sisters. It's been supported by performers, who are like cousins, and by the general public. Not that the public exactly clamors for it!

**RD** Just several last names, rather familiar ones. Barber?

**NR** Barber's hyperthyroid sentimentality was never my cup of tea; but when I hear his music, now that he's dead, I'm sometimes very moved. As a man, he had a love-hate affair with me. We were the only song composers for a long while, and I was younger than he. I'm impressed by all Sam's music except his songs.

**RD** Bernstein and Copland?

**NR** Underrated. At his best, there's nobody doing what he's doing. Lenny is worth twenty Sondheims. His non-Broadway element can be persuasive, too. The *Jeremiah Symphony* is in part quite overwhelming. He's a bit inclined, like Milhaud, to wreck everything that he does. Very few long pieces hold together from start to finish. *Kaddish* and *Mass* are half unique masterpieces and half blush-making junk. His Achilles' heel is his need to be hip. But Lenny is a major composer. He's underrated precisely because of his fame. Aaron's more than a mere landmark; he is the Rock of Gibraltar. His main influence on me is that he taught me not to be afraid of leaving notes out. . . . Of American composers over forty-five, I like David Diamond, Barbara Kolb, John Harbison, Lou Harrison, and Paul Bowles. I like Lucia Dlugochevsky and George Perle, and I like David Del Tredici, or used to. I don't know what he's doing now. Perhaps he's gone stark mad. Let's add my old friends Louise Talma and Miriam Gideon. I like Virgil Thomson the man profoundly, but I have mixed feelings about his music. I don't know where we're going, but the last thing I'll say is this: although we began on a pessimistic note, that the huge outlets for good music no longer exist, there are paradoxically more proficient young composers around than ever before. So God must want them. They are going to have to make their own rules, start their own publishing houses, their own recording companies, their own chamber music groups. Nuance and tragedy have been squashed by pop. Therefore, young, serious composers will have to find their outlet, and they will.

# Narrative III

The ICE concerts in 1960 with the New York Philharmonic and the Philadelphia Orchestra, which took place one week apart, marked my first trip to the East Coast. This, along with the subsequent national and European tours, brought me in contact with a very broad spectrum of contemporary music and composers, many of whom I began to meet personally. I felt as though, previously, I had only been looking in the "store window." Now was the moment to go in and "buy" something for myself. I think this was the real beginning of my own exploration into new music. I was totally fascinated by it all; but what was coming out of Europe in those days particularly piqued my curiosity. In the broadest sense, the '50s and '60s seemed like one gigantic "investigation" that encompassed all of the arts.

Because of a new exuberance, new emotions, the multiplicity and contradictions of artistic directions and questioning (from the *technical* to the *philosophical*), I would label the entire epoch a "romantic" hotbed. Not only was "today" being sifted and filtered before my very eyes, but both the immediate and the more distant past were also under a microscope. All of which left me, as a conductor and musician, more than challenged and perplexed. What "ground" was underfoot? Where could one step without stepping into something that one would learn to regret?

Music had become a labyrinth of expression and examination. I didn't become a conductor because it would be easy or comfortable to take refuge in a "cult of the new" at the expense of traditional music, which I loved. I became a conductor because I was overwhelmed by the problems of inventiveness that I faced. Also, I was certain that by beginning to solve these phenomenal problems, I would look more clearly at the repertoire of the past. Just playing the clarinet was not the answer, nor was just "specializing" in conducting the music of the eighteenth and nineteenth centuries.

In June 1–11, 1961, the First International Los Angeles Music Festival took place. There were more than a dozen composers present from a variety of countries, and they included Karl-Birger Blomdahl (Sweden), Werner Egk (Germany), Lukas Foss (U.S.), Blas Galindo (Mexico), Iain Hamilton (Britain), Roy Harris (U.S.), Kara Karayev (U.S.S.R.), Tikhon Krénnikov (U.S.S.R.), Darius Milhaud (France), Walter Piston (U.S.), and Igor Stravinsky (U.S.), who was to turn seventy-nine on June 17. With the exception of the two Russian composers, all the others conducted their own music. Krennikov was elected General Secretary of the Union of Composers of the U.S.S.R. in 1948 under Stalin. In 1961, he was also the President of the Music Section of Soviet Societies of Friendship and Cultural Relations with Foreign Countries. Such formidable titles, such average music! In fact, with the exception of Foss's *Time Cycle,* most of the music heard in the festival for the first time on the West Coast was extremely bland.

However, on June 5, Stravinsky conducted his Concerto in D for violin and orchestra, with Eudice Shapiro, violin, and once again the *Symphony of Psalms.* Only this time, because the ICE was performing *Time Cycle* on June 3, I was allowed to watch Stravinsky rehearse. There I saw what I had only read about before; namely, how Stravinsky dealt with the stylistic performance problems of his own music; how he insisted on articulation and rhythmic diction in order to gain nuance and syncopation; how he achieved articulation through separation; how he focused on one of his pet aspects of articulation—*sf* ♪ 𝄾𝄾𝄾 instead of ♩, which he maintained was something that he endeavored to teach musicians for over fifty years. What seemed astonishing was the ease with which he negotiated, even at seventy-nine, a series of variable meters with small precise gestures, while the orchestra exploded and erupted in complex and virtuosic passages.

Robert Craft conducted the remainder of the program, which was comprised of the American premiere of Schoenberg's *Four Songs* for voice and orchestra, op. 22; and *Die glückliche Hand,* which was receiving its West Coast premiere. Such an I.S./A.S. concert caused some conjecture about A.S. rolling in his grave, since it was said that the two men had not exactly gotten along in the past.

1961 was also the year that Lukas Foss was invited to direct the Ojai Festival, then in its fifteenth year; and to my great pleasure, he asked me to become Festival Coordinator. This meant budgetary planning, hiring of personnel and soloists, playing and conducting, as well as brochure and program development. We had been performing together for five years and had planned many interesting programs at UCLA, but this was my first attempt at running an important music festival.

In previous years the Ojai Festival had distinguished itself in many ways. Stravinsky had conducted *Les Noces* for the first time in America, and Aaron Copland was in charge of the festival in 1958 and 1959. The ensuing three seasons, from 1961 to 1963, were therefore a challenge for Foss as Music Director-Conductor and for me as Festival Coordinator.

Over these years, the repertoire ranged from Monteverdi, Palestrina, and

Bach *(The Passion According to St. John)*, to the present. In the first season, for example, "A Day of American Music" began at 11:00 a.m. with the music of Carter, Dahl, Kirchner, Ives, and Smit; and it ended with an evening concert featuring jazz and non-jazz. André Previn, Shelly Manne, and Red Mitchell improvised on Bernstein's *West Side Story* and the ICE performed improvised pieces. Previn also played the "Masque" from Bernstein's *The Age of Anxiety.* The final festival concert concluded with Stravinsky's *Pulcinella* (complete) and contained the music of Schoenberg, Webern, and Foss's nine-minute opera, *Introductions and Goodbyes* (libretto by Menotti).

As a festival innovation, a select group of outstanding young musicians from the universities, colleges, and conservatories of California were invited on scholarship to take part in "informal meetings" with renowned artists, including Gregor Piatigorsky, an Honorary Member of the Ojai Festival.

This innovative touch was extended to the second season when a group of twenty-two young composers and performers were again invited to the festival, which in 1962 had as a theme Mozart and Contemporary Music. Of all the composers of the past, Mozart seemed to keep good company with 20th-century music, or vice versa. "Four Days with Four Composers" was presented as a prelude to the festival, featuring Berio, Babbitt, Schuller, and Foss, in addition to music by Maderna, Nono, Ligeti, Cage, Bussotti, Arel, Davidovsky, Kagel, and Stockhausen. All-Mozart programs alternated with programs of contemporary music. A composers' symposium was held at midnight after an evening concert. Critic Alfred Frankenstein moderated. On the last program, the "Bedlam Scene" from *The Rake's Progress* and the U.S. premiere of *A Sermon, a Narrative and a Prayer* by Stravinsky were coupled with excerpts from Mozart's *Idomeneo,* and brought the festival to a fitting close.

By the completion of the third festival season in May 1963, I had conducted my first *L'Histoire du soldat* with outstanding Los Angeles musicians, including Eudice Shapiro, solo violin; and I was greatly impressed by two exceptional singers, Marni Nixon and Cathy Berberian, with whom I had worked. The lists of standard works and the contemporary composers performed were long. The Argentine-born composer Mauricio Kagel had given the U.S premiere of his *Anagrama.* Unbeknownst to me at that time, I was being musically prepared for many things that were yet to come.

I list certain features, programs, etc., not simply to document historically what occurred in Ojai but rather to emphasize what was awakening in me: namely, how to build a festival, how to build programs, how to present the old and the new, and the responsibility of considering younger performers and composers.

# *Mauricio Kagel*

BORN DECEMBER 24, 1931, BUENOS AIRES

**April 9, 1987, May 9, 1988** **Cologne**

My first talk with Kagel in 1987 was the day after speaking with Stockhausen and was at 10:00 p.m., the night before Kagel departed for the Zagreb Biennale. His study is extremely well organized—like a medieval library. Our second talk in the spring of 1988 was much more concise. When I left I asked about some flowers he had growing in his garden. He said, "They are 'don't-forget-me' flowers." I said, "You mean, 'forget-me-nots.'" He said, "Yes, that's the problem of translation." Left Cologne the next morning by train at 8:18 a.m. for Paris to speak with Pierre Boulez.

RD Can you speak about yourself as a composer?

MK I'd like to point out right at the beginning that composing, for me, is an all-comprehensive activity. Whoever has been trained and educated as a composer is conditioned to work with time—musical time, for sure, but this conditioning enables him to transfer his experiences with time to other media. Thus I have made films, and I am still making them: mounting pictures, arranging the sequence of scenes: all this has eminently to do with time. And I notice again and again that everything we do is dominated by rhythmic proportionality. And who would be better attuned to this than the composer, who had to learn from the beginning to deal with durations and rhythms? I wouldn't even claim that I am a film-maker, because I am not interested in having two professions. But as a composer I have worked at what I have called the painless extension of the métier—an extension into other dimensions, including, of course, the visual dimension.

RD You express yourself in a very practical sense about construction and proportion, etc. Could you talk about somewhat more subjective material, in terms of the contents of your pieces, or your compositional process? Does it involve subjectivity, and to what degree?

MK I don't think that anyone can be totally objective—I doubt that objectivity exists at all. At any rate, my profession (of composer) is based on a process of introspection. This holds true for all artists. I have to journey deeply into the "within" of me to discover how I can make my subjectivity "objective"—objectivity is attained at the moment of artic-

*Universal Edition*

ulation. Articulation always includes a degree of clarity, of communicated intelligibility: in this way the subjective becomes objectified. This most variable process goes on all through one's life. As long as one is very young one has a reserve of spontaneous power which distracts from this journey into the "within." It takes time to discover how to deal with whatever one has to say. I strongly believe in the almost proverbial "three periods" every artist has to go through. Only the mature artist has the courage to leave out the unnecessary, to concentrate on the essential, and this indicates an ideal degree of objectivity as well as communication.

RD This doesn't explain necessarily, Mauricio, how others look at a person's work. You obviously feel that objectivity and subjectivity (as realized in your work) belong to you, to your soul. And you are looking at it back and forth. When people, however, observe somebody's work, they might say: "The work has a subjective quality." Or, "it has an objective quality."

MK Objectivity and subjectivity, to me, are not such polar opposites; rather, they complement one another. I don't believe in perfect objectivity. Think of something as seemingly "objective" as railway or airline schedules—once you read them, they turn into mental phenomena. Numbers always veil intentions. And if music impresses us forcefully as "objective," it is nonetheless replete with subjective impulses and with feelings of subjective nature. Think of the famous "All the news that's fit to print" of the *New York Times*—but someone has to decide subjectively what's fit to print!

RD I would like to know your feelings about your particular sense of chronology, be it in your music or otherwise.

MK I have noticed that nobody really listens to music in a very concentrated manner, not even those who think of themselves as professionals. Our listening, in fact, is like the cicadic motions of our eyes, which locust-like jump around. We listen to music, but at the same time we are preoccupied with our inner world, perhaps with very trite things, the university, traveling, eating. Our thoughts go on, unrelated to the music to which we are listening. They never stop. The result is a fascinating mixture of life and musical perception, and this mixture is never the same. Sometimes we concentrate, but only for short moments. Suddenly there are holes, *lacunae;* then again one comes back; concentration lapses, and then one is "with it" again. All this happens very quickly. The new music of the second half of this century suffers—most fatally—from the lack of exposure. Most people hear new pieces only once. In this way no enriched memory can form. To give an example: we listen to a work of the past, let's say, Beethoven's "Eroica". Most people know this work quite well, or so they think, and they focus on those parts which they remember and love. Love and knowledge determine the perception of the piece. Knowledge in particular plays an enormous role. First hearing almost overtaxes the listener's capacities, because everything is new to him. He can't really listen in peace: he is often nervous, tense, or too curious. Only if the listener has the chance to hear a new piece several times—which is rarely the case—can he combine the remembered with the new, and this leads to understanding in depth, to "classical" listening. One wishes for this situation: tired listening is unhelpful to the cause of new music which is often rejected because people literally don't know it. Apparently these remarks have nothing to do with my way of composing, but often I think about the entire context, and my music reflects some aspects of it. For instance: I have now composed my Third String Quartet—for the first time a four-movement work. And I am using shadow-reflections of classical string quartet forms—structural props, as it were, isolated elements of scenery which I tie together with the help of themes. It is an attempt to integrate the process of desultory listening into the structure of the piece. I haven't heard the quartet yet, but perhaps this work will show me that I progressed in fusing the remembered with the new.

The subtitle of my piece ("In Four Movements") shows me now how profoundly I was engaged in adapting the multi-faceted form of the string quartet to my purposes. Writing for the classical ensemble of the string quartet is still a challenge! One cannot deny that it has become accepted practice during the last decades to write music without movement signatures or movement profiles. I, on the other hand, think it exciting to organize and construct a coherent movement. This enables me to avoid those questionable transition periods which are found so frequently in music that tries to achieve the semblance of unity over long stretches.

The thoughts expressed by sound have an existence of their own: they don't depend on music for their realization. But we experience them through music, and that, in our memory, endows them with an anecdotal coloring. I think we need music which can prove its independence by its inherent abstract nature. The string quartet is a perfect tool for the careful testing of musical ideas; it enables us to auscultate them, to exhaust all their possibilities. First and second themes, development sections, and recapitulations might factually and realistically evolve, but they might also be mere shadow images of their original functions. But the poetics of the form have always to remain in focus, even in instances of "form-semblances" playing the role of mere stage-props.

RD This particular principle of chronological displacement—do you use it also in your instrumental theater, or in your music theater?

MK I have tried in various ways, utilizing varying approaches. For instance, in *Aus Deutschland*, an opera based on German texts which Schubert,

Schumann, and Brahms also used. This was intentional and therefore the subtitle of the work is a *Liederoper*. The idea was to set these texts to music again and also to *stage* them. The words and situations described in the *Lieder* become living theater. Liszt called Schubert's songs "operas in miniature" and recognized the dramaturgy of these songs. The combination of something already well known with the unknown evokes strongly conflicting feelings in the listener: he hears texts of *Winterreise* and *Dichterliebe;* at the same time he sees on the stage the characters as they are outlined in the songs. This combination is exactly what I wanted to achieve. The listener remembers, and he also faces the confrontation with the new. What I offer him forces him to redefine and to complement and expand what he knows. I composed the opera ten years ago, much earlier than the Third String Quartet. But it was also an attempt to operate with musical culture in a stimulating manner. All of us, those who listen and those who compose—we all carry within ourselves a reservoir of fragments of musical culture and a repertoire of pieces we love. No matter if we have a general orientation or if we are specialized, perhaps in being one-sided musicians; we all compare that which we know with the new "input." Nobody can ignore his musical memories. It is not merely comparing interpretations of the past and today—we perceive at once the differences in interpretation and work.

Culture was never chronologically determined by *use* or experienced in a linear way, step by step. The same is true for music: there is no need to begin with Perotin and end with "now," in order to understand better the musical process. We all have jumped constantly within and between the centuries. And the sum total of all these experiences is a non-linear, a-chronological musical culture. This kind of musical world is perhaps the one in which we feel most comfortable being able to be entertained by an enriched experience. Misunderstandings and conflicts are generated often because people erroneously believe that the music they know is identical with music history as it happened. There is no chronological understanding of music history or objectivity in judging what was. There is only a subjective concept of the past, of the musical events of the past. By the same token we cannot judge what is history today. I hope we go on jumping for a long time. . . .

RD Mauricio, is this chronological aspect of history susceptible to analysis?

MK This depends on what you want to analyze, and which chronology you have in mind. Every piece of music has its own natural chronology, related to its overall form. It surely would be desirable to hear a piece of music in this manner. But, the sequence of before, now, after, as given by the composer, is not really understood and worked out by the listener—why not? Because, as I said before, the listener is distracted (held up, as it were) by music which holds particular interest to him;

partially because of the impact of what he heard before, he listens less concentratedly to the rest. This happens so often when we are listening to classical music; let's say, to the Adagio in the Fourth Piano Concerto of Beethoven: we are so deeply moved that we are almost not ready for the Allegro when it comes. A part of the Allegro gets lost; we are not attentive, "not quite there," as it were.

Analysis! We have to differentiate between various forms of analysis. It's essential to know what one expects to *find* when one analyzes. During the last two or three decades a cult of analysis has sprung up which aims at changing it into a way of composing. *What* one analyzes seems less important than *how* one analyzes; which procedures are used, which logical processes, which subterranean connections are to be unveiled. Analysis as a substitute for composing does not strike me as quite legitimate. Sometimes the results are evident only on paper; they cannot be heard or sensually perceived. Analysis which has emancipated itself from its object cannot be called of interest for composers and therefore used legitimately. Webern's music at nine o'clock in the morning does not sound the same way to us as at eleven o'clock in the evening. Musical analysis claims that chronology maintains itself independently, without change, and is part of the "objective" world. One forgets that people are exposed to an infinity of influences which change and vary their perception (and the intensity of their perception) of music.

RD Could we reflect on the post-World War II period called the Darmstadt era? You taught summer courses there. There was an analytical attitude in those days in Darmstadt. Is that correct?

MK Of course. And it was very much what I tried to describe to you. Some of the neurosis of the analytical approach towards music originated and developed there. Also the strange need to analyze the inaudible. In their field there is today really no difference between Europe and the USA.

This might be the explanation: one tried, at that time, to reformulate and to expand the fundamentals of craftsmanship. That made it necessary to talk about technical questions. Questions which emerged out of the confrontation of the composer with notepaper, not with the practical realization of musical thoughts. There are explanations, which became examples of historical importance, such as the analysis of Schoenberg's *Gurrelieder* by Alban Berg. Many teachers today will not accept this kind of "naïve" approach. The idea of enhancing listening experience by means of explanation goes back to the German concert guide. The tradition of the concert guide came into being during the 1830s and 1840s; it's alive in Austria as well as in Germany. Do you find comparable booklets in America? The concert guide assembles knowledge about the compositions, it explains what happens thematically and structurally. Alban Berg did this with the *Gurrelieder,* and he has shown how the themes are interrelated, and everything is unified and connected. Schoenberg dealt with his Kammersymphonie in the same way in *Style and Idea.* All these analytical efforts try to prove that musical masterpieces do not contain meaningless, isolated elements; all the details, no matter how disparate they appear, are connected and form a closed universe.

Fortunately, most of this indigestible analysis today is hardly ever read. Dissertations have to be written, musicologists have to keep busy, musicological journals need articles. Remember the American "publish or perish"?—this keeps musical analysis alive, on our side of the ocean as well. But: I prefer to perish.

RD In a general historic way, how do you look upon Darmstadt from the distance now?

MK It was important and exciting for those *who were there.* It is easy to judge something negatively from the distance. But those who were in Darmstadt during the important years really experienced what the Chileans call *"olla podrida,"* a melting pot. And then historical logic imposed the moment to finish a golden era—it is correct in that way. All important cities in music history became focal points only for a definite period. Like Vienna from 1780 to 1880—for a century an essential place, thriving also because of its immigrants. Neither Beethoven, Brahms, nor Mahler were Viennese. Musical life in Vienna was determined almost exclusively by non-Viennese people. The people in Berlin have a saying: "the best Berliners are the Poles and Saxons." This was the extreme case with Darmstadt. Today Vienna's contribution to the world of music is insignificant—it reproduces: that's all. Of course there are composers everywhere, but Vienna is so much less important than Paris or London. Interestingly, these musical centers were not always capital cities—think of Mannheim during the Mozart era; much happened in Hamburg or Munich and now in Cologne. We should not judge Darmstadt by what came afterwards; one has to look at Darmstadt as it was during the period when it prospered. You cannot hold the Bauhaus responsible for what came afterwards, even if this is, to some extent, the historic truth. But the Bauhaus was immensely exciting as long as it functioned.

RD I would like to remind you of the Ojai Festival in 1963, when the U.S. premiere of your work *Anagrama* for four singers, speak-chorus, and chamber ensemble was given. You conducted.

MK Ojai was a very pleasant experience. It was the first time I gave a concert in the American countryside. We worked hard and it was a good performance.

RD *Anagrama* was one of many works widely discussed at that time as a work of great innovation, especially in the area of using language—the word spoken or sung. I couldn't imagine anything more advanced at the time you wrote it, which was 1957–58. Did this precede some of your colleagues' work in this direction, let's say in Darmstadt, or was it on the same rail?

MK I was interested very early in language as a sound source and wrote most of the piece in Argentina, before coming to Europe. At the same time I was fascinated by the semantics and complexity of words. Language as a material of the same importance as pitches: clusters of consonants, tone-rows of vowels, etc. As a matter of fact, I studied literature and philosophy regularly in Buenos Aires; and read the palindrome on which *Anagrama* is based in 1953 in Argentina. A palindrome is a sentence which can be read in both directions and you have the same words: IN GIRUM IMUS NOCTE ET CONSUMIMUR IGNI. English translation: "We circle in the night and are consumed by fire."

RD What source did you find that in?

MK It was an article about the Middle Ages and Renaissance mannerisms in literature. The author of this paper said the palindrome was by Dante, from the *Divine Comedy.* But in the middle '60s I read it was not. In any case, the *Divine Comedy* is a marvelous example of the use of hidden meaning by such symbolic techniques. I was immensely touched by the idea of composing a piece on such a magic sentence. The original palindrome is a very special version of an anagram.

RD May I read the program note you provided in 1960? "The spoken and sung texts in my work have been formed from the vowels and consonants of this palindrome (anonymous from the Middle Ages). Hence the title *Anagrama.* The four languages used are German, French, Italian, and Spanish. The Latin appears only in the original sentence. Utilizing the literal translation and an artificial language, I have created several metamorphoses, some derived from the meaning of the palindrome and semantically correct, others constructed from words whose letters are borrowed from the palindrome. Far from being variations on the text, these metamorphoses are autonomous forms. As they are superimposed, they begin to 'relate.' The words and music are allied by a vocabulary which suggests a parentage of categories and common meanings. . . . To 'understand' speech is still the most appropriate form of communication. One could also understand the spoken in the music when mankind begins communicating logical meaning only through sung words—Utopia?"

MK There are various categories of anagrams. For example, this palindrome is an anagram because you don't change the order of the vowels and consonants. There are other kinds of anagrams. For example, with vowels and consonants of a sentence, you can build up other sentences. You have the same number of consonants and vowels, but instead of six words, you can make a sentence with nine words. These techniques are very old magic techniques related to theology, to alchemy, but also related to music.

RD By virtue of what? Their code? By virtue of their meaning? Because we're avoiding meaning here.

MK No, we're not avoiding meaning. I think there is a constant in the human thought—a constant manipulation with the logic of the illogical. Or the meaning of the meaningless. I am not the only composer who has reflected on these things. I remember Anton Webern with the "magic square."[1]

SATOR
AREPO
TENET
OPERA
ROTAS

Translation: "The sower Arepo keeps the work circling." This is a famous anagram which very much influenced his thought and inspired him to build series, chords, etc. He was interested in this topic, of course, because he studied the music of the Middle Ages and Renaissance.

RD There are a number of composers who continue to be interested in various musical developments of the Middle Ages; for example, the structures of isorhythmics. Why are we going back? What principles did these people have that were so important in rediscovering that a 20th-century man would want to return to? I expect a 20th-century man maybe to discover the moon! Why this return, Mauricio, to what *was*? Why is that so fascinating?

MK I don't think this is a return, but on the contrary, an expression of continuity. I don't feel that we are in an exceptional period of music history, because we dispose of a great number of heterogeneous elements which can be used in any possible way. Richness or poverty: that's not the question. Composition today has not become easier than in the Middle Ages, without electricity or gasoline! I'm not prejudicial about our time; but I really see no difference between using isorhythmics and operating computers. It remains a matter of quality and goodness of music. Computers are interesting as long as the composer is exceptional. When the resulting music is of average or poor level the computer can be understood as a tool even more "old-fashioned" than an isorhythmic motet.

1. *Anton Webern: The Path to the New Music*, edited by Willi Reich (Bryn Mawr, Pa.; Universal Edition: London, Wien, 1960), pp. 56–57.

RD What strikes me now is that some people say that as a composer, no label fits you neatly. You make music; you make films; you make radio plays. You are a theorist and a philosopher. Your works vary in their thrust. They vary in their framework of organization, theatrical or not. Sometimes the instrumentation is weird. . . . I don't know how many composers are writing for a group of zithers or a group of accordians or strange folk instruments or street organs. And it all ties in with the little bit of what I know of you. I understand many of your musical perceptions and you have used material from these perceptions that ranges from humor to God!

MK There is perhaps no difference between humor and God. One of the most extraordinary sects in Jewish religion—the *chassidim*—try to get very close to God by joy and laughter. Humor is absolutely essential to be serious. And I'm very suspicious when somebody defines himself as a serious personality. I don't think this is always true. Buddha laughed all the time. He had arrived at that high point in which laughing and smiling are the evidence of happiness and of perfect balance.

RD You have this enormous perception of humor. I see it in the way you talk and think. But then, I see it in the works, too. There it almost appears tragic at times, which is about as serious as you can be.

MK There is much misunderstanding about humor. Humor is not only irony or sarcasm but rather a mixture: a very special way of understanding and misunderstanding, of contradiction and affirmation, and of the specific comedy in so-called "normal situations."

RD Could you define your musical genetic code?

MK I can speak about some principles, but not about *the* genetic code. One principle is: I am not afraid of adding apples, chairs, and shoes, working with very disparate elements in trying to accomplish a unity. From the early beginning I saw no aesthetic difference between pure noise and pure sounds. I do not reject something because it is not pure or is not "beautiful" enough. I don't think something can be so ugly that you have to reject it. Ugliness as a definition does not exist for me. I open my mind for things which perhaps other composers reject because they don't fit into their composition code. The catalogue of my pieces is rather heterogeneous because of that.

RD Last night I asked Stockhausen what his musical genetic code was, and he said that he invented it. In other words, he formulated what he uses as his genetic code. I said, "It has nothing to do with past individuals or past times?" He said, "No." Now where are you?

MK I will say . . . I created my code in reflection of the experiences of the past.

RD That's already allowing the past to enter your life. In his case, he didn't allow that.

MK If I tell you that I am the first composer of music history and the last one, I am lying! Do you remember Mark Twain? "There are three ways of lying: lying, big lying, statistics." Forgive me if the quotation is not textual. My memory loves to compose.... Neglecting the past, or neglecting the sources, is very easy. Much more difficult, really, is to allow them. I am someone who was born in the early 1930s. That means that when I was fifteen or sixteen, at least two very big revolutions in contemporary musical language were accomplished: first the Impressionist one, and second, the Expressionistic one of the Second Viennese School. Both are very important for me. By the time I was fifteen, in 1946, the fight was Expressionism versus Neo-classicism. Of course, I was involved in this aesthetic battle like many composers around the world. But at thirteen, I invented my twelve-tone theory. It was perhaps primitive and naïve, but it was already a serious attempt to serialize my material.

**RD** What is it that allows composers to discover these things? At one point here, at another point there ... independent of each other. It's "in the air," and yet you wouldn't have discovered it one century earlier.

**MK** This is something which also fascinates me. Say, every twenty years there appears a new generation of composers. They don't have the possibility of knowing all the music of the past; but they feel the necessity of expressing themselves. And the continuity is not based on deep knowledge, but on an intuition of what to do at a certain moment in music history. We are witness to a secret mixture of the rational with the irrational and a strong sense for the further evolution of musical language. A quite extraordinary mechanism, which is absolutely unpredictable.

**RD** Could we touch upon what you are composing now?

**MK** It's really difficult to say, because I am always writing simultaneously two or three compositions. I have done this since I discovered that in this way the pieces turn out to differ sharply, and I don't like to repeat myself. If one works at an instrumental piece in the morning and at a theater piece in the afternoon, one is compelled to observe strict demarcation lines, but one also creates a dialogue. As things are, I should talk to you about the piece which I finished just now. The piece is entitled: *Quodlibet* for a woman's voice and orchestra. I am using French texts, chansons from the 15th century.

During a concert of old music I came for the first time upon the old Spanish term *ensalada* for "Quodlibet"—a term doubtless apt to stimulate one's appetite! The corresponding French period term, *fricassée,* on the other hand, has no vegetarian connotations, and the German *Bettlermantel* suggests an impoverished patchwork of melodic fragments and textual choice items—nothing "culinary" at all! It is known for certain that the Latin term *quodlibet* (whatever you like) entered musical nomenclature around 1550, and that the *quodlibet* is the oldest form of musical collage. Research has found a direct connection between the *Disputationes de quodlibet* at the Sorbonne and "mischmasch" as musical form. I find it exciting that the shaping of musical thought was, as so often, influenced by an extra-musical model—and that, in this case, it took its cues from rhetorics. Renaissance composers created compact anthologies of well-known melodies. One of the most astonishing examples—anonymous, of course—unites fragments from thirty-nine chansons within only fifty-eight bars—a "ragout" in a four-voice setting, an aural *Reader's Digest.*

For my *Quodlibet* I used exclusively French chanson texts from the 15th century. However, I did not use original melodies. For the textual montage I availed myself of two sources. In terms of origin and character the chansons are either "courtly" or popular. But the borderline between the two is in flux. Contents and themes are not interchangeable; while they appear and reappear all the time, they are also constantly varied. The Renaissance affords generous space to lifestyle and custom, to questions of what is "seemly" and "unseemly," to love as etiquette and measuring tool of politeness, to chastity and its opposite, to everything which has to do with the business of the heart. We recognize relics of the love poetry of the troubadours in the way sensual feeling is presented, in the description of sensuality contained by virtue, in personal life histories. Politics and war are not excluded. They set the scene of the knightly life.

I have tried to illuminate in the vocal parts some of the ambiguities of the texts, redeeming them, as it were, into outspokenness; on the other hand I interpreted instances of drastic licentiousness in a "broken" and many-layered manner, making them in this way "ambiguous." For this reason a woman interpreter would have to assume occasionally a masculine voice. Clearly delineated monologues change imperceptibly into dialogues. One of the most striking characteristics of these texts is the sincere—sometimes also distorted—need for undisguised, candid statements. Yet they stay away from the confessional box. In this kind of poetry the partner, distant or close by, is mute but always present. This role then with good reason is given to the listener, who should not be left excluded like an acoustical *voyeur.*

A short while ago I wrote *Dance School (Tanzschule),* which will be performed in Vienna in September 1988. This composition is based on Lambranzi's textbook, *Neue und curiose theatralische Tantz-Schul,* which for the first time appeared in Nürnberg in 1716. This Venetian dancing master introduces us to about seventy different ballet scenes or ballet situations, which are supplied with short commentaries, pictures, and the corresponding dance tune.

The pictures and melodies of this collection form the starting point of a *ballet d'action,* which consists of eighteen numbers. The finale reverses the sequence of the motives of the preceding seventeen numbers: 17, 16, 15, . . ., 1. Thus "Entrée" (No. 1) becomes "Sortie."

In my own way, I have tried to be faithful to the original concept of the *Tantz-Schul,* and I don't think it necessary to try for historic authenticity in the execution of the dances. Lambranzi stresses the elements of theatrical dance and "scenic" significance inherent to any kind of human action: this makes his work seem to be new and attractive even today. Court ballets are analyzed and valued in the same way as folk and peasant dances, and the characteristic stylized gestures of craftsmen and artisans interest him as much as danced society games in bourgeois surroundings. We meet, in his *Tantz-Schul,* the classical characters of the commedia dell' arte (Pantalone, Pulcinella, Scaramuzza, Scapin, and Dottore); besides, he familiarizes us with the entertainers of the period: fools, magicians, exorcisers, conjurers, gypsies, necromancers, jugglers, men with two faces. There is the satyr next to the reeling peasant-drunkard, the Turk next to the fastidious courtier; ship-captains, Bacchus, hunters, soldiers, sculptors, Moors, gardeners, people engaged in racket games, slaves, masked ghosts, tumblers, shoemakers, witches, coopersmiths, and an untold number of others.

Anything that surrounds us can be turned into dance and scenic events. This book represents the first systematic attempt to achieve an all-comprehensive dance theater. In consequence, one might consider it to be a *guide to a total collage,* with frame structures so porous that further discoveries (from the immediate surroundings, from history, fiction, literature) can be inserted, "mounted in." The renovation, rejuvenation of the material, in this instance, is part of the formal principle of invention.

The choreographer of this newest *Tantz-Schul* has a difficult task. On the one hand he will be preoccupied with 17th- and 18th-century figures which, although constantly varied, have never ceased to enrich the repertory of mimes, actors, and dancers. On the other hand there is the challenge to color and interpret present-day actualities in new ways and provide them with new connections and ties. Present-day figures, professions, and models might be used, as well as new sports, characteristic dance impersonations, roles from the film or the theater. This is like a kaleidoscope of the typical which ties together that which is timeless with present day actualities: a theater of danced actions with independent props, with many strains of drama and also with humorous pranksterism: a dance piece the subject of which is the dance. I am creating a kind of ambivalence—and, I believe, that ambivalence is really one of the principal themes of all my work.

RD You see, I asked you before about your musical genetic code. What is in your blood? Ambivalence . . .

MK My blood is black. I am the son of a printer, and my blood is ink.

RD You're getting very cryptic.

MK And when I have no pencil, I can write with my finger.

RD I see. It's cheap.

MK It's indelible!

# Narrative IV

Shortly after the Ojai Festival in May 1963, I was invited to become Assistant Conductor of the Buffalo Philharmonic. Again, the invitation came from Lukas Foss, who had just been appointed Music Director. In the four seasons of my stay in Buffalo, 1963 to 1967, I participated in and helped develop what could be called a pioneer model for the presentation of the arts on a regional level.

The Buffalo Philharmonic under Foss provided an exceptional base for programming the old and the new. Many American and European contemporary composers were performed. Many composers appeared in Buffalo, sometimes conducting their own work. For example, Karlheinz Stockhausen conducted the North American premiere of his *Momente,* with Martina Arroyo (March 1964). My own responsibilities included subscription, youth, and pops concerts, which broadened my orchestral repertoire considerably.

Along with all of this, something very innovative occurred in Buffalo during the Foss regime. In October 1963, a proposal for the establishment of a Center of the Performing and Creative Arts at the State University of New York (SUNY) was drawn up and presented to the Rockefeller Foundation for support. The proposal was funded, and as a result an international group of composers and performers was formed in the fall of 1964.

This group, called the Creative Associates, was made up of brilliant performers; and some of the composers in residence over a three-year period included George Crumb, Richard Wernick, Sylvano Bussotti (Italy), Cornelius Cardew (England), Vinko Globokar (Yugoslavia), Carlos Alsina (Argentina), Niccolo Castiglioni (Italy), Fredric Myrow, Don Ellis, Stanley Silverman, Frederic Rzewski, and David Tudor. A concert series, Evenings for New Music, was planned for Buffalo and New York City.

In the spring of 1964, I found myself organizing the roster of participants, corresponding with composers all over the world regarding possible premieres and other suggested repertoire, and ultimately making a survey of concert halls for a New York home for the "Evenings." We decided on the Carnegie Recital Hall; and the series, in conjunction with the Carnegie Hall Corporation, was launched in New York in December of that year. The program: Cowell, Mayazumi, Myrow, Xenakis, Kontonski, and the New York premiere of Brown's *Novara.*

I conducted the group in addition to my position with the Buffalo Philharmonic. All of this activity reached its peak in the Buffalo Festival of the Arts Today, February 27 to March 13, 1965, when the various components were pulled together. The Philharmonic performed the music of Feldman, Varèse, Boulez, Penderecki, and the world premiere of *Traces* by Berio. Evenings for New Music presented Babbitt, Pousseur, Cage, Ligeti, Foss, Kagel, and *La Passion selon Sade* by Bussotti. John Cage, Merce Cunningham and Dance Company were present, as well as poet Gregory Corso, painter Larry Rivers, and art critic Harold Rosenberg. The Albright-Knox Art Gallery held an exhibition: "Art Today—Kinetic and Optic." There was a film series that included Andy Warhol's *Eat,* and four plays by Eugène Ionesco were given their American premieres: *The Leader, L'Impromptu, Foursome,* and *Bedlam Galore.*

In May 1965, we took the Buffalo Orchestra and Evenings for New Music to Washington, D.C., to participate in the Inter-American Music Festival. I conducted a program of world premieres by South American composers: Santoro, Tauriello, Gandini, and Paz. The "Evenings" produced the music of Orbon, Shapero, Kagel, Foss, and the first Washington performance of Crumb's *Night Music II.* Buffalo became an artistic "crossroads" for the new. My involvement with new music intensified as I conducted and experienced a plethora of compositional styles, approaches, and techniques.

In my first season with the orchestra, SUNY Slee Professor of Composition George Rochberg shared the podium with me in conducting his *Time-Span.* Later that first season (February 1964), Buffalo had the good fortune of receiving an illustrious, unexpected guest: Maestro Leonard Bernstein came to conduct his Symphony No. 2, "Age of Anxiety" (Foss—piano), on an "All-Bernstein Program." Foss conducted the Overture to *Candide* and *Symphonic Dances* from *West Side Story.* I conducted *Songs* from *The Lark,* a premiere.

# *George Rochberg*

BORN JULY 5, 1918, PATERSON, NEW JERSEY

**April 9, 1986** **Newtown Square, Pennsylvania**

RD George, I think we met for the first time in Buffalo, New York.

GR That would have been in the first half of 1964. I was Slee Professor at the State University of New York (SUNY) for one semester. I think I had all of two students. Of course I worked, but also went to a lot of dinner and cocktail parties! I also gave three lectures, as I recall, and a number of concerts. It was a very strange time for me.

RD Although I saw you on occasion after Buffalo, I really didn't conduct your music until you were able to come to Aspen in '72.

GR Yes, I was there the same year that Roger Sessions and Jacob Druckman were there. I'm glad to be reminded.

RD Do you recollect that stay in Aspen at all?

GR Certainly do. That was a very pleasant time . . . the preparation and the performances of *Music for the Magic Theater* and *Tableaux*. Remember that? *Tableaux* is a piece I still retain a certain fondness for. But I have to confess there are things about it which distress me. Not because they are badly done; but it's what's being said. I'm referring specifically to the absolutely violent moments. . . .

RD When I met you in Buffalo you were involved with the control of serialism. I've asked a number of composers about this . . . how they arrived at it, why the attraction, and had it anything to do with post-World War II?

GR For me very much, because one of the most powerful impulses toward twelve-tone, serialism, whatever you want to call it, was my reaction to my war experience which began to take over after the war. You see, I fought as a second lieutenant and was wounded and all of that . . .

RD In the army? In Europe?

GR In the army. I was in France, I was in Germany, I was in Belgium, I was in Luxembourg and all of the last business of the war had to do with the Battle of the Bulge. I wasn't in Bastogne; but the Third Army, Patton's Third Army, was brought up from the south to begin to break the hold the Germans had on that situation. What I experienced . . . in terms of my wound, came earlier, after we came out of the hedgerows on the way to Verdun. As a matter of fact, do you know what I found

*Associated Music Publishers*

out recently? I was wounded just outside a town named Mons. That was the town Josquin des Prez was born in! Isn't that crazy? I didn't know it then. At that time I hadn't even heard of Josquin des Pres. That was in 1944, September of '44. . . . After the war, and after I began to feel I had my feet on the ground musically, and in other ways too, the drama, the darkness of that whole experience . . . really had rooted itself. It didn't show itself right away, but it started to make demands on me emotionally. And that's what started to push me into a kind of atonal world. . . .

RD Really? You felt that consciously?

GR Yes, it was very, very powerful. In other words, I didn't adopt a severe chromatic palette out of any intellectual or musical interest per se. It came out of a deep emotional need to express what I felt had happened, what I'd been involved in and what it meant to me. And so, that all began to emerge in the First Symphony. By then I had already had my first experience with Schoenberg, which was very traumatic. It came in the form of his Fourth Quartet, and I had a real hate-love experience with that work and with Schoenberg. But I sensed that I had to go that way. Only years later did I realize that there were only two or three ways that a composer could go in relation to Schoenberg. It was as though you're traveling in a huge land and you come to an enormous mountain range. What do you do? You either tunnel through, and that takes forever, or you go over it. If it's not small enough, you can't go around it. Now you can avoid any of those actions, you can just simply stop and pitch your tent and just look at that mountain forever or move away—just remove it from sight. I climbed over it, to use a simple metaphor, that's what I did. So that from about 1948 to 1951, I really wrestled with the whole question of how do you make this kind of music? Because what Schoenberg was doing in actual musical terms: melodically, harmonically, not so much rhythmically, but melodically and harmonically . . . a total mystery. I could hear it, I could sense the intensity of it. The work that really convinced me, the thing that really pushed me over the line, was the Violin Concerto. That's the work that I really felt was a burning, intense, hyper-emotional piece. And I loved it for that. My reaction to music, then and now, is still very much related to my perceptions, my emotional responses. The intellect comes in later. It's always a matter of trying to figure out what this all means in itself, what it means to me and what I'm going to do about it or with it. And so, I think it was the experience of the Violin Concerto that really convinced me that Schoenberg had something important for me and for my time.

RD What was your reaction to the early Schoenberg piece, *Pelleas and Melisande,* that you heard me do with the Philadelphia Orchestra recently?

GR Well, that's interesting. You know, we all have our particular and peculiar characteristics and preferences. Mine as a musician are really quite simple, but really blunt. If I'm attracted or drawn to a piece, or if I dislike it intensely, but think there is something there that I have to know, I will study it. And I've done that with a lot of stuff that I didn't like before I studied it and liked even less after I'd studied it. With Schoenberg, it was a question of grasping the world he had made for himself. For some reason, which I can't explain to you, when I first heard *Pelleas,* I didn't really respond to it. And it's curious . . . I heard it recently on the radio, and I still don't respond to it. I like the sound, in a general way, but that's not enough for me. I need ideas. What draws me to a composer, what draws me to a piece, is the nature of the ideas. Now the ideas could be melodic, they could be rhythmic, they could be textural, they could be coloristic, they could be harmonic, what have you. But if they are not specific enough in the sense that they really attract me, I hold back. It's for those reasons, basically, I've never studied that piece.

RD In the case of *Pelleas,* it's very leitmotif oriented. It's very Wagnerian in that sense; and I suppose if the musical "psychodrama" doesn't appeal to the ear or the story isn't of any interest the piece seems a vagary of some sort.

GR Well, I'm not so much involved with the story as with the question, what is there musically? It's the Wagner side of it that repulses me. I've always thought of Wagner as a fantastic composer. . . . I recently suffered through *Lohengrin* when I saw it on TV the other night . . . I cursed him from the Prologue right through to the end, but the guy's fascinating! The thing that always puzzled me about Schoenberg was that he had a profound relationship, a human, personal relation with Mahler. I suppose he knew Strauss. He would not have known Wagner . . . maybe he knew Bruckner, maybe he didn't . . . of all of those composers, the one that he did not seem to take into his musical world as a composer was Mahler. He couldn't compose the kind of tunes that Mahler could . . . he didn't enter that world. But he did enter the world of Wagner, the world of Strauss, and, by association, the world of Bruckner. It's paradoxical. The one he seems to admire and love most is Mahler, yet he doesn't enter his world. You can't hear Schoenberg's music and hear Mahler. It's very odd, a very contradictory thing, very anomalous.

RD Some of the European composers, one generation or so younger than you, that went through the Darmstadt experience have felt after a while that the administration of Schoenberg's technique, the twelve-tone carried to extremes, left them a little too cold to use. These people I'm speaking of wanted to have a system that is outside the composer, so to say, so they can use it—but not so that the rationale of number and tone predominated. Did that ever repel you?

GR This is what I wrestled with for years until I was forced to give it up. You know, in 1955, I discovered for myself, in my own clumsy, non-mathematical way—because I'm not mathematically oriented—the secret of the hexachord. Now that, for me, that was a tremendous event; because during those years, I read everything I could get my hands on. There was not very much published about the actual, theoretical side or the manipulative side, the rational side of how Schoenberg made the relationships; just the business of one through twelve or zero to eleven, but that was in itself very simple, not very complicated, not very interesting. When you look at a work like the Trio, the very last works, a work like the Fourth Quartet, the Violin Concerto, he's working with hexachords. I don't think he knew a hell of a lot about the geometry of it in the sense of making musically spatial relationships: the tune—the first six notes ... the accompanying structure—the other six notes. Then you start playing with these possibilities. I discovered it for myself. So I published a little monograph called "The Hexachord and Its Relation to the Twelve-Tone Method." What I never published, of course, was a whole set of other ideas which came rushing out, all within the space of a couple of weeks, which had to do with the harmonic side of it. All I published was the basic data ... it was like a manual. I was saying, "This is a problem I had to solve. I'm laying it out for you so that anyone who now wants to deal with a hexachord, which is a much more subtle, sophisticated way of dealing with the whole twelve-tone method ... anyone who needs that or is interested, here it is. Here is the explanation." Of course, it was reviewed as a clumsy, unsubtle way of explaining something. It wasn't sophisticated, it wasn't mathematically elegant. Well, how can you expect someone who isn't mathematically oriented to be mathematically elegant?! In those years, I started to write to George Perle and Milton Babbitt. I didn't know either of them at the time; I'd never met them. But I couldn't get anything from them, not a word. They were very tight-lipped. As it turned out later, Babbitt had made some sort of a mathematically oriented theory of the whole thing. Which was just a different way of saying the same thing I had said in my simpler, more blunt way. But that didn't bother me, because it served my purposes completely. The point is ... take for example my Second Symphony.... It's a strictly hexachordal work, and I think that's where it derives a lot of its strength; because the harmonic relationships in the Second Symphony are right out of the interlocking world of the hexachords and how they multiply. It's like something that has a kind of genetic capacity and it starts multiplying. It's never the one form. It's always exfoliating. But because I never like to repeat myself (I've had a horror of it all my life), having done that work—and I knew it was good—I moved away from it.

RD Would you call that a quest for originality?

GR NO. That's something that never bothered me . . . I mean, it never bothered me in a psychological, emotional sense. I figured if I had something to say that was musically worthwhile, that was enough for me. Whether it was original, according to whatever standard of measure or not, that was okay with me so long as I was writing what I felt was important to me.

RD Do you have a view of any 20th-century composer whom you consider original?

GR Well, of course Schoenberg, for one; Bartók, for another . . . very original. But maybe not in the sense that you might mean. You see, I feel there are original natures. If you talk about an artist and you say he has an "original nature," he can't help but say things in his own way, so clearly, so indelibly, that it really comes out unlike anyone else's. He can't help himself, but he's not thinking about it. That's very important. I think the thing that puts me off entirely about the whole business of originality is the conscious *search* for originality. It's when the composer or painter starts monkeying around with all the possible devices and things that can be manipulated, in order *not* to do what X has already done or what Y has already done; but somehow if there's an open space between what X and Y have done, or are doing, that's for him. I don't like that sort of thing. I find it arbitrary, because most of these things turn out to be dead ends. They might turn out to be good for one or two pieces, but not for a lifetime.

RD When you began to veer away from a Schoenbergian method of organizing tones, what came to your mind?

GR Tremendous doubts, a lot of anxiety about what to do and how to proceed. You know that in 1950 I met Luigi Dallapiccola. As a matter of fact, he was the first musician to hear anything of my first twelve-tone piece. I had been in Italy on a Fulbright and at the American Academy in Rome, and we really hit it off. So when he came to Tanglewood in 1952, I played it for him. The last time I saw Dallapiccola was in 1963 at a conference on "Music in the East and West," which took place in Jerusalem. There were composers from the States, Europe, and Russia; and the problem we discussed was what was going on, what were the directions, what was happening, who was doing what, and what it all meant, if anything! I had some talks with Luigi, and by then, I had just finished my Piano Trio, which turned out to be the last twelve-tone piece I wrote. At that point, I was obsessed with doubt about continuing with twelve-tone. I communicated this to him by saying, "What's going to happen to music after twelve-tone, after we give it up, lay it aside?" (words to that effect). He said, "But, my dear, this is our language!" There it was, absolutely clear conviction, belief, certainty, that not only had we arrived at something solid, but there was no reason to question it, no reason to doubt that it was capable of continuing on through the generations. So, in a sense, I suppose he had accepted Schoenberg's idea of a new basis for music that would go on and on; but as we know, it was a very short-lived thing.

RD I have conducted Dallapiccola's opera *Il Prigioniero* on two occasions, and I remain impressed by how beautifully he used the system in a lyric and harmonic way.

GR Right . . . because of his Italian heritage, people said, he was imbued with the whole spirit of Monteverdi through Verdi. So I suppose it was inevitable for him. But you know, it's curious that people should have claimed that the Germans or the French or any other non-Italians were more suited to the twelve-tone system because they were less melodic . . . that doesn't make any sense. There isn't a German composer starting with Bach who wasn't supremely melodic at his very best. And if you take the French. . . . Who wrote some of the most gorgeous art songs, if not Fauré and Debussy?

RD What about the ultimate totalitarian, serial aspect which later developed and which might be compared to a kind of totalitarian socialism?

GR The whole Marxist idea applied in any direction, especially music. It was pretty deadly when you come down to it. It's a question that lies between two worlds. One where you are controlling all the elements—that does somehow fit the analogy of the totalitarian political system—where you have the bureaucratic mind that sits on top of the pyramid of the whole society and works out how everything will be done. It ensures its security and manages that nothing from below is going to upset that. So you are in absolute control . . . same thing in music. That was the conviction. The other side of it, the other world, is: What can you say through this methodology? Because if you repress, if you suppress, if you oppress human beings, eventually what you get is a dead society. What happens to music if you do that? I'm not going to make the simple-minded transference and say, "Well, you get a dead music." But it's always possible to leave that as an option, that something meaningful can still be said with it. But what did happen musically—and this is totally different from what happens with human beings—is that it ended up being indeterminate. You didn't know what it meant, really. And the ear was left out. That was the most serious shortcoming . . . that the ear was left out. After all, when you're asked, "Is music manipulation or is it expression?," I must opt for expression . . . always.

RD What about Cage, who opts for non-expression? I mean, that's the other polarity.

GR Yes, of course. And what he set up was the opposite pole of all the old certainties. And now, I feel the really desperate need for certainty.

RD I remember Saul Bellow said, "Modern character is inconstant, divided, vacillating, lacking the stone-like certitude of ancient man; also deprived of the firm ideas of the 17th century—clear, hard theorems."

GR That's a lovely phrase—"clear, hard theorems." Because that is I-IV-V. That's the musical counterpart of clear, hard theorems. You see, one of the things I learned very, very quickly from Schoenberg was you couldn't make a cadence, harmonically. And you know very well, in twelve-tone music, the only thing he could do was to make a *ritenuto* and then start again. Well, you know what that does. If every time you come to some sort of a major cadential point in the music, it slows down, then you have to build it up again. That's quite different from making a clear cadence and not monkeying with the tempo. That was one thing we lost. Another thing we lost, and this hit me like a ton of bricks, I remember . . . I don't know when it happened . . . must have been in the late '60s . . . we gave up scales. We gave that up! We gave up the arpeggio. We gave up all the time-honored things.

RD What about your dilemma: What to believe? What about your use of musical "quotations"?

GR It started in '63, '64. 1965 was the year I wrote *Music for the Magic Theater* and *Contra Mortem et Tempus,* and for me those are the first two important "quotation" pieces. But the reason I chose to go that way was not to quote, per se. Again it was not mechanical.

RD I recall you had used Mahler, Beethoven, Mozart, and jazz quotes, for example. Was there a connecting thread in your mind?

GR There is a thread that tied all of those sources in the *Magic Theater,* tied them all together for me, and this was the element of the three-tone, descending, half-step progression. All of the "quotations" contained this element, which covers a very wide range, from Mozart through to Miles Davis! And almost at the end . . . Varèse and Webern and so on. . . . The interesting thing is that what tied the "quotations" together in the *Contra Mortem* was the opposite half-tone direction! So that every one of the choices I made had to do with whether that element was present. Of course, there were many other elements present. I became fascinated by the fact that perhaps it was true after all . . . this old Jungian idea of the archetype. That we are much less individual than we like to think . . . that we are all individuals and have uniquely individual minds, is part of our democratic upbringing. But actually, there is something in the unconscious that is non-individual and nevertheless affects all individual directions in terms of belief, images, thinking, and conviction. And that's the archetype. I began to do some non-scholarly research . . . because I'm not a scholar regarding the archetype. It really came out of those two works. Yet, in some very interesting, strange, and intangible way, my feeling was then, and remains, that this helped to make those two works cohere, internally. So that it didn't sound like a loose assemblage of a lot of different chunks of other people's music, sort of interrelated and combined with connections that I made, things that I wrote to bridge the gap between here and there. That was the first step *away* from twelve-tone. The next thing that happened was, of course, the growing realization that I was going to have to find a way back to tonal thinking, to writing tonal music.

RD That brings up, again, the quest for a certitude. The Czech writer Milan Kundera has reflected upon the literary worth of ambiguity and relativity existing in the West. Does this provide you with any creative advantage?

GR Ambiguity combined with relativity. . . . While it's very attractive from a certain point of view, as a general idea, to *live* ambiguity or relativity is very hard, if not impossible, in one's daily life . . . and more specifically as a musician, as a composer, because you're trying to say something. I keep coming back to that . . . I have to say something. I can't write unless I have something in my head that is clear and carries with it the freight of feeling and emotional charge that is extremely intense and very real for me. If I don't feel that, I can't compose. So that to live and compose ambiguities is for me virtually impossible. And those composers who have, let's say, dealt with ambiguities and relativities, other than in a quotational mode . . . I'm a little hard-pressed to think of who that may be other than John Cage and those around him . . . I have never been able to make sense of what they are doing.

RD I asked Cage what he thought of the music of Reilly, Reich, and Glass. He thinks, "They bring out a very convivial feeling in the audience. People who enjoy the music are all turned, so to speak, into a group." Does this "group" or "herd-like" instinct relate to your concept of archetype?

GR The "herd-like" thing . . . I'm not sure that's exactly what Jung meant. It's certainly not what I mean by archetype. The archetype, for me, is not some kind of animalistic instinct, like that which drives a school of fish, for instance. They all flick in one direction and then you see them all flick in another direction, and they *all* go . . . instantaneously. I think the archetype really means something else. It's something deeply unconscious, which rises to the mind in the form of an idea or an image, and though not individual, per se, can only rise in an individual. At this point, it can potentially mean the same thing to different individuals. Whereas, for instance, what the minimalists are doing . . . some of it is charming and attractive . . . convivial is a nice word. . . . But it's curious, I get bored with it very quickly . . . and I don't mean after a half-hour, I mean after a minute, a minute and a half! Immediately, the means are obvious. You know exactly what is involved. It's very simple-minded, and while there is a kind of playful charm about it, that's not enough. So, if people enjoy it, that may have something to do with "herd instinct"; it may have something to do with strong needs these days to relate to art experiences in a collective way . . . I don't know. I think it's a very short-lived movement; and if it's not finished yet, it will be soon. The thing that bothers me is that nothing lasts. Every new possible idea or movement is hailed ... no one rejects anything anymore. If it's innovative, if it seems fresh, unlike anything that's happened before, immediately a kind of hurrah goes up and everyone goes that way, like the school of fish. But then, over here, something else pops up, and that's all forgotten.... And so now they are on the something else. It is not only a sign of our vacillation, to use Saul Bellow's word, it's a sign of an inability to settle into any kind of conviction about anything. It's as though the whole society is simply playing. There is serious play, and there is non-serious play. I can see serious play, because, in a way, that's what music really is in the end. That's one of its great joys. It's a wonderful form of serious play. But I can't see non-serious play, where A would be just as good as Z ... they're interchangeable ... just as long as you're distracted, as long as you occupy those few hours with whatever it is and on to the next thing, and the next.... That's what bothers me, that's the nature of our time. I feel that we move too rapidly, there are too many things that we have to deal with; there are too many things zinging in and out of our lines of sight and perception. A lot of it, I think, when you get to be mature enough, you simply ignore. You realize that it's just a momentary thing and nothing to be taken seriously ... and you go about your business. The curious thing about minimalism is ... I've heard Steve Reich say that he has to get away from what he's been doing to something more solid. And that's something that refers back to what I said earlier; that I see the faint sign of desire here and there to get back to more solid things; in other words, back to certitude, back to conviction about something, a belief. But in order to believe in something, you've got to have *something* to believe in. We don't have it yet. That's the problem as I see it. For example, even though I have played with the juxtaposition of tonal and atonal, there are times when I find even that totally unsatisfactory. When I wrote my Fifth Symphony, which is my last big orchestral work, there is no play back and forth between the one and the other. It's almost entirely chromatic, but no longer in an atonal fashion. While it has a slight edge of atonality, if you listen intently, you'll sense the tonal pull underneath. In other words, there are gravitational points which are drawing the music along. I think that's essentially what I believe is possible now. So it's no longer a matter of speaking, or trying to speak the language of Beethoven ... which I did in my Third Quartet. Some people absolutely delighted in it and some absolutely abominated it, because I spoke directly in the language of the 19th century. I can't do that anymore. When I did it, I believed it was possible. Now, I don't believe in its efficacy any longer ... in its rightness ... and if I have that feeling, I must go on.

RD Once a thing is done, some people like to just leave it be.

GR What I'm referring to is not the work, per se, because I don't like to repeat directions as such. I'm speaking of a certain cast of mind, a feeling and an attitude, where you are able to go in any direction you like because the core is solid. I've always been pleased when, on rare occasions, I've been likened to Picasso. Because, of all the modern painters, and there are many I enjoy, I enjoy him the most ... and for a very specific reason. He was not afraid to do many different things at all stages of his life. Every one of the things he touched turned to gold.... The pre-Cubist stuff, the Blue, the Pink periods, the Greek period, I call it the Classic period, with those huge monolithic, almost statue-like female figures sitting on the beach; and the various combinations that he made later....

RD Well, he and Stravinsky were both a kind of embodied "pluralism," weren't they?

GR In a sense.... He was much richer and much wider in his range of interests than Stravinsky. He had ferocious power and passion in him ... you see, what it comes down to in the end is the question: Can we compose again with passion? That's what I miss in almost everything that's written these days. I want to feel the intensity of experience, where music is concerned. I must feel I'm in the presence of a passionate voice, a passionate nature ... someone who knows what he is doing ... and someone who knows, consummately well, how to project what he wants to say. Not some kind of huge outrush, wash of unorganized, unthought-through feelings, a kind of musical finger-painting.... That, I want nothing of. But passions, real passions, something that's said with tremendous power.... That's the only thing that's going to save music ... the emergence of passions, again. That's why the minimalists ... they leave me cold. They're too tame. They are very cool. I don't like cool art. I want a strong, clear art, a kind of directness and simplicity of choice among the things you have to say. And you say it, and you leave out all those other things that might be included, because they obscure the essential thought.

*Stephen Smilak*

# *Richard Wernick*

BORN JANUARY 16, 1934, BOSTON

**April 9, 1986** **Media, Pennsylvania**

RD What brought you to Buffalo in 1964, the first year of the Creative Associates, and the year we first met?

RW I'll bite my tongue! What brought me there was Alan Sapp, that's what brought me there! Believe it or not, I was actually supposed to come there either as an assistant professor or a lecturer . . . it was some beginning-level position at the university. It was supposed to be a purely academic position; and as Alan put it, it was supposed to have a *small* amount of administrative work connected with it. And of course what happened was I ended up hardly teaching at all, because there was no time for it. I was putting in some sixty or seventy hours a week on the Creative Associates project.

RD Where did you go after that?

RW I went to Chicago. I remember I gave a lecture in Buffalo, as part of the Creative Associates project. They asked me to give a lecture on John Cage, because nobody up there knew anything about him. So I just put a few notes together on 3 by 5 cards, and I gave a lecture on Cage. It was not improvised, because I did prepare notes, but it was done from notes, not from a prepared speech. It was just a few weeks later that I got a call from the University of Chicago, that they were looking for a composer-conductor. . . . It was Ralph Shapey who had recommended me. They asked me to come out. And one of the things they asked me to do was to give a lecture. On my C.V. I had listed my lecture on John Cage, and they said they wanted me to deliver the Cage lecture! Well, no such thing existed. So for three days I sat and hacked away at the typewriter, preparing this lecture, which turned out to be a pretty good one, as I look back on it twenty years later. And it was on the basis of that lecture, plus meeting people, and so forth, that I got the job! By then, I was ready to get out of Buffalo. I could not take that tension anymore. It really was making me crazy. First of all, the hours were nuts. And being caught in the huge vise, this kind of trap between Alan and Lukas on the one hand, and being trapped between different musicians; and trying to keep all these forces reconciled . . . no one else was doing it. No one else had the capacity or really cared about the human side of it. It was falling apart at the seams, and it all devolved on my shoulders. The Chicago years were very pleasant. I didn't leave there at all under the same circumstances under which I left Buffalo. I had a very good schedule, because Chicago, like Penn, had built into it what they call research time, which was the composer's composing time. So part of the academic load was built-in composing time. I taught two courses and conducted the University Orchestra, and the rest of the time I was able to compose. It was a good situation. I met people like Neva Pilgrim there—she was singing there—I guest-conducted Ralph's group a couple of times; I did a piece for Neva which I conducted at Hunter College in New York—it was a very good experience. Ralph and I solidified our friendship, which has lasted a very, very long time.

RD After Chicago, you were invited to the University of Pennsylvania, where George Crumb and George Rochberg were.

RW Penn wanted me to come. George Crumb invited the family for a visit in 1967; and it became apparent that he wanted me to meet George Rochberg, who was chairman of the department at that time. Mel Strauss was the conductor of the University Orchestra and was leaving Penn. What were then called the Penn Contemporary Players were essentially Arthur Weissberg, Ronnie Roseman, Sam Baron, Gil Kalish, the New York crew . . . they came down for concerts. What Penn wanted was someone to take over the University Orchestra, to teach, and set up a real contemporary group in residence at the University. And along with that, the big attraction was a promise that after one year they would promote me to tenure. Chicago wasn't willing to match that; and there was no way of having my own group at Chicago because Ralph's group was entrenched. So I decided to accept the Penn position.

RD And essentially is it the same as when you arrived or has it changed?

RW Well, the department has changed enormously. Not so much in composition, but George Rochberg is retired now; Jay Reise is here and Chinary Ung is here, and we have a slightly different composition department. But we have a very different musicology faculty. A much more active one, so that the whole department is a more distinguished one than when I arrived. When I arrived, I found myself in a crazy situation in which the new chairman was summarily thrown out. They brought in a new chairman to replace George, who lasted nine months. And after petitions from the faculty and the students and all kinds of unpleasantness he was forced to resign, and I became acting chairman the second year that I was here. And then I became permanent chairman for three years. So I was chairman of this department for four years.

RD That time in the music world was the time for groups, wasn't it? They were starting to form. Now it seems a very normal thing for groups to start up, but those were pioneering times.

RW Well, they were. I think you have to remember also that they were easier to start. Organizations like the Rockefeller Foundation and the Ford Foundation were still giving money for that sort of thing. The Ford Foundation has almost completely backed away from that now. The Rockefeller Foundation is still doing a lot, but not specifically that sort of thing. They're doing things like supporting the Meet the Composer/Exxon Project, putting composers in residence; but the Buffalo project was started with Rockefeller money; the Chicago project was started with Rockefeller money. When I came into this group, it had Rockefeller money. I came in at the tail end of the Rockefeller money. I guess that's one of the main reasons they wanted something to be homegrown, because the Rockefeller money was just about to run out. And it's been supported by the University ever since. It's marvelous on the one hand, but it's anomalous on the other. I don't have any place to perform. We don't have a building! I have a group, but I don't have a building! So for at least ten years now, I've been performing at Swarthmore. I worked out a very nice quid pro quo with them, where they give me their hall, piano, *pianos* if I need more than one, percussion equipment, whatever they have around, logistical support, in terms of stage help, lighting people . . . in exchange for which we give them the concerts. All for nothing.

RD Some time ago, I remember speaking with you before a panel. I asked questions about your becoming a composer. You had a great deal to do with commercial work of certain kinds. Am I wrong about that?

RW No, no, no. You're right. When I was a freshman in college, I decided not only to major in music but to minor in theater. I was always interested in theater. The head of the theater department at Brandeis at that time was a man named Elliot Silverstein. He was a *brilliant* director. I guess his greatest claim to fame is that he directed *Cat Ballou.* In 1951 he was a very young, brilliant, but unknown director. I began to work with him not only at Brandeis but also at the Wellesley College Theater on the Green which had a professional company. And that was a very distinguished group, had some very good professional people and did some marvelous productions, including an uncut production, if you can imagine such a thing, of *Man and Superman.* Five and a half hours of Shaw! And Elliot engaged me to do the music. I did some very interesting things for him. I did the first performances of the expanded version of Arthur Miller's *A View from the Bridge.* It was a one-act play, he expanded it into a full evening and I did the music for that; and I did music for the *Bourgeois Gentilhomme* of Molière, in which the whole orchestra appeared in costume on the Green itself, which was the stage, which was loads of fun. And through Elliot I met people in New York, and decided that, rather than pursue an academic career, I would pursue a commercial career. So I did quite a few theater productions . . . not musicals, but theater with incidental music, and from there I met people involved with dance and I did a couple of things with John Butler and Glenn Tetley. Because my name had begun to circulate a little bit, film directors began to approach me. I did a considerable number of industrial films . . . documentaries. I wrote the music for the fiftieth anniversary celebration film of the Girl Scouts of America! That kind of stuff. Nationwide Insurance, I did a couple of small featurettes for them. I never did a big feature film. By the time I was getting close to thirty, I decided this was not the kind of life I was meant to have. It was too stultifying musically; I was not writing any concert music. I wanted out, so I got out.

RD Compositionally, from whom did you learn and what direction did you take in terms of your method? What attracted you?

RW Well, coming from Boston and studying at Brandeis I was maybe in slightly more insulated circumstances than other young composers. The only new music I knew as a kid studying composition and learning music was the prevailing system in Boston: Neo-classicism. I was Stravinsky-Copland-Piston and down through the next generation of Fine-Shapero. As a matter of fact, a couple of years ago, Milton Babbitt asked me, "When did we first meet?" And I told him the story of when he came to give a lecture at Brandeis University when I was a junior. That's the first exposure I ever had of any kind to let alone, serialism, but set theory as well. It was the first time I had any real exposure to twelve-tone music. I knew a little bit of the music of Schoenberg and Berg and Webern, but not very much. Because Boston was the center of Neo-classicism. I knew everything of Stravinsky cold. I remember hearing tons of Alexei Haieff's music, which has not lasted terribly long. And of course the Boston Symphony under Koussevitzky was doing a tremendous amount of American music; but it was mostly the neo-classical and what would then have been called a neo-Romantic music. It wasn't until I got out to California in 1955, when I started to study with Leon Kirchner, that I really got exposed fully to the other side.

RD Did you take well to that? Are you talking about serial writing?

RW Yes. Serial writing, and what I would consider chromatic writing, although I did a lot of work in serial music. I had a lot of catching up to do in learning something about how that other side of the world worked. A curious thing just came up with *Time* magazine a couple of weeks ago. The question was raised, "How would you characterize yourself?" And that's very difficult for me to do. Most people who don't listen terribly carefully and don't study the music or know the music tend to make an assumption that my music is serial. Or twelve-tone.

Because it's very chromatic. But there's no serialism, there's no twelve-tone-ism. I've never really done that sort of thing. I guess my music comes out the way it does because I was very highly attracted to the harmonic texture of chromatic music. But the way I compose, the way I put music together is pretty much Neo-classical—that aesthetic. Not with diatonic structures over it but chromatic structures over it. That produces what I guess you would call my music.

RD So that's your approach?

RW Yes, but it's not a terribly conscious one. It's one I've only become conscious of because I've been asked so many times in recent years, "What camp do you put yourself in?" I'm not in a camp. When the Violin Concerto came out, I thought it was very clever. Bernard Jacobson, who wrote the Philadelphia Orchestra program notes, referred to me as "a cat who walks by himself." I am told that my music is individual in that respect, that it doesn't fall into a camp. But that can be said of a lot of composers. I'm probably in the first generation that just missed being directly involved in the Second World War, and also just missed the essential conflict between Stravinsky and Schoenberg. I went to college in 1951. The point I'm making is, the generation just prior to me, I think, grew up having to make a commitment to one school or another school; and I think that even if they didn't wish to make that commitment, they still felt the pressure. I didn't feel that pressure, and I don't think people of my generation felt that pressure. I think that as you come down to the present time, the kids in their twenties, who are writing music today ... they're feeling *nothing!* They are feeling no pressure from anybody ... every man for himself ... everything and anything goes!

RD We did have a discussion last summer in Aspen regarding that idea of pluralism; and I think it's on the one hand valid to say there is a multiplicity of style and direction today. But the concept of pluralism isn't a new one, is it?

RW No, but it's on a much greater level. I think that the 20th century in many respects bears a resemblance to the 17th century, which also had a kind of a pluralistic cast to it. There was still Renaissance music being written in the 17th century; yet there was early Baroque music being written also. You had Monteverdi on the one hand and Schütz on the other hand ... the music is very, very different. The styles are totally different. You've got Italian schools and Flemish schools and German schools.... But nothing like today. I think the reasons are clear. Exponentially, there are many more composers; and with mass communication the amount of information that's circulating is enormous. And not just in our Western culture, but the influences that have come after the Second World War, particularly from the Eastern cultures. I'm talking about the real influences, not what I consider the not-so-quite-authentic Zen that was proliferating in this country after the Second World War. I think that this country had a very misconceived notion of what Eastern culture and aesthetics were, mainly through people like Alan Watts, who wrote very interesting books; but I think if you look back on those things now, they were not really a true picture of Eastern philosophy and aesthetics. I think this picture became clearer when the actual people from the East began to arrive in the U.S. or their music became available on records; and you had Japanese composers and Korean fiddle players and Cambodian composers, like Chinary Ung, for example, who's in his mid-forties. He's not a kid, he's a mature composer. He writes so-called Western style music; but he is also in the forefront of preserving Cambodian musical tradition. Because of the political situation in Cambodia, he's desperately trying to hold this thing together. I think it's through people like that that the real Eastern influence began to be felt more in the '60s and '70s, rather than the late '40s and '50s.

**RD** Did the music of Cage and Varèse make an impression on your composition?

**RW** I'm not sure that it ever made an impression on my composition, at least I can't see it. I've always been a tremendous admirer of Varèse. I've never used any of that kind of three-dimensional, spatial concept of writing. He composed different musics that go at different speeds and work in harmonic blocks against one another. I've never done that, although I appreciate it. It's interesting that you mention it, though, because when my Cello Sonata was done in New York last year, Milton Babbitt was at the concert, and he said that parts of it reminded him of early Shapey. I said, "Milton, I can't believe that you're saying such a thing!" And I began to think about it, and actually it's not Shapey that Milton was hearing in that music, but there is, in certain sections of that music, a kind of a background-foreground overlapping of static harmonic textures which could be a Varèse influence. It's more Varèse than anything else. Cage's music I've never felt had any influence on my music at all.

**RD** The idea of the liberation of sound which both men somehow claimed in different ways; and in another area, not just directed toward them, the freedom of the performer . . . did those two things cross your mind in terms of composition?

**RW** They did. Some of the music I was writing from the mid-'60s into the '70s, for about ten years, had some of those aspects. I tried solving it in a different way. I tried writing aleatoric music, but when I did it, it was more controlled, rather than less controlled. I've always been reluctant to give up control in composition, not necessarily to the extent where a performer or a conductor or a singer couldn't interpret, however.

That, I feel, is still terribly important. One of the things that we almost lost in a lot of the very rigorously notated music was the fact that you can get to the point where you can't interpret anymore.

RD So that much freedom you were interested in giving the performer.

RW I think that performers should be able, within reason, without disturbing the music, to have the freedom to interpret it.

RD Hence the "boxes" that show up, of notes that are given that are freely interpreted?

RW I never went quite that far . . . I never wrote "box" music. As I say, my music is principally controlled. There was a period in the mid-'60s to mid-'70s in which the concept of having music sound very free interested me a great deal; and I wrote music that would sound as if it were aleatoric, some of it. And yet I could never quite give up that whole business of control.

RD Now, just to take a side step . . . performers. Can you name a few performers whom you regard highly in terms of performing your music?

RW Of course the first person who comes to mind is Jan DeGaetani. She's probably performed more of my music than any other single person. Not only performed it, but I think she has been responsible for more of my music being composed than any other single person. In terms of singers, I would say that she's extremely important. Among other individual performers who have done my music, I'd certainly have to include violinist Greg Fulkerson. I was reminiscing with Greg about when we had first met. . . . I don't know if you remember the circumstances. . . . Greg played the viola in the world premiere of *Introits and Canons,* which you did, at Juilliard in 1978. He played viola. He was a graduate student then.

RD And he played your concerto recently.

RW He not only played the concerto, but in between he did the *Cadenzas and Variations II,* which is one of the pieces he played to win the Rockefeller–Kennedy Center Competition. Although he didn't actually commission the concerto, he kept sticking the spurs in my side! To get myself a NEA Grant. So I wrote the concerto for him. In terms of instrumentalists, Greg would have to be extremely important, as would others like Lambert Orkis, who recorded my Piano Sonata on my first compact disk.

RD In 1976, I was able to ask you to come to Aspen for the first time. Do you have a recollection of that visit?

RW Oh, that was a marvelous visit, I have tremendous fond recollections of that. First of all, I thought it was extremely interesting to have two British and two American composers there that year. It was Ollie Knussen and Max Davies and Jacob Druckman and myself. It was a fascinating summer in that respect. The whole year before was fascinating because it was, of course, through you that I wrote *Visions of Terror and Wonder,* so you are directly responsible for my winning the Pulitzer!

RD I think it's fair to say that Jan was at you, too ... you were writing it for her. But I remember being on the phone with you many times about how we were going to finance that for you and how it was all going to fall into place. And you had copying costs, you had printing costs ... it was almost endless. And little by little, somehow, money began to piece together.

RW Well, the NEA supported the composition of the work, but they turned down my request for copying funds. It was also one of those fortuitous circumstances in that I had just come out of my chairmanship. I had finished my term as chairman in 1974, and you asked me to write the piece in 1975, and since I had not had any leaves or any other "perks" since I had essentially saved the music department from going under, my successor as chairman, Larry Bernstein, did wangle the money to extract the parts.

RD Anyway, it did get done, and we did the world premiere, and it seems to me the very next year you won the Pulitzer Prize!

RW That's right. But the fondest memories of that summer were the rehearsals, not just the performance, but the rehearsals of *Visions;* of hearing that piece for the first time. It was the first big orchestra piece I had written since leaving Chicago, so it had been about eight years between orchestra pieces. And I felt that, since I had come to so-called serious composition late anyway, that was a very important transitional stage for me. It was essentially a student orchestra, wasn't it?

RD The quality of players, even though students, was very high that year.

RW It was very high, because that performance was excellent! I remember the scrambling around you had to do in order to get some extra rehearsal time. And I remember Jan's first reaction when she came in and you were rehearsing the last movement without voice. Jan came into that barn-like place ... the Music-Hall. She had a look on her face, the look of absolute terror, and she said, "My God, Dick's written a piece for Birgit Nilsson!!" I remember it being a *very* special summer. The kids were there, Adam and Peter were there and they had a marvelous time, and Bea and I had a marvelous time ... it was just one of those extraordinary experiences.

RD Well, I want to talk a little more about the piece, *Visions of Terror and Wonder.* I just want to remind you, however, we did *Cadenzas and Variations* too that year, Steve Clapp did it; and we did the *Kaddish Requiem,* which Jan DeGaetani also sang. Speak about the piece ... the conception of it.

RW It came to be because of your phone call and having the opportunity to do a piece for Aspen and for the bicentennial. I remember discussing whether it should be an "American" piece . . . and I remember you said something to the effect that you didn't want it necessarily to be limited or constrained . . . it might be nice, but let's see what happens . . . it was the piece that was more important. . . . I remember looking through what seems to me now like a hundred texts, of course it was nowhere near that number, but it was endless, and I simply could not zero in on something that was American. And I had had ideas for this kind of thing on a much, much smaller scale for quite some time; and I mean by this kind of a piece, a piece that involved, on a fairly high, poetic level, the whole problem of the Middle East, the whole problem of Israel and Jerusalem. I had written a piece for Jan, as you know, called *A Prayer for Jerusalem.* It was a theme that was very important to me. I had the idea of the Hebrew text and the Greek text of *Visions.* Of course, they are naturals. They had gone together for a long time. I had had those things stuck away in a drawer. And as I began to think about a much larger piece, I began to put together in my mind the notion that there must be somewhere in Islamic scripture something parallel to this idea. So I got myself a Koran, in English of course, and read it, and came up with the Arabic verses . . .

RD The Vision from Mecca . . .

RW The Vision from Mecca, which of course is very close . . .

RD And the vision of Paradise . . .

RW Yes. That was more like a coda, but the vision from Mecca, the business of the scroll, you know, the world being rolled up like a scroll. . . .

RD Well, you say, "On that day, we will roll up the heavens! As one rolls up written scrolls."

RW Which is parallel to the idea of the new heaven and the new earth, and the destruction of heaven and earth and the re-creation of the new Jerusalem. That appears as a prediction in Isaiah. It appears in the past tense in the Book of Revelation; and it appears also as a prediction of sorts, although Jerusalem is never mentioned in the Arabic text. And so what I worked out was a way to combine these texts in their own form, which then dictated the musical form. I took the Hebrew text and the first Arabic text, and although they were two separate movements, I made them Part One because they were the vision of the future. The Vision of the End had the sense of having been seen already, so I decided to make that the second part. And because I had two rather large movements in Part I and only one large movement in Part II, the ending, the Coda-Recitative I call it, the vision of Paradise, was added for mainly musical reasons, because I wanted to make the reference back, at the end, to the Arabic portion of Part I. Even though the first movement and third movements were very, very different from one another, there would be a reference in Movement 4 back to Movement 2, which would round out the piece. And in doing that I realized that having done the Vision from Judah and the Vision from Mecca, which were the prophecies, and then the Vision of the End, which was a revelation of the prophecy, to do another prophecy of the same sort would be poetically anticlimactic. So what I had to do, since I had a musical sense of how the piece was going to end, was to find another text which would be appropriate to the ending. The one thing that I'll never forget, and which I was going to put in the notes, but I left out, . . . was that at the very end, you have the last Arabic line which ends on *Salaama Salaama,* and then it ends on *Irene-Shalom,* which I have in parentheses there. The piece originally ended with *Salaama Salaama.* I remember the precise day that I finished the piece. It was November 11, 1975, and the reason I remember is because it was the day the United Nations voted that Zionism was racist. And I said to myself, "God damn it, this piece is not going to end in Arabic." And I took out the last pages, and I rewrote the music so I could add the Greek word for peace—Irene—and then the Hebrew word for peace—Shalom. I wanted it to end in Hebrew. And I rewrote the ending of the piece the next day to make it work out that way. The extra irony here is that my mother's name is Irene.

RD Well, I think that should be part of the notes. That's a very personal story.

RW It is. I think, nine or ten years later, if the piece were to be done again I think maybe I'd revise the notes and put that in. I think it's even more telling now than it was then. Now that we have a few years of removal. But you know, that's how music gets written . . .

RD . . . and of course we did the piece again in New York with the American Symphony. Has it been performed a number of times?

RW Well, Jan did it here with Dennis Davies in 1981 or 1982. And I did it in Jerusalem.

RD What was the reaction?

RW Well, it was a terrible performance since the Jerusalem symphony was simply not up to the piece. Christopher Keene was supposed to conduct three performances in Jerusalem. And Bea and I were going off to Cairo for four or five days and then coming back to Israel in time for the rehearsals, and then planning to spend a couple days after the performances enjoying ourselves. Two o'clock in the morning the day before we left, the phone rang. The manager of the Jerusalem Symphony saying, "We have a small problem." And when I hear the words "a small problem," I know we're in *big trouble!* He said that Christopher had canceled, because he had had a recurrence of hepatitis. And I had two choices . . . I had three choices, actually. We could cancel the performance, in which case they would still be happy to have me come and be their guest and be in Israel for the week; or I could let this other conductor, Alun Francis, who had been doing most of his work in Switzerland, conduct the piece, because he had been engaged to conduct the rest of the program; or I could conduct it myself. Well, if it had been a conductor I knew, even by reputation, as having some dealings with contemporary music, I would have agreed to let the other conductor do it. I didn't want to cancel, so I said, "Okay, I'll do it myself." So I threw that great big score into a suitcase and went off and spent my evenings in Cairo studying the score. Most people think that just because a guy composed the piece he can conduct it just by waving his arms, and it's all going to come out! I had to study the damn thing! I got to Jerusalem, and it was a shambles. The orchestra is full of Russian string players who can't count. They simply can't count! It is the most undisciplined group of musicians I ever saw in my life, and it was a misery . . . an absolute misery! Three performances.

*Charles Abbott*

# *George Crumb*

BORN OCTOBER 24, 1929, CHARLESTON, WEST VIRGINIA

**May 19, 1986** **Media, Pennsylvania**

After driving in from Connecticut on a hot afternoon, I appeared on George Crumb's doorstep. He greeted me, proudly showing me the stone fence that he had built outside his front lawn. As we entered the house, he ushered me past his extensive library with two pianos into his private study, where his attractive scores were on display.

He sat in his presumably private chair, behind which was an air conditioner that bellowed at the level of a small jet airplane. And in his soft voice he began to speak with a quiet spookiness that belied a fierce sense of inner determination, betrayed externally by his intense, wide-open gaze. I had learned of his compassion for and his sensitivity to life through his music, which he produced in a slow but constant flow, like a gentle mudslide, egged on by nature's silent force.

RD The first thing I want to establish is exactly when we met for the first time. I think it must have been when you came to Buffalo.

GC Yes, that would have been in 1964.

RD What was your position there?

GC They called us "Creative Associates." I had been teaching in Boulder, Colorado, for several years and was delighted when I was offered a grant so that I could participate in the Buffalo project. After the year in Buffalo, I came directly here to a position at the University of Pennsylvania. That was partially a result of meeting so many East Coast performers and composers while in Buffalo. I became excited about the general scene and wanted to be a permanent part of it.

RD I know that your early years were spent in West Virginia. When did you leave and where did you go?

GC I left West Virginia in 1951 for graduate study in Champaign-Urbana and Ann Arbor, Michigan. My years in Ann Arbor were interrupted by a year in Berlin as a Fulbright grantee. After finishing in Ann Arbor I spent one year at Hollins College in Virginia and then went on to my first permanent position in Boulder.

RD You were teaching composition there?

GC No, I was on the piano faculty! I did manage to have a few students in composition and gave one or two analytical classes in comtemporary music. But I was always nominally a member of the piano faculty.

RD What tipped the scales?

GC It was never my intention really to move in the piano direction! It just so happened that at the time there was a dearth of positions in my own field. But it all worked out fine—the piano faculty had some very dynamic people. That's where I first got to know David Burge.

RD But in terms of composition—had you been pursuing this, too?

GC Oh, yes! In fact, it was during this period that I wrote my *Five Pieces for Piano* (which David Burge premiered), *Night Music I,* and *Four Nocturnes* for violin and piano. You might recall that Paul Zukofsky and I did the premiere of *Four Nocturnes* during the Buffalo year.

RD I remember when your name first came up. It must have been about 1963. There was a whole series of composers' names in the air at the time, and I remember Lukas listening to one or another. I was helping out in the planning stages for the new music project and of course the selection of composers for the project was of paramount importance. In fact, I talked to your colleague Dick Wernick quite a bit during the planning stage, and of course he was involved with the administration of the project during the first year. I imagine that both you and Dick have recollections of the Buffalo year.

GC Oh, indeed! Dick writes about that year in a very humorous way in the book [*George Crumb: Profile of a Composer*] we were just now looking through. Incidentally, Dick and I were both students at Tanglewood in 1954 although we really didn't get to know each other then. So it was only in Buffalo that we developed a close association—and here we are now, colleagues at the University of Pennsylvania!

RD In these formative years, did you have to face the problems of choice and chance, of control and serialism?

GC I feel that every composer these days has to work his way through the jungle of technical possibilities, and I was certainly no exception. During a period in Ann Arbor I must have carefully analyzed dozens of twelve-tone works, but somehow I was never comfortable with this approach when it came to writing my own music. I was impressed by many works of Berg and Webern and was also very fond of Dallapiccola's music. I think especially his *Quaderno Musicale* influenced my own work.

RD What did you see in it?

GC Interestingly, the pitch structure per se, the rigorous serialism, was not primary for me. I was more intrigued with his concept of piano color, even though it has nothing to do with extended piano resources. *Quaderno* shows an incredible ear for sonority and texture. Also the symbolic apparatus of the piece is so beautifully worked out.

RD I continue to be struck by the lyric quality also, especially in *Prigioniero.* And yet everything is so beautifully organized.

GC I think you've pinned it down. Despite its advanced chromatic language, it preserves all that marvelous Italian lyricism.

RD I remember, and I'm sure you do, too, when Dallapiccola was held in pretty low esteem by adherents of "hard-core" serialism.

GC That's right. He was considered not to be "uncompromising" enough. I think the objection really had to do with his intensely lyrical mode of expression.

RD Let me now pursue the other aspect of my question relating to technical procedures: were you ever much influenced by the concept of "chance music"?

GC I was initially intrigued by the concept. And I certainly had a chance to actively perform (as pianist) in many works which included varying degrees of the chance element—you recall that a lot of the programs in Buffalo were featuring works of this sort. But my own application of the principle has not been very extensive. The original version of my *Night Music I* included several passages where I asked for free improvisation; but later I went back to the work and made a new performing edition—I wrote everything out! I know that in earlier years you were very much interested in improvisation and even performed in the Improvisation Chamber Ensemble. But I'm sure that you would agree that, as exciting as improvisation can be, it is rare that the quality of musical inspiration can be sustained equally be several performers over a longer segment of time. (The legendary improvisations of Bach, Mozart, and Beethoven were of course solo affairs!)

RD However, there is a certain amount of freedom that you did then, and still allow to the performers, isn't that so?

GC That's right. I do frequently allow a certain freedom in respect to the vertical coincidence of parts. That might be termed a modified or limited use of the chance principle.

RD Did your thinking in this area have any influence on the marvelous visual appearance of your scores? I'm thinking of the varied ways in which you "bend" the staves in order to achieve symbolic visual patterns—circles, ellipses, crucifix shapes, and so forth.

GC Yes, it did indeed. I feel that visual symbolism can communicate something very meaningful to the performer. And, practically speaking, there are passages in my music—for example, in *Eleven Echoes of Autumn* or in *Star-Child*—where the relaxed vertical alignment is more clearly expressed by the visual "opposition" of different notational modes.

RD Have you run into problems with musicians trying to read your scores?!

GC The performer usually solves the problem by photocopying the passage and rearranging segments of bent staves into a more or less linear form!

But these are small inconveniences—the visual symbolism of the page is very important to me.

RD I recall that the mid-'60s period was highly charged with graphic experimentation of all sorts. Which piece of Sylvano Bussotti was that—*Rara?*—which had that really incredible calligraphy?

GC I remember several of his scores which might well be hung in museums as surrealistic drawings!

RD You recall that the Buffalo project was largely sponsored by the Rockefeller Foundation, and that one of the paramount goals of the project was to bring composer and performer together, or composer-performers and other performers together with each other. Do you think that this collaboration was successful?

GC I think so, yes. I enjoyed very much being involved in several of the performances as a pianist. In addition to pieces by Xenakis and Bussotti, I performed in Kagel's *Sur Scène*—an immensely difficult piece, not in terms of the actual notes, but in terms of the highly structured theater aspects. We must have put in fifty hours of rehearsal time on *Sur Scène!* I also served as conductor for my own *Night Music I.*

RD There were some wonderful performers during the early time in Buffalo, and I imagine that you have kept in contact with many of them. You have already mentioned Paul Zukofsky—who were some of the others there who may have had some influence on your musical thinking?

GC Well, certainly the percussionists John Bergamo and Jan Williams were immensely interested in developing new percussion techniques. I learned quite a bit from both of them. And then, too, I worked a lot with Carol Plantamura; she alerted me to many interesting new vocal techniques, some of which I incorporated into later scores. There were so many excellent performers of all kinds, and of course they were chosen principally because of their interest in new music.

RD So in that sense Buffalo was a kind of workshop for the composer.

GC It was immensely stimulating for me. I have always felt that there must be a very close relationship between composer and performer—I feel that composition really is a kind of extension of performance. My experience in Boulder centered around the exploration of piano idiom; Buffalo represented for me a more expanded sense of instrumental possibilities.

RD Many of your works show a whole new range of piano techniques, which go far beyond the conventional manner of treating the instrument. Did you have any model for unconventional techniques—for instance, Cowell?

GC At the time I wrote *Five Pieces* I had not heard any of Cowell's piano pieces. I had heard some of John Cage's early piano pieces, but of course "prepared piano" is not a device that I have ever used, strictly speaking.

I was always more interested in the possibility of *constant* change of color rather than in a predetermined and unchangeable kind of thing.

RD I spoke with John Cage recently and asked him what prompted him in this connection. He tried a pie plate on the strings and it jumped; then he went for screws, and so forth. Do you have a parallel story of how you found the inside of the piano?

GC Nothing so dramatic, I fear. Since I am a pianist, I simply experimented on my own. Of course my final goal was to somehow integrate all these beautiful sounds with the conventional keyboard sounds. And finally, to integrate timbral elements with all the other equally important elements of pitch, rhythm, form, etc. Nothing is so uninteresting as a purely gratuitous use of color.

RD I know that you have collaborated frequently with Jan DeGaetani, George. What sort of an influence did she have on your vocal writing? It occurs to me that roughly half of your compositional production is in the vocal area.

GC I got to know Jan during my first year at Penn. The first works I did for her were the *Madrigals, Books I* and *II.* I was astounded by Jan's flexibility and her command of an infinite range of nuance. She is able to go far beyond normal *bel canto* singing. Another important element in Jan's make-up is the incredible musicianship—she is able to control very difficult rhythmic configurations, and her sense of ensemble is infallible. She represents for me an absolutely new type of singer. Her performances of my earlier pieces permanently altered my sense of vocal idiom.

RD In addition to your interesting and original contributions to the piano and vocal literatures, your music has also over the years shown an intense involvement with percussion. It's probably not so important who was actually first, but you certainly had people bowing tamtams and vibraphone very early on.

GC Here again, I have to confess that there are very few things that I actually invented myself. Most were in fact borrowed from percussionists I have known over the years. You probably recall the sense of excitement John Bergamo produced when he first hit on the possibility of "bending" vibraphone pitch! But of course some of my ideas came out of my Appalachian background—I'm thinking now of the hammered dulcimer, the musical saw, the banjo, the jug. I feel that my obsessive interest in percussion carried over into other areas by perhaps honing my ear vis-à-vis timbre generally.

RD You have also put some heavy demands on orchestral musicians from time to time. I've not yet had the privilege of conducting your *Echoes of Time and the River*—but tell me, how do the musicians react to that piece?

GC I always seem to have to compromise when performances of that work are involved. There's a kind of built-in conservatism among professional orchestral musicians and, thus far, aside from a few student orchestras, only the Boston Symphony has done an "*Ur*-text" version of the piece! In all other cases, the musicians rebel: they don't want to shout, they don't want to pick up a second instrument without getting double pay, and they resist the idea of moving about the stage (in processional style) while performing. So I've pretty much decided that this work is more suitable for younger, less inhibited players. The level of technical difficulty is not that high. I've heard several excellent performances by university orchestras.

RD I can just imagine the temperament of the experienced "pro" when confronted with your *Time and the River!*—"I'm not moving about! I have to sit when I play!" Well, anyway, I had a lovely time putting *Ancient Voices* together for the Aspen performance in 1971. That was the year I was able to invite you and Xenakis and Wuorinen.

GC Yes, I remember very well your beautiful performance of *Voices.* I also recall a very stormy panel discussion that took place that week.

RD I frequently like to ask composers what they feel about the concept of originality. Do composers ever steal?

GC Oh yes, they steal! You recall Stravinsky saying that good composers steal rather than borrow, or something to that effect. I do feel that originality is very important in music. All the composers of the past that I admire were highly original.

RD Originality is perhaps a difficult concept to articulate. Do you have any theory as to how it comes about?

GC I doubt that one ever sets out to become original. I suspect that it's something that simply develops by itself, unconsciously. Perhaps originality really means being true to oneself?

RD That year you visited Aspen, Nicky Nabokov was still alive. Do you remember him?

GC I remember meeting him, yes. He came up after the *Ancient Voices* performance and gave me this enveloping Russian bear hug. He seemed quite moved. That certainly was a very strong performance.

RD I do remember him crying and applauding afterwards, and then coming back and telling us that he knew Lorca personally. Just how did you become interested in Lorca, George?

GC I got on to his poetry about the mid-'50s, when I was a student. But my first Lorca work—*Night Music I*—dates from 1963.

RD What was in the poetry that excited you as a composer?

GC The images must have corresponded with my own musical ideas. The simplicity of the language, the fact that the poetry allows so much space for music—some poetry is so dense, all the chinks are filled in. There is certainly a pronounced musical quality in Lorca's poetry—did you know that he was himself a musician, too, and made some arrangements of traditional Spanish folksongs? He was quite well versed in musical notation.

RD What were some of the images in Lorca that fascinated you?

GC Lorca's images are very basic, one could say "primeval." They coalesce around the ideas of life, death, love, nature . . . just what songs have been written about all through the ages. Then, too, his work is overlaid with a kind of ancient Spanish mysticism. You know, too, that he borrowed from folk sources—there is lots of Gypsy imagery in Lorca's poetry.

RD Did this ethnic sense in the poetry influence your music in any way?

GC Yes, in certain respects. For example, my use of the bolero rhythm in *Voices;* or again, the prominent use of guitar in *Songs, Drones.* Also, Moorish culture had exerted much influence on the Spanish, and there are several passages in my cycle of Lorca works which suggest a kind of Moorish character. But on the general subject of ethnic borrowings, my music also owes a lot to Asian influences. I really do feel that music is at present undergoing a process of synthesis on a grand scale.

RD I think that's certainly true. Even the music of Boulez has not been untouched by certain exotic sounds—certainly his use of percussion instruments, but also his treatment of the human voice suggests this. But as regards your own treatment of the voice: is there something specific in Lorca which led to your extended sense of vocal idiom?

GC I feel that it's as if Lorca's emotive power can become so strong that the voice just simply has to break out of any traditional mode. Lorca's language implies a wide gamut of utterance; it may erupt into overt shouting and suddenly scale down to whispering. It may suggest *parlando* style, or again, a melismatic style of singing—it is always changing. I have also made liberal use of microtonal bendings, glissando, and vocal-percussive sounds—all these devices seem to me to be appropriate to Lorca's surrealism. Ultimately, a composer has to find that poet he can treat musically.

RD I've talked with many composers who have widely divergent views on treatment of a text . . . whether or not to repeat a word or phrase, whether or not to excerpt portions of an existing longer poem, whether or not to break words into their phonetic components. Where do you stand on this?

GC One refers to "poetic license," but I feel that composers should have license, too! First of all, I feel that one doesn't have to treat the whole poem—I have sometimes excerpted only a line or two. I feel that words can be repeated, if musically justified. I can even imagine cases where the actual sequence in the original poem may need to be changed around. One can even combine lines from separate poems. I have sometimes treated words as pure sound by stressing certain phonetic components. Composers have probably offended many poets—I wonder sometimes what Lorca would have thought of my settings. But when one does combine music and words, one is dealing with a synthesis, a composite. In this case, new laws are in effect.

RD Have you ever set the poetry of a living poet?

GC No, I haven't. I've set some Walt Whitman and Edgar Allan Poe, and also some medieval Latin texts. It might be interesting to have that experience one day.

RD Music and dance would be another kind of synthesis, wouldn't it? I know that many of your works have been choreographed. Perhaps in this case, the shoe is on the other foot—now we have the choreographer adapting a piece of *your* music to *his* purposes!

GC You're absolutely right! *Ancient Voices* has been used maybe twenty or twenty-five times for different dance productions. There is a different title for each production. I've seen a few of these and always have the feeling that the music has lost part of its own identity in the process. But this new thing—this synthesis—can be interesting in its own right.

RD This reminds me that I conducted one of your pieces with dance at Aspen—what's that piece where I have to wear a mask?

GC You mean *Lux Aeterna.* Did you do the theater version with the robes and special lighting?

RD We couldn't do the special lighting, because it was a daytime performance in the tent. But we persevered and used the lighted candle which you specify in the score! And then—I don't know whether you were there—we did your other piece where the performers have to wear masks . . .

GC That would be *Vox Balaenae* for flute, cello, and piano.

RD Yes, that was it. That was done in 1974. On another occasion we did *Makrokosmos I*—David Burge was the pianist.

GC I was there for that performance.

RD That's right. You were composer-in-residence along with Elliott Carter and Aribert Reimann. We had all three Boulez sonatas performed by Claude Halffer, and of course there were a number of Carter pieces, and at least two of yours. I guess you must have visited Aspen several times, all told. What are your memories of these visits?

GC Even without the music, Aspen is a wonderful place to visit! But I am very grateful for those fine performances—after all, that is what composers live for. I remember that David's performance of *Makrokosmos I* acoustically filled the whole tent. Whatever might be said of tent acoustics generally, one advantage of amplified sound became evident to me!

RD Incidentally, George, I have a recollection of doing *Voices* on two other occasions when you were not present. One performance was in Holland with Jan. We also made a studio recording for the Dutch Radio. That may have been the first European performance. After that, I was able to invite Jan to Vienna, and we did *Ancient Voices* again on Friedrich Cerha's series called Die Reihe. Of course the main problem in doing the piece is finding a good boy soprano. In the Dutch performance we used a young lady, because there just wasn't a boy soprano to be found. In Vienna, however, the Mozart Knabenchor produced someone for us.

GC There's quite a tradition for that sort of thing in Vienna! It's quite a hard part, you know, involving lots of tritones!

RD How about moving to musical criticism? How do you react to critics who imply that any extensions of vocal or instrumental timbres are mere sound effects? I'm sure that kind of comment can come along.

GC I'm very philosophical generally as regards criticism. It would bother me, though, if the purely musical aspects of my work are ignored. Any special technique or timbre should only serve the musical meaning. If a critic hears only a disconnected series of sound effects, then my music is not projecting the way I want it to project.

RD Or else the perception is a bit faulty.

GC Either it's my fault or it's the critic's fault! I wonder what the critics in Beethoven's time thought about his use of the *una corda* pedal effect or the *sul ponticello* passage in the C sharp minor Quartet? This is not a new problem!

RD Or consider the music of Varèse, with all those special percussion instruments—the lion's roar, the anvil, things of that nature. Of course, in his case we are told that he was intuitively searching for a new kind of "instrument," which he finally achieved by electronic means. Your music seems different, however; there is no sense of a groping for some ultimate electronic solution.

GC That's right, I've only used simple amplification of instrumental sound. My music seems to focus obsessively on the live performer. I think it's the "danger element" in live performance which intrigues me so much. There's the sense that live performance can collapse at any moment. The cellist, by sheer bravura, has to sustain that fragile harmonic that's just on the verge of breaking. Or the soprano has to make that leap of two octaves and land pianissimo on that high note. Or the trumpet player has to play in the clarion register without being a quarter-tone flat. These elements in music are very important to me.

RD I know that you consider Bartók to have been an important influence on your development as a composer. Are there others? For example, Varèse?

GC I respect Varèse's music very much, although I doubt if his music has influenced me very much. I do sense stronger affinities to Messiaen and Berio.

RD In Berio what would that be?

GC That Italian lyricism again, the marvelous sense for texture, his enormously inventive use of percussion. Also, his ability to synthesize widely contrasting elements.

RD How about today's music? You have had a great number of students over the years, and I suppose you still observe the scene?

GC A few years back, on a visit to Taipei, I met a student of a student of a student of mine! In other words, sitting at one table were all of us—a former student of mine from Taiwan; his student, a very talented young man; and that young man's student, a very young lady. I suddenly felt very old. But apropos your question, Dick, the compositional picture now is very confusing, but also very dynamic. Just about everything is going on. Much of it may not endure, who knows? None of us knows the future, but the vitality is certainly there.

RD There's a lot of talk around New York about the "uptown composer" and the "downtown composer." Do you perceive that basic dichotomy?

GC Yes, it's there, in microcosm. The same basic division could be traced all over the world. I do very much like this sort of musical pluralism. Just about everything is being questioned, don't you think?

RD I do. In fact, musical pluralism was an important topic of discussion in Aspen in 1985. But I feel that we've always had some form of pluralism in music, wouldn't you agree?

GC I'm sure you are right. The postwar serialists wanted to believe that their kind of highly structured music was the one True Way, but even at that time John Cage and other composers were doing their own thing. I'm delighted that there are now *no* certainties; I think it's always a sterile situation when people think they have the truth all wrapped up. I think a good case in point is Russia. I visited there recently and was delighted to find something entirely new and very potent in several recent Soviet works. I feel this could only have come about through a growing spirit of openness.

RD Tell me a little bit about the musical scene there.

GC I met two very impressive composers—Shchedrin and Schnittke. I heard recent ballets by both of these composers at the Bolshoi Theater. I feel that their music finally makes contact with so much that has been going on in the West recently, and this is something unexpected in Soviet music!

RD I recall Nicky Nabokov mentioning to me that even many years back a number of scores by Western composers were somehow being circulated in Russia. But it's interesting that the impact is now finally being felt. Did you have performances of your music while in Russia?

GC Not on this occasion. But my music has been played in many Russian cities over recent years. There is even a recording of *Ancient Voices* by the Bolshoi Ensemble in the works—it will be released on the Melodiya label.

RD So your visit was a sort of musical good-will mission?

GC I was invited by the Soviet Composers Union to participate in their annual meeting. There were several European composers present, but I was the only American. I imagine that I was invited on account of those recent performances of my music there.

RD Do you mind talking about colleague-composers? If I were to mention a few could you give me your impressions?

GC No, I don't mind at all. But I must tell you that my critical sense is not very acute. I'm not very analytical, generally—I tend to see only the good things in the music of other composers.

RD I'm not looking for bad things! You've already mentioned Berio, but how do you feel about Boulez's music these days?

GC I'm especially fond of the earlier works—to me these are the most challenging. I'm thinking now of *Le Marteau* and the *Mallarmé Songs*. Boulez has such a fine vocal sense.

RD And Stockhausen?

GC In the case of Stockhausen, I would have to say again that the earlier works impress me most. Works like the *Gesang der Jünglinge* and *Refrain* have many beautiful qualities. In the cases of both Boulez and Stockhausen, much of the later music has seemed to me too abstract and not as potent musically. Can it be that there is such a thing as a "20th-century-composer syndrome" which results in the equations: early is good, and late is less good? I must confess that I also prefer early Stravinsky to his later work; I feel the same about Schoenberg and Webern.

RD How do you feel about the Polish school—Lutoslawski, Penderecki?

GC I like them both. I guess I am more impressed by Lutoslawski's music because there seems to be a wider expressive range. But both composers handle the orchestra beautifully and Penderecki especially can invoke the sense of epic scale. This is rather rare, don't you think, in a time when most of us tend to think of orchestra as a larger chamber ensemble?

RD Let's skip over the Channel to Max Davies. What about his chamber works and also his symphonies?

GC I think he has done very impressive things in the chamber area. I must say that I do not know the symphonies, with the exception of the first, which I feel is not truly representative of his mature style. He has a marvelous sense for theater—I recall your performance of *Mad King* on one of the Aspen programs. I also admire his references to medieval music and his talent for synthesis.

RD Apropos American music—what are your feelings about, say, Elliott Carter and Jake Druckman?

GC Carter's mode of construction would be altogether foreign to me, but I very much admire the expressive power in his music. I feel that the *First String Quartet* is his overwhelming masterpiece—by the way, I heard the premiere of this work when I was a student in Champaign-Urbana. I like other pieces of his, too, but my favorite is that early quartet. In regard to Druckman—I think Jacob is an enormously gifted composer. What impresses me most, I suppose, is his incredible sense for virtuoso instrumental writing—especially his treatment of wind instruments. His handling of the orchestra is brilliant.

RD Are you working on a new piece at the moment?

GC Yes, I've recently been sketching a piece for two pianos. I just recently finished a work for amplified flute and three drummers which I call *An Idyll for the Misbegotten.* I wrote the piece for the Canadian flutist Bob Aitken. He'll premiere the work in Toronto this coming November.

RD How about the public for new music these days? Are things getting better in your estimation?

GC I think there is a developing audience for new music, and that helps all of us. Progress seems slow, but I think there has been progress. There are so many fine recordings of newer music now available. I've done a lot of traveling abroad over the last dozen years or so, and I see interest everywhere in new music.

RD I know that you've been to the Orient on several occasions. I imagine that the public there would be very interested in your music.

GC Well, in Tokyo they wanted to make me an honorary Japanese composer; and in Beijing, an honorary Chinese composer! They must pick up certain things in my music that they can relate to.

RD How about the performance situation in Japan?

GC The Japanese performers are quite excellent, virtually on the level of the finest American performers. In China the performers lag a bit behind, but they'll catch up. After all, I can remember very well when there were not very many excellent performers for the newest music here in the United States—I'm thinking of the early '60s. How many pianists were there like Gil Kalish or David Burge? How many singers like Jan? Really only a very few. It seems to be a very gradual process.

RD I'm looking at the book about your music—*George Crumb: Profile of a Composer*—which C. F. Peters Corporation recently brought out. And on the first page I see a statement: "I believe that music surpasses even language in its power to mirror the innermost recesses of the human soul." Signed: George Crumb. That's a rather definitive statement, George!

GC Sometimes, in weak moments, I think aphoristically! But although this is a statement that poets might not like very much, it seems to me to be true. I really do believe that music can cross that threshold . . . it begins almost where poetry leaves off.

The 1963–64 Buffalo season hadn't ended before the New York Philharmonic engaged Lukas Foss to become Artistic Director of their first summer festival in the newly built, air-conditioned Philharmonic Hall at Lincoln Center. I was appointed Assistant to the Artistic Director. Lukas and I began working on the project during the summer of 1964.

Motivations for festivals sometimes have unusual beginnings. In this case, the fact that Robert Casadesus and Zino Francescatti were not only willing to take part in orchestral programs but would join forces in chamber music events was a germinal idea. Casadesus would also write a Concerto for Three Pianos, and he would be joined in the premiere by his family, Gaby and Jean Casadesus.

This led to the idea of a French-American Festival, planned to take place during three weeks in July 1965. While the festival fell during the period of Music Director Leonard Bernstein's sabbatical year, Bernstein agreed to take part in it, conducting his *Age of Anxiety,* with Philippe Entremont, piano; and his *Serenade,* with Zino Francescatti, violin; plus the world premiere of *Chichester Psalms.*

It seemed natural to find out if Charles Munch was available and whether Aaron Copland and Darius Milhaud were interested. With some schedule adjusting, Munch became available for *The Damnation of Faust,* Berlioz, and *Daphnis and Chloé,* Suites 1 and 2, Ravel; Casadesus (Ravel, *Left Hand Concerto*) and Francescatti (Saint-Saëns, Concerto No. 3) were to be his soloists. Copland agreed to conduct a concert version of his *Tender Land.* Milhaud, for health reasons, could conduct only half a program; his Symphony No. 10 and *In Memory of John F. Kennedy* were scheduled.

By this time, Carlos Moseley, Managing Director of the Philharmonic, had enlisted the cooperation of the French government in presenting a festival devoted to French and American music. Composers, conductors, singers, and instrumentalists of both nations would participate in a French-American Festival in New York.

My first major task was to find out if Duke Ellington was interested in conducting his music in the festival with the New York Philharmonic. Ellington and his band were playing at Basin Street East in New York in August 1964. After contacting his manager, I flew in from Buffalo and went down to hear the band. In his dressing room, over Chinese food during intermissions, I spoke with Ellington and Mrs. Ellington about the festival. Our conversation was frequently interrupted by his friends or musicians from the band, such as Cootie Williams. We discussed the possibility of performing his music, and he gave me his new album, *Symphonic Ellington.* He was definitely interested. I spent the entire evening at Basin Street East and thoroughly enjoyed hearing Duke Ellington and his great band. Eventually, we programmed his *New World A-Coming* and the premiere of *The Golden Broom and the Green Apple.* Ellington also agreed to narrate Copland's *Preamble for a Solemn Occasion.* Other American orchestral works scheduled were by Schuman, Harris, Ives, and MacDowell and the world premiere of Wuorinen's *Exchanges.*

With most of the orchestral programs planned, we focused on the four chamber programs: the first, a program of 18th-century French music with Jennie Tourel and Leopold Simoneau; the second, Casadesus and Francescatti with Debussy, Fauré, and Chausson. For a while we entertained the idea of bringing members of the Domaine Musical from France to perform Boulez's *Marteau.* The idea was to juxtapose the Domaine with the Buffalo group (Creative Associates). After a lengthy corrspondence with Pierre Boulez and members of the Domaine, plus discussion about a program called "Boulez and the Second Generation"—in which the music of his students Eloy, Amy, and Méfano would be performed—we had to drop the entire program because of projected costs.

Instead, we asked Rudolf Serkin if Marlboro would be interested in presenting a program of American chamber music. They were: Diamond, Ives, Foss, Barber, and Kirchner were planned.

This left one last chamber music program, and the choices were: Varèse, *Hyperprism;* Boulez, *Improvisation sur Mallarmé, No. 2,* with Bethany Beardslee (both of which I conducted); and Carter, String Quartet No. 2, plus the world premiere of *Variations V* by John Cage, choreographed by Merce Cunningham with Carolyn Brown and the Cunningham dancers. Cage said of the work, "We shall use electronic and film means, audio and visual. Our intention in this work is to implement an environment in which the active elements interpenetrate, sounds affecting images, images affecting sounds, etc., so that the distinction between dance and music may be somewhat less clear than usual."

I had corresponded with Varèse previously about what he would like to have performed. At first, he suggested *Amériques,* which he pointed out

". . . has not been given in this country since Stokowski performed it with the Philadelphia Orchestra in 1962, or, if *Amériques* requires too large an orchestra, then *Arcana.*" We also considered *Equatorial* and *Nuit,* the latter a work that he was trying to complete and which was based, apparently, on words taken from Anaïs Nin's *The House of Incest.* I was quite startled when Varèse showed up in Philharmonic Hall for rehearsals of *Hyperprism.* That incredibly imposing face bristled with intensity and expectation! It was the kind of face that made you freeze. Much to my relief, he was pleased by what he heard and only fussed a bit about the quality and timbre of some of the percussion instruments. At the end of the final rehearsal he said to me, "Where did you come from?" The performance was on July 23, 1965. Varèse died that November 6.

As it turned out, the festival was an example of how one can have enough variety and make it appeal to both "popular" and "intellectual" tastes, maintaining a basically serious tone.

My association with the New York Philharmonic continued shortly after this event in September, when I participated in the New York Philharmonic Conductors' Seminar under William Steinberg. In the concluding concert, September 24, I conducted two works: Bartók, *The Miraculous Mandarin,* Op. 19, and Copland, *El Salón Mexico.*

Upon returning to Buffalo, I looked forward to yet another wonderful event. The New York State Arts Council had arranged to tour the New York City Ballet, and I had the great pleasure of working with Maître George Balanchine in Buffalo. Rehearsals with him were a joy. He was so totally musical, so poetic, so kind. I personally don't enjoy conducting for dancers, but he made it an entirely different artistic experience. He would play passages on the piano (even though his left forefinger was partially missing), to illustrate his intentions, the musicality of a phrase, the exactitude of movement.

On November 26 we produced Stravinsky's *Agon* and excerpts from *Petrouchka,* as well as works by Tchaikovsky, Borodin, and Minkus, with dancers such as Suzanne Farrell, Arthur Mitchell, Patricia McBride, and Edward Villella. By comparison, the 1965 to 1966 season in Buffalo was relatively calm regarding new music. However, very glamorous soloists appeared with the orchestra, and they included Rostropovitch, Rubinstein, and Schwarzkopf. I had the pleasure of conducting Segovia in Boccherini's *Concerto for Guitar.*

What didn't let up was the influx of visiting composers and new scores to examine. Then the New York Philharmonic wanted Foss and me to produce another festival in the summer of 1966. Lukas had the brilliant idea of making it into a Festival of Stravinsky, and we proceeded furiously.

*Jeffrey Aaronson*

# *Earle Brown*

BORN DECEMBER 25, 1926, LUNENBURG, MASSACHUSETTS

**March 30, 1986 Rye, New York (overlooking Mamaroneck harbor)**

In the 1960s, I looked upon Earle Brown as a pioneer in the truly American sense. He had taken his music and thoughts to Europe. Although I had met him for the first time in New York in 1966, I began to know him better after seeing and talking to him in Paris in 1967 and after inviting him, for the first time, to Aspen in 1970. It was Easter Sunday, and I drove down from Southbury, Connecticut.

RD I want to first consider a bit of history, the history surrounding post-World War II, which will necessarily amount to a brief recollection of Darmstadt. I'd like to know when and why you were invited to Darmstadt.

EB The first invitation was in 1958, but we might even go back to the first invitation of my *music,* or the first appearance of my music in Darmstadt, which was in 1953 or '54, when David Tudor was invited to Darmstadt. He gave what were called *"Nacht Konzerte,"* which were apart from the regular official schedule of Darmstadt, because they didn't really know much about David or about us American "radicals" at that time. David took with him scores of mine, and of John Cage and Morton Feldman and Christian Wolff; and they were really a kind of startling revelation for the European "Darmstadt composers" like Stockhausen, Nono, Berio, Maderna, Kagel. So the first appearance of my music there, which obviously led to my being invited and commissioned later, was in those very early '50s. That's when David Tudor presented my *Folio* works (1952) and *Twenty-five Pages* (1953), which I consider to be the first open-form piece as most people do, and graphic scores, new "proportional" notations, and other scores. The first real invitation I got was to write a piece for Boulez's Domaine Musical, which Boulez was to conduct first in 1958 in Darmstadt. It was a piece called *Pentathis,* in standard notation (normal metric notation), closed-form, and basically "normal" in its scoring. Although I had done a lot of experimentation in graphic and proportional notation and open-form before that, I didn't really want to write a piece for Boulez and the Domaine Musical that would throw him into a situation with which he was, to a large extent, unfamiliar. So I wrote it for Boulez, but that was the year that the Donaueschingen Festival commissioned Boulez to write a piece for two orchestras and tape. He subsequently rejected his piece, because he didn't think it was a good work. As he frequently does, he writes, rejects, revises, and transforms his works. His piece *Poésie pour pouvoir,* however, was performed, and I heard it at the Donaueschingen Festival in 1958. As a consequence, Pierre was not able to conduct the piece that I wrote for him at Darmstadt, because Heinrich Strobel would not allow Boulez out of Baden-Baden until he finished his piece. So it turned out that Boulez didn't go to Darmstadt at all that year, and Bruno Maderna conducted my piece at Darmstadt in 1958. It was one of the first of my pieces that Bruno conducted, and that was my first Darmstadt invitation.

**RD** So you got there. What did you do?

**EB** I just attended my rehearsals, but there were many, many discussions and arguments about my notations and open-form concepts.

**RD** What was the atmosphere like? What made it so special that everybody seems to remember?

**EB** For me, the atmosphere was fantastic. Darmstadt premiered a lot of very important pieces. The director of Darmstadt at that time, Dr. Wolfgang Steinecke, was a wonderfully open-minded person. I remember hearing performances of Ravel and Bartók as well as of Stockhausen, me, Nono, Berio, and Cage, etc. The important fact is that Steinecke wanted to make a very wide open, world-wide festival of all kinds of musics. And unlike many people who make festivals, he was willing to go as far out in the avant-garde as was necessary, or as was going on at that time; as well as going back to the roots, like Bartók, Ravel, Schoenberg, Webern, Varèse, Berg, etc. Anyway, what I remember as the atmosphere at Darmstadt was open. Other composers would come to my rehearsals, and I would go to theirs; and it was a really vital, creative atmosphere. I felt that I was really part of a *community* of extremely interesting composers. That first year I met René Leibowitz and Rudolph Kolisch. I remember the Parrenin Quartet, who were playing in my piece, kept breaking strings on their instruments . . . they were having particular difficulty with pizz, and snap-pizz. I showed the score to Kolisch and said, "Am I stupid to be writing such a thing for the strings?" And he said, "No, that's perfectly possible." To have Kolisch say that that's a perfectly possible and proper thing to write for the violin was a *very* important thing for me. The first time my music was published, it was published by Schott, because Luigi Nono heard *Pentathis,* which is for nine solo instruments. Nono really liked it a great deal and said to his publisher at the time, Schott, "You must publish this composer." Which they did. They published two piano works and took *Pentathis* as a rental score.

**RD** None of your other works were performed?

**EB** Not at that time. No, they were still very far away from what Darmstadt was interested in doing at that time. As I said, David Tudor had done those *Folio* pieces on "Night Concerts" at Darmstadt. The concerts began at about 11:00 p.m. and ended at 2:00 a.m. But those early graphic works were not included in the main body of Darmstadt performances in 1953, '54, '55. The first time I did anything at Darmstadt that gave the performer any latitude—and by performer I mean conductor—was 1961. As a result of the success of the 1958 piece, *Pentathis,* Darmstadt commissioned me in 1961 to write a piece. The piece turned out to be *Available Forms I,* which was the first open-form score for orchestra, and it was in proportional notation. Bruno Maderna conducted it. I have the review upstairs, and it's a very interesting document. But that work started a whole controversial and influential atmosphere . . . that really was the time when people started paying attention to my music and to the innovations of my open-form work and scoring and notation techniques.

**RD** Who seemed to dominate the scene? Was it Stockhausen, was it Boulez? Or were they all pretty much equal?

**EB** It didn't seem to be a dominated situation at that time. Later it came to be that Karlheinz dominated the Darmstadt Festival like crazy. But at that time, Karlheinz was one of the composers, along with Boulez, Maderna, Berio, Nono, and Pousseur. It was a very well-balanced thing, and that's because of Dr. Wolfgang Steinecke. He was a very quiet man, but he was a very strong man, and he wouldn't be pushed around. It was, as I remember, a very equalized situation. I have a lot of the early Darmstadt programs upstairs. You look at them, and you see that Varèse, Debussy, Bartók, Ives, Babbitt, Wolpe were performed there. The so-called Darmstadt "School" was really a misnomer, because Steinecke's policy was *much* broader than the reputation that has come down. The ones who attack it usually were not there. And that is probably why they attack it!

**RD** Would you consider it as unique as we now have it described to us? Was it unique as an international situation?

**EB** Absolutely, I think it *was* unique. The composers that we now know as major composers in the world had first performances there of pieces that are now very important and influential. It was really a kind of clearing house for the avant-garde. I don't know of any place in which so many significant works and/or aesthetic attitudes were allowed to flower and mix. One of the things that happened around the time that I got there in 1958–'61 was the collision between aleatoric ideas and pure serialist thinking. They were really the polarities, and it was really where they significantly bumped into each other. And as I've said in an article, they were such extremes. Serialism is such an extreme of control; and aleatoric music at its worst, or at its most extreme, was almost carte-blanche improvisation. And those two things—total mechanical control and total intuitive, inspirational, improvisational aspects—were polarities which *had* to come back together as far as history is concerned. Some of us really went to the extremes of disorder and order within a very short period of time. What I think has happened, and I think it was inevitable and necessary, is that we have come into a balance between these two things. Of course, I was writing graphic and open-form music in this country long before I went to Europe. What made me write music the way I wrote it, and still write it, is more connected to my experiences as a young artist, influenced by Kenneth Patchen's poetry and James Joyce and Gertrude Stein and the Abstract Expressionist painters and sculptors. That all happened to me before I ever went to Europe. What really happened was that those influences which were totally American *really* influenced Europeans. I've always said that I could document the fact that this was the first time that musical influences went from this country to Europe, rather than the other way around.

**RD** Do they recognize that?

**EB** They are *very* reluctant, but more and more they recognize that they *must* or they will be seen as musicologically *dishonest!* A lot of Europeans will not recognize it. For instance, Boulez writes a lot of books, and he writes for and against things without ever mentioning sources or the names of anybody. I first showed Boulez my graphic scores and the proportional notation in New York in 1952, when he was here on his first trip to America (he was the conductor of the Jean Louis Barrault–Madeleine Renault Theater Company). And he also saw my early twelve-tone serial piano works which he liked a lot. And then I showed him and talked to him about the *Folio* pieces, and he said, "Oh, no, no, no. Composers cannot do that. *We* are the ones who know, *we* are the ones with taste, we are the ones who know the way it should be." And soon after that he wrote the Third Piano Sonata and became more and more "open."

**RD** What led you to writing serial music? Was it the Schillinger experience or what?

**EB** Yes, it was Schillinger and also studying with Roslyn Brogue-Henning. She was a twelve-tone composer in Cambridge, Massachusetts. What led me to write twelve-tone serial pieces was a combination of her teaching me twelve-tone writing, à la Berg, and my own other studies, at the same time, of Schillinger. Schillinger has an arithmetical, or mathematical approach to the organization of pitch groups, rhythmic series, etc., which strangely turned out to be quite parallel to Messiaen's teaching. What he and Boulez call cellules, Schillinger called rhythmic groups. They are almost exactly the same thing. In my early twelve-tone serial pieces are Schillinger-influenced manipulation of rhythmic groups and tone rows: *Three Pieces for Piano* (1951), *Music for Violin, Cello, and Piano* (1952), and *Perspectives for Piano* (1952).

**RD** Are those performed much?

**EB** No, they're not.

**RD** Do you want them to be performed?

**EB** Absolutely yes. They are very tough. They will be performed, but it's going to be after people come around to realizing that they have to look at the *roots* of things in order to study and understand the *leaves.*

**RD** What made you switch from that process and go to the opposite pole?

**EB** What made me become dissatisfied with Schillinger-based, twelve-tone serial constructivist kinds of pieces was my jazz background. None of my music sounds jazz-like, but I played trumpet in jazz combos and big bands; and I have always had a lot of respect for spontaneity and improvisation. And that's where the influence of Jackson Pollock comes in; because Pollock, in a sense, was improvising and performing his paintings, in real time, to a large extent. Other things were added later of course, but he was not sitting there and saying, "Well, I'm going to structure this up here and this over there." He was a very intuitive painter. And jazz is a very intuitive musical art, as you know, and I came out of that. Neither Cage nor Feldman ever had, to this day, any interest in jazz. But my background in the spontaneous creation of music with other people, without scores, was very important; I would not have been able to make *Folio* and graphic scores and collective improvisational scores and a notation which allowed flexibility in 1952 if I had not had the experience of trusting musicians.

**RD** And you knew Pollock quite well?

**EB** Not quite well, but I knew him, and I saw him frequently. He was a very difficult person to know well. He was usually "drunk and disorderly"! I knew his work better than I knew him. The look of his work and what I knew about his process of composition, let's say, was a tremendously big influence on me. It gave me the nerve to write music which was more direct and immediate than one would say about Webern's music, for instance, which was so calculated. So I was a schizophrenic about the balance between intuition and impulse, and, on the other hand, with Schillinger and calculation, the idea of making plans.

**RD** Do the words "ambiguity" and "relativity" make any sense to you as a composer?

**EB** Sure. I've always been interested in what I call *creative* ambiguity; ambiguity as a creative function of the activity of a listener, conductor, or musician. I've always considered that, as a composer, I was a primary creative input; to create situations which then went on to stimulate further situations, in which ambiguity would be a creative part of the outcome. In other words, this was a very strong thing that I felt from literature, from Joyce and Stein especially; the fact that we didn't know *exactly* what they were talking about, *exactly* how they wanted us to interpret it. This allowed my mind, as a reader, to become a part of the actual creative process. You know my music very well—you can see the parallel between what I do as a composer who allows you as a conductor to have a creative input. You might do something that I would never think of doing. A composer who was very strict or determinate about what his score should sound like would tell you, as a conductor, that you did it wrong. Whereas I might say, "Hey, Richard, that was fantastic! I never expected that to come out of that situation!" But I, as the composer, created the situation out of which it could come, through *you!* This is so deep. It's almost like I create a program (as in computer "program"), which is *Available Forms I,* for instance. In that work there are twenty-nine musical events that I've written, and that's a basic "palette" of material.

**RD** So "ambiguity and relativity" you feel are artistic virtues.

**EB** Absolute virtues. And the fact that I would be surprised by certain things that would occur in your conducting of *Available Forms I,* or Bruno Maderna's conducting of it, I consider to be a *positive* effect, rather than a mistake. I mean, you can make a mistake, I can make a mistake, you can conduct Mozart and make a mistake. I haven't taken anything away from the fixative of traditional writing and conducting. What I hope I have done, and what I was consciously trying to do, was to inject a degree of ambiguity and relativity into a score by Earle Brown, which was conducted by you or Bruno or Boulez or me. I have a lot of examples; for instance, of how Arthur Weissberg, with his background as a conductor, conducted a performance of *Available Forms I* that was very Stravinsky-like, and it was very good. The important thing is that I wanted that "personality" to come into the life of the score. Ives said a similar thing. He said, "If I write it down and I make it absolutely fixed and accurate, it will take the life out of it." In other words (my words), the more I solidify it, the less life of its own it has. Now, whether it's dumb on my part, or idealistic or romantic, I really felt that I wanted to make something which could have a "quasi"-life of its own. It's like a quasi-organism. Every time *Available Forms I* is done from 1961 on, it will be the same piece, but it will have a different ambiance and formal manifestation. And you as a conductor are a part of its environment. Seiji Ozawa and another conductor did *Modules I* and *II* in Japan and sent me the tape of it (it is for two conductors and large orchestra). Seiji and the other Japanese conductor conducting *Modules I* and *II* will perform the same piece, but with quite a different feeling from *Modules I* and *II* conducted by myself and you, for instance. The difference between the two realizations would upset a lot of composers. But to me it is my inclusion of a second degree of creative expansion . . . it's an injection of ambiguity and relativity as a creative potential.

RD You mentioned Pollock. Now, of course, we know that Alexander Calder influenced your thinking tremendously; and certainly here was a man who considered ambiguity and relativity in space in a wonderful way. Was that a different influence as opposed to Pollock, or were there similarities?

EB No. I think the first influence was Pollock. I remember in 1949, I saw a *Life* magazine article about the Abstract Expressionists in New York. Pollock started his characteristic painting in 1947. It took *Life* about two years to get around to recognizing that there was a de Kooning, Rothko, Kline, Pollock, etc., in New York. And I have the magazine. I was studying in Boston at that time; and I remember seeing those painters, and I was very impressed by all of them, although more by Pollock than by the others. I was studying basically twelve-tone polyphony with Roslyn Brogue-Henning. Pollock is a visual representation of tremendously intricate polyphony; the webs of lines and splashes, and bursts of color. It *looked* like what I wanted my music to *sound* like.

RD And you see nothing figurative about it?

EB No. I've always been drawn to sound as an abstract material, rather than writing a piece about mountains or the ocean. I've always been drawn to sound as an abstract material for construction and creation of sonic objects . . . *objets sonores.* Pollock was the first thing I was really drawn to because of the intricacy of the textures. Calder was the one who gave me the aesthetic base for making open-form music. It was about the same time. But I remember looking at Calder's work at the time, and I have quotes and things in my notebooks from Calder. You know, Calder makes a mobile of fifteen different metal objects, each designed and created by him, and put into a relationship. But he didn't make it *one* relationship; he allowed it to float. So this Unit #1 and Unit #12 were sometimes overlapping, and sometimes you saw #12 on the left and #1 on the right. And sometimes you saw it the other way. That was the key that gave me the whole basis and confidence to make an open-form score.

RD So it was a visual cue that sent your mind going toward the aural.

EB Yes. I thought, wouldn't it be marvelous, for instance (putting it in traditional musical terms), to make variations for orchestra without making them go 1-2-3-4-5-6-7. Why can't I make a score which would be a "variations for orchestra"? But I have worked out a way for the conductor to be able to control the sequence and juxtapositions in a completely spontaneous and different way each time.

RD The difference might be, however, that a variation usually has a theme, and you don't have a theme.

EB No, but each of the twenty-nine units in *Available Forms I* is self-descriptive, to the extent that they could be considered twenty-nine themes. In a sense, if you or a musicologist looked at it from an open point of view, and I said to him as I've said to you, "What I've made is 'variations for orchestra' which can be constantly juxtaposed in different ways," what's wrong with that?

RD So that was somewhat the atmosphere for you in Darmstadt and a little after . . .

EB Well it started earlier than that, with 1952 and 1953, with *Twenty-five Pages* and *Folio. Twenty-five Pages* (1953) is the key-work of open-form. It is a score of twenty-five unbound pages, to be performed in any order or inversion of pages. And that is a definite thing. Doing the *Available Forms* in 1961 was my first opportunity to be commissioned to write an *orchestra* piece, which used the same principles.

RD Now, I'll repeat a bit the idea that you had introduced . . . "giving the conductor a certain latitude in performing your music." And you found that exciting. How about today?

EB Well from the beginning there were composer/conductors who loved it. Bruno Maderna loved doing it, but Bruno was also a composer. I know for a fact that it scares the wits out of some conductors, because they are put into a situation, by the open-form score, of making decisions which they are not used to making. And they worry about whether their decisions are going to be as good as someone else's decisions. So I recognize that a lot of conductors shy away from these scores. Also, I have a reputation of being able to do them very well. I've really inadvertently created a kind of competitive situation.

RD I remember the first piece I did of yours was *Novara*. I must say, I had a kind of queasy feeling, too . . . am I going to invent? Am I going to be creative enough?

EB But you see, that's my idealistic and romantic feeling that *everybody* is creative.

RD Do you still believe that?

EB Well, I still think that everybody is creative, but I now know that a lot of people don't know it, in the sense that they don't have the confidence. I mean, a composer has to have a lot of guts, whether it's open-form or closed-form, but basically, talking about closed-form . . . to write a piece and say, "This is my statement," and get knocked on your ass by Henahan and/or other critics, takes a lot of confidence. Number one, it takes a lot of confidence to *be* a composer. I have a lot of closed-form pieces, if someone wants to check out my ability to fix a form. But the open-form pieces, I really thought, wouldn't it be marvelous if I could create this "environment" and invite other people to make music *with* me.

RD You found that a necessity then. Do you still find that a necessity today?

EB I find it a necessity, but I am more circumscribed, not convinced anymore that so many people can do it well. I think they can, but they frequently lack confidence.

RD You've mentioned Bruno Maderna many times; and of course, we all remember him as a marvelous musician and a wonderful, spontaneous person in his nature and in his life and in music, too. What about conductors? In terms of your music?

EB Bruno was the best, but as to conductors in general, relative to my work, it's very strange. Two of the best performances of *Available Forms II* (for ninety-eight performers and two conductors) were in Paris with Charles Bruch. He's not a composer, and he was a relatively old guy when we did it. But he had no problems with it. I think it's a psychological thing. If somebody is basically nervous, they have difficulty doing the pieces. Bruch was not a composer, but he had no trouble entering into it. And Bernstein, with whom I conducted it with the New York Philharmonic, introduced the piece in the 1964 Avant-Garde concerts . . . and he said, we have a very interesting and unusual piece. It asks the conductors to *form* the music which the composer has written, differently every time, from performance to performance. He said it was one of the most difficult things he had done, but he was a composer and he could do it, and he would do it as a creative act of the moment. He really explained it very well to the orchestra, with great respect, and I know he liked doing it. He said it was one of the most challenging things he had ever done.

RD As opposed to the conductor, and in comparison—to what degree do you give freedom to the *performers?*

EB First you have to understand that there is not a great deal of music of mine that gives the *performers* a great deal of freedom. The early *Folio* pieces do—and you know; I got a call from the *New Grove Dictionary of American Music.* They wanted to have the right to have a copy of *December 1952* reproduced in my bio. And I said, "Everybody wants to reproduce that piece. It's as if I hadn't written anything after *December 1952.*" It is this famous/infamous graphic piece, which is simply horizontal and vertical lines in a spatial field, graphically indicating high, low, long, short, loud, soft.

RD Well, by now it's a postcard. I have one.

EB But anyway, that's the most famous and most far out and most notorious piece that I or anybody else ever did, except for Cage's *4′ 33″* (4 minutes and 33 seconds) of silence. But they both were extremes, and people don't realize that I have written perhaps eighty-five pieces of music, most of which don't give any freedom to the performer at all. Many give the freedom of *forming* my composed material to the conductor. Since 1952, I have written some pieces that give freedom to the performer. For instance, the piece *Centering* for violin and orchestra, commissioned by the London Sinfonietta for Paul Zukovsky. Now that is a balanced score, between closed-form and open-form. With the violinist in one of the cadenzas, there is improvisational possibility based on my notes; I give him a stream of notes and give him a verbal construction as to how to play it and what he can do with it . . . his parameters within which he improvises. And that works extremely well. It has been done by many, many people. I think of these open-form compositions as being in the hands of the conductor once I've written all the material. The conductor is improvising with the orchestra as the instrument.

RD Then, one step further. We worked together on those two-conductor pieces, which is another form of engagement that I find very, very challenging, when you are trying to complement or offset what you are hearing from the other side.

EB The idea of two conductors conducting the same work came right out of *Available Forms I.* When Bruno did it in Darmstadt in 1961, he immediately got me a commission from the Rome Radio Orchestra; and that became *Available Forms II,* for ninety-eight instruments and two conductors, because I was jealous of all the fun Bruno was having up there doing *Available Forms I.* So, in effect, I sort of built myself into the piece. And the first time I ever conducted was in *Available Forms II* at the Fenice in Venice in 1962. That piece I consider a sonic, conversational collage. It's one piece created by the interaction of two creative and sympathetic people.

RD The other piece, *From Here,* involved voices . . .

EB A chorus of sixteen voices. That I wrote on a commission from a New York organization, and it turned out that I was going to be able to have Alvin Lucier's chorus from Brandeis University . . . that was in 1963. So I built that . . . it's a different relationship from *Available Forms II* . . . you know, because in Aspen you conducted the orchestra, and me, in effect. Whereas I was the chorus conductor, you could cue me in or out according to what you wanted the final, overall form to be. That's exciting. I'll never forget when we did *From Here* and you started the performance with the orchestra, and then I brought the chorus in (the chorus was sitting right behind this blonde violinist); and because it's open-form and we don't do the same thing in performance as we do in rehearsal, I brought the chorus in with a big "YAH!" And she jumped about three feet off her chair! That was a terrific performance.

RD I believe you had four formal invitations to Aspen, and I think you also just showed up once.

EB The time Gilbert Amy was there I showed up just for fun.

RD Your first appearance as composer-in-residence was in 1970. I had programmed three areas; Music Theatre, Alea, and The Transcendental. And in Alea, we did *Event: Synergy II.*

**EB** I remember Nicholas Nabokov being there and liking it very much, and I was very flattered that he was so impressed by it.

**RD** In 1975, the American Bicentennial Part I, "The Americans," you were there with Copland and Del Tredici. The pieces included *Centering*, with Paul Zukovsky, *Times Five* and *Syntagum III*. Then you were there in 1981. I had as a theme "The Liberation of Sound and Forms Freed," subtitled "Sound Works." That was the year you brought the Calder mobile and we did *Calder Piece*. You and Philip Glass were composers-in-residence. And of course, last year, which was unfortunate, because you had to give up your rehearsal time for other things, and we never got to do your work *Available Forms II*. Do you have any overall picture or recollection about Aspen? How did it figure in your musical experience?

**EB** Some of the best performances of my music happened there. One of those years (1975), I was three weeks at Tanglewood and three weeks at Aspen; and I'll never forget that. Theodore Antoniou and I were to do *Synergy II* in Tanglewood, and we did a lot of rehearsing before I went to Aspen. Then I went back to Tanglewood to perform it; and one of my rehearsals was cut out. I was *incensed* that they had cut three hours out of my rehearsal time. I recall that especially because, when I was at Aspen, there were three or four works of mine done by you and/or me; and the performances were incredibly better—better musicians, better concentration, better performances overall than anything that I had had in Tanglewood. That's why, when I became a director of the Fromm Foundation and Paul had moved it out of Tanglewood, my first feeling was—since I had had such fantastically good performances in Aspen—that's where the Fromm Festival should go after Tanglewood. But then, as we know, the performances in 1985 were not that great, because we were not given the best performers, and that was a great problem for us. But apart from that, I love the environment, the musical-orchestral environment, and your devotion to the pieces.

**RD** Aspen really benefited by the appearance of composers like yourself. I did feel the excitement in the performances when we had good people. It was unique in that sense. In '85, we discussed the idea of the New Romanticism. Do you have any overtones from all of that discussion, now that there has been some time in between?

**EB** About New Romanticism? I think that Jake Druckman came up with that title. It's still going around. I've always considered my music romantic. The early piano pieces are very severe, but most of my music since then I have always thought to be very romantic.

**RD** What makes it romantic?

**EB** Well, it has a lot of color in it and delicacy and rhythmic intricacy, and it basically has a long gesture. The open-form pieces produce a sort of continuity which is not typical of what you might call romantic continuity; but from moment to moment, I always think that . . . I guess any composer thinks that his music is beautiful. And I do think it is beautiful and romantic. I don't write ugly music, and I don't write "angst-full" music. I have never been wildly involved in middle European angst or expressing terror. But as I said earlier, I think of sound as an abstract material, like the Abstract Expressionist painters thought of paint . . . line, color, texture, density, intensity as abstract materials. And when I write a piece, I think of all of those things.

**RD** There is another word, in addition to the discussion we had about ambiguity and relativity, and that is the notion of irreducibility, which was a concern of some painters. Musically speaking, it could be something that the composer calls "basic." Like Webern is "basic" and sparse. Did that enter your mind as a concept?

**EB** As economy, yes. René Leibowitz's book *Schoenberg and His School* always talked about Webern as the epitome of economy, using the simplest materials to get the maximum effect; and I think I try to do that, too, but from a totally different point of view. Very often my scores, like the orchestra piece that was played just now in Cincinnati, *Time Spans,* are all written on three pages. Now that is a kind of paginational economy, which I'm not talking about. It's also restricted to one single chord: sonority. After a certain point, which I express in the program notes for *Cross-sections and Color Fields,* I tried to strip away and to simplify the tremendous complexity of *Available Forms II* into simple chordal structures. *Modules I* and *II* are like that. *Time Spans* and *New Piece: Loops* are like that. *Cross-sections* is like that. . . . To do the maximum with a simple amount of material, and to limit the material that I compose and still have the maximum of interactional complexity. That's characteristic of a lot of recent scores. I never thought about irreducibility as a concept, but I know what you mean.

**RD** There are two words that can cause a lot of confusion. One is "chance" and the other is "choice." The term *chance-music* has often been used, perhaps incorrectly, to describe anything that involves improvisation or the spontaneous act and result of performance. Composers of so-called *chance-music* are indiscriminately lumped together, in terms of their compositional input. The word "predetermined" can heighten this confusion about "choice" and "chance." In your mind, can *chance* be predetermined?

**EB** "Predetermined chance" is completely self-contradicting, and intrinsically irrational as a philosophic point. There is either "predetermination" (choice) or "*other*-determination" (chance). I don't use chance! Do you think Indian music is *chance-music?* Do you think jazz is *chance-music?* Nobody ever called it *chance-music.*

RD If a player is improvising on a tune or just on chords, is he taking a *chance?*

EB No! He's not taking a *chance,* he's playing what comes next in his own mind. He *determines* what will come next. The only way a jazz musician could be anywhere like *chance* is if between one note and another he flips a coin. And they don't do that. *I* don't do that. When you conduct my open-form pieces, you're not doing it by chance. You're doing it because you want the next thing to happen. Because you think it's right. And that's what an improviser does. It's what a composer does who writes closed-form music: but he does it in his room upstairs, rather than doing it on stage. But still, you can't conduct my open-form pieces by chance unless you do *Event 1,* and then you flip a coin which gives you twenty-nine possibilities, and the chance flip comes up 13, so you do *Event 13.* There's a *huge* difference between improvisation (spontaneous decisions) and chance. Chance really has to be an *exterior,* objective thing. You go to Las Vegas, you have a machine . . . the one-armed bandit. It's got a chance thing going on inside it that they structured into it, right? Or you play "21," the way the cards come up. But those are exterior to any subjective decision. Whereas in my open-form pieces, it's all interior and subjective decision making. In my opinion, John Cage is the only composer I really know who is a chance composer.

If two people are conducting *Available Forms II* and I conduct one event which is strings, your mind is immediately going to say, "I want 'X' event with that." That's not a chance thing; it's a *decision.* In other words, there is a real difference between *deciding* and "chance" being allowed to operate. That distinction is very important to me, because I have never written *chance-music.* When Lenny did the 1964 Avant-Garde thing, he put Cage, Feldman, and myself on the program and called us the Chance Composers. I won't fight what he wants to call it; but musicologically it's got to be clarified. The difference between flipping coins as John Cage does, true chance operations which are to separate himself from the event that is going to happen, is morally, intrinsically, and in every other way different from my putting you as a conductor in the situation where you can make a *decision.*

RD What do you have to say about the word "original"? Do composers have to be original, or do they steal?

EB Well, they're influenced. Some of them steal. This is an interesting question. I've written things in my notebooks about this. It's just inconceivable that composers would not be original. I mean, who wants Mozart rewritten? If you take the opposite of original, it is *un*-original, and who respects that? "Original" has come in for a lot of badmouthing. "All he wants to do is be original!" Like LaMonte Young or whoever it was, Nam June Paik, burns a violin onstage, or destroys a piano by pushing it off the stage . . . that's not original. A very interesting thing has occurred to me, in terms of Harry Partch for instance, whose music I don't like, but he was original. My question is, "What is the difference between being original and being merely eccentric?" In my feeling . . . somebody can be original because nobody ever wrote a piece for four motorboats crashing in Mamaroneck Harbor. It might be an original idea, but it's just eccentric. It has nothing to do with the mainstream of the history of the development of art. I consider myself in the mainstream. What I've done will affect the history of music tremendously, and it already has.

**RD** How many ideas do you think a composer has in the course of his lifetime?

**EB** It depends. How many ideas did Picasso have? How many ideas did Bob Rauschenberg have? I mean, those are two visual virtuosos.

**RD** I mean, is an idea repeatable or non-repeatable?

**EB** That's the thing about "original" or "innovative." An innovation or an original move that an artist makes is only important if it's not repeatable. "Repeatable" is important, but only if somebody follows. That's another thing about "original" or "eccentric." In my opinion, there can be all sorts of people calling themselves avant-garde. And other people calling them that, too. But in the real sense of avant-garde, it means the person who's out ahead of the troops, in effect. Like in the old West, it's the scout. He goes out and finds the path through the rocks, the mountains. And if he finds a path, the wagons follow. That's avant-garde. You may be "avant-garde" and nobody follows. Then you're *not* avant-garde, you're just eccentric.

**RD** Well, now, who was original?

**EB** Cage was certainly original. But I think I was original, and I think Feldman was original. I think Boulez was original, out of Messiaen. But you see, I'm original out of Calder. Calder gave me ideas. And Pollock. Nobody is totally original. But within music I do things under the influence of Pollock and Calder that nobody else thought of doing.

**RD** So there's a little thievery involved. It's not a bad thought.

**EB** No, but it's a very negative and un-philosophical, insensitive way to put it. Did Picasso *steal* from African masks? I don't think it's thievery.

**RD** Borrowing?

**EB** *No . . . influence!* I've been talking about Calder and Pollock as big influences on me for years. And twenty years ago, in writing about it, I came upon a quote from Baudelaire which I wrote in many articles and continue to use as a quote. Baudelaire said that "the arts tend, if not to complement each other, to lend one another new energies." And that's what Calder, Pollock, and many others did for me. That's across the arts; Stein, Duchamp, Joyce, or whatever. They really lend each other new energies. And Calder lent me an energy which went into music for the first time, in the concept of open-form mobility. And that was an energy. . . .

RD And Bucky Fuller lent you a name. . . .

EB *Synergy!* Sure. That's not stealing. It's a valid and indispensable fact. We're all in this culture together. As Gertrude Stein once said, there's no difference from one century to another, except the *lighting* is different. It's true. From one thing to another. And I have a lot of quotes . . . Varèse used a quote about nothing changes from century to century except the focus . . . or something like that. It was almost a paraphrase of what Gertrude said. I love that idea. . . . The lighting is different; and between Phil Glass and Steve Reich and myself, Feldman and Cage, they put a spotlight on a different aspect than we did. Nothing comes from zero. It all comes from sensitivity.

I have always been very sensitive to all of the other arts. And for some reason, more so than a lot of other composers, I have been very influenceable by paintings, sculpture, literature, etc.; basically the nature of compositional *process!* That's the difference between how I was . . . it goes all the way back to childhood or something. What made me be dissatisfied with Webern? You can't answer that. What made me be dissatisfied with jazz and start to write arrangements? What made me start to be dissatisfied with arrangements and start to compose, while Quincy Jones is still composing and doing arrangements? Which is not to knock Quincy or anything, he's a brilliant guy and a great musician. There are different things that one puts one's lights on. I put on a spotlight for myself, as an artist, on Calder and Pollock, which nobody else did in the same way that I did. I see the development as a totality, a Gestalt thing. If you're sensitive to it. I think if you sit around as a composer and simply write music because of the music that you hear from other sources, then you're not going to write music of any cultural significance; because all you're doing is making variations on what came before. Which may show great talent, but it's not creative. There are a lot of composers whom I will *not* mention who are not sensitive to these exterior influences to their own art; they are writing in a vacuum. It may be interesting momentarily, but it won't be culturally significant if it doesn't change the way music is thought of poetically. I never set out to be self-consciously trying to say, "What can I do that will change the history of music?" But what I did has made a big difference to the way people write scores these days. . . . The open-form thing is a kind of activation of a stream of consciousness. And maybe, contrary to Boulez, I have been very interested in the subconscious mind . . . what *your* background is feeding you when you are confronted by *my* score. Your history is going to come into it if you let it. And what a lot of conductors do is freeze. They say, "I don't know what's best here." They shouldn't worry about what's "best" here. They should worry about what sounds best, next. . . . If I've done my composing well, and my conditioning of the sonic environment in which I request that you enter my composed but flexible sound world with musicality and sensitivity—not to worry—enjoy!

# Narrative VI

"A Festival of Stravinsky—His Heritage and His Legacy" was the official title of the 1966 New York Philharmonic summer festival that ran from June 30 to July 23, in Philharmonic Hall, Lincoln Center: Lukas Foss, Artistic Director; Richard Dufallo, Assistant to the Artistic Director. It was the first major festival in America to pay homage to Igor Stravinsky.

Since Stravinsky would be present and would conduct, it was obvious that he should bring the festival to a close. It was equally obvious that Music Director Leonard Bernstein should begin the festival, and he agreed to the subtitle "Stravinsky and American Music" for the gala opening. His program began quite logically with Stravinsky's arrangement of *The Star-Spangled Banner,* something Stravinsky wrote shortly after he took up residence in the United States at the outbreak of World War II. He had applied for citizenship and in return said, "Searching about for a vehicle through which I might express my gratitude at the prospect of becoming an American citizen, I chose to harmonize and orchestrate as a national choral the beautiful, sacred anthem *The Star-Spangled Banner.*" The score is dated July 4, 1941. Much later, in 1944, after a performance of the anthem conducted by Stravinsky in Boston, a police commissioner evidently appeared in Stravinsky's dressing room admonishing him and advising him of the existence of a Commonwealth law forbidding any "tampering" with national property.

Bernstein chose to end his program with *Le Sacre du printemps* (1911–13). What a beginning! In between *The Star-Spangled Banner* and the *Sacre* appeared Samuel Barber's *Capricorn Concerto,* Aaron Copland's *Dance Symphony,* and Silvestre Revueltas's *Sensemaya.* Stravinsky chose to conclude the festival with his *Symphony of Psalms* (1930). What an ending!

Foss quite rightly gave each program a subtitle. "Stravinsky and Russian Music" needed a Russian conductor, and Kondrashin was available. "Stravinsky and French Music" needed a French conductor, and Pierre Boulez was asked, but he was not available. However, the Swiss conductor Ernest Ansermet was available. Elisabeth Schwarzkopf was engaged for "Stravinsky and the 18th Century" and "Stravinsky and Song." For "Stravinsky and the Dance," George Balanchine choreographed two short pieces: *Elégie* and *Ragtime.* "Stravinsky and Italian Music" needed a special cast for *Oedipus Rex;* and Jason Robards was engaged as narrator, as well as Shirley Verrett (Jocasta) and Ernst Häfliger (Oedipus). The painter Larry Rivers was to make a special visual presentation. "Stravinsky and Recent Years" was a program to be chosen by Igor Stravinsky. That was the framework, but the juxtapositions in each category proved very enticing.

Stravinsky's love affair with the 18th century could not be more aptly framed than with *Pulcinella* (1919–20), after Pergolesi, and his opera *The Rake's Progress* (1948–51). Schwarzkopf had been chosen by Stravinsky to create the leading role of Anne Trulove in the first production of *The Rake.* In this program she not only sang the "Bedlam Scene" but also sang in *Pulcinella* and offered two arias from *Cosi fan tutte,* an opera with which she was closely identified, and one in which Stravinsky had steeped himself while writing *The Rake.*

Mme. Schwartzkopf's song recital nicely complemented this vocal emphasis with Stravinsky's early songs and those of Moussorgsky, Tchaikovsky, and Rachmaninoff.

Of the three nationalistic areas to be explored—Russian, French, and Italian—Russia and France also represented citizenship for Stravinsky before he came to America. In considering a program of "Stravinsky and Russian Music," certainly there would be two important elements: first, his teacher Rimsky-Korsakoff, and second, his admiration for Tchaikovsky.

The *Fireworks, Fantasy,* Op. 4, is rich in historical significance. At twenty-five, Stravinsky composed it as part of a celebration of the marriage of Rimsky-Korsakoff's daughter. Days after sending the score to Rimsky-Korsakoff, Stravinsky learned of his teacher's death and soon after received a package containing the score marked "Undelivered on account of death of addressee."

It was because of *Fireworks* that Diaghilev invited Stravinsky to collaborate with his Ballets Russes in Paris. Therefore, it is not surprising that *Fireworks* (1907) and *Petrouchka* (1910–11) were chosen as representative works. The not-so-familiar Piano Concerto by Rimsky-Korsakoff, Op. 18, performed by André Watts, provided a gentle historical reminder of the Russian period and Stravinsky's heritage.

His early association with Debussy could be regarded in programming "Stravinsky and French Music." Debussy dedicated *En Blanc et noir* for two pianos to the young Stravinsky. He, in turn, dedicated *Symphonies of Wind Instruments* (1920), written in France, and the short, enigmatic cantata *The King of Stars* (1911–12) to Debussy. The latter work, also known as *Zvedoliki* or *Le Roi des étoiles* is rarely performed because of its brevity and the size of instrumental and choral forces required. It was written between *Petrouchka* and *Le Sacre.* The dedication had little to do with the music of either composer. These two works plus *Perséphone* (1933–34), a melodrama by André Gide, also written in France, made up the Stravinsky part of the program; Poulenc's Concerto for Organ and Machaut's *Hoquetus* completed it.

In the program "Stravinsky and Italian Music," the late Renaissance music of Gesualdo, Gabrielli, Monteverdi, and the *Te Deum* by Verdi provided a proper reflection of what attracted Stravinsky to the musical traditions of Italy. This music was also an appropriate prelude to the opera-oratorio *Oedipus Rex* (1926–27), after Sophocles, written with Jean Cocteau.

The visual presentation by Larry Rivers had its problems. In his statement Rivers said, "Oedipus is a leader, yet he is everyman who lives to see that his destiny is out of his control, whose grim fall and oblivion is waiting in the wings. My conception of what to do with this not so very Neo-classical score is to surround it with flags, skies, and plagues. To erect a bright stadium. Packed bleachers. Points of power."

The chorus and orchestra were arranged in such a way that a "stadium" or "boxing arena" effect was apparent, complete with flags. Rivers suspended painted clouds above the stage, but they seemed too small in scale. The tenor Häfliger had the most trouble in that he was required to wear a "practice" boxing helmet which made him sweat profusely during the performance. Toward the end, when Oedipus gouges out his eyes with his mother's (wife's) golden brooch and the chorus is bidding him farewell:

> Vale, Oedipus,
> miser Oedipus, noster,
> te amabam, Oedipus,
> Tibi valedico, Oedipus,
> tibi valedico

Rivers had large, blinding klieg lights turned into the faces of the audience. The reaction was negative and cast a sour note on what otherwise was a stunning performance under Foss.

The title "Stravinsky and the Dance" was stretching the point a bit. This was a chamber music program, and no full-scale ballet was presented. However, Balanchine's choreography of the *Elégie* (1944), for unaccompanied viola, with Suzanne Farrell, coupled with *Ragtime for Eleven Instruments* (1918), including Hungarian cymbalom, which I conducted with dancers Farrell and Arthur Mitchell, constituted a small "dance" offering. This, plus three excerpts from *Le Baiser de la fée* for violin and piano (Laredo, Votapek) and *Piano-Rag Music* (Votapek), all preceded a performance of *L'Histoire du soldat* with three American composers reading the text: Aaron Copland, narrator; John Cage, Devil; Elliott Carter, Soldier.

After the performance, Stravinsky came backstage and sat on the couch in the conductor's room, and was besieged with admirers, well-wishers, and others seeking a glimpse of or an autograph from the great man. The crowd droned and buzzed around him much like bees around the queen.

I stared at Stravinsky for a while, absolutely paralyzed. I didn't dare ask for his autograph. Eventually, I did muster my courage and asked him if he liked my performance of *Ragtime.* He looked up and smiled and said, "Eet vos batter zhan zhe premiere!" I stepped back and recalled our private car trip nearly ten years ago in California.

The works chosen by Stravinsky on the chamber program, "Stravinsky and Recent Years," were framed by two short works of his: *Fanfare* for two trumpets (less than one minute), and the first New York performance of *Introitus,* "T.S. Eliot in Memoriam" (four minutes). The other works included the New York premiere of Boulez's *Éclat;* Carter's *Études and Fantasy;* Foss's *Echoi,* in which I played the clarinet; the world premiere of Webern's *Kinderstück* (1924) played by a very young pianist, Caren Glasser; and the American premiere of Webern's *String Trio* (1925); and Varèse's *Octandre,* which I conducted. During the intermission Milton Babbitt's *Ensembles for Synthesizer* was performed on the Grand Promenade beneath the Lippold sculpture *Orpheus and Apollo.*

I tried to observe as many rehearsals as possible throughout the entire festival. Watching Bernstein prepare the *Sacre* was an overwhelmingly exciting experience, and he did what he always does—produced a *great* and inspiring performance. Ansermet, on the other hand, seemed bland and unfocused as a conductor. In contrast, Kondrashin was nearly violent on the podium, lunging forward and back like a panicked slave driver, producing extreme contrasts in the music of *Petrouchka* that struck me as artificial. That he needed an English/Russian interpreter in rehearsals was one thing. But that he used the interpreter in order to relate the story of Petrouchka, blow by blow, to the members of the New York Philharmonic, was rather tedious and condescending.

As we approached the final concert, tension did begin to mount. This was the third occasion that I would hear Stravinsky conduct his *Symphony of Psalms,* and the second time I watched him rehearse. After presenting more than twenty works by the great master, the atmosphere at this culminating point in the festival was electrifying.

At the finale, Robert Craft conducted the *Symphony in Three Movements* and the first New York concert performance of *The Flood.* After intermission Igor Stravinsky, eighty-four years old, stepped onto the stage and received an outpouring ovation of love, admiration, and respect that I shall never forget. We settled down, finally, and Stravinsky's performance of the

*Symphony of Psalms* transcended anything I had heard before. All of the performers literally hung on his every gesture and responded with a kind of spirituality rarely seen or heard. Needless to say, at performance end the place was nearly torn apart . . . *laudate dominum.*

I returned to Buffalo spiritually elevated by the summer and also actually elevated to the status of *Associate* Conductor of the Buffalo Philharmonic.

*Charles Abbott*

# *Bernard Rands*

BORN MARCH 2, 1934, SHEFFIELD, ENGLAND

**July 10, 1986** **Boston**

RD I don't remember when we met for the first time. Do you have any recollection?

BR It was Buffalo, in the spring of 1967. I visited because Lukas Foss had suggested that I might like to consider becoming a Creative Associate there.

RD Can you give me a summary of how you grew into becoming a musician and what led you to become a composer?

BR Yes. It's a simple story. I was born in England in 1934, into a so-called "working-class" family. In fact, it was one that was essentially deprived economically. My father was a school janitor but was a "natural" musician. He had never been trained, but he had a fantastic ear. He played the piano and the flute. My mother had a wonderful contralto voice . . . a voice which, when I was a little boy, I was slightly embarassed about, because it had such a low register. It was an extraordinary voice; but she too never received any training, nor indeed barely any formal education. Together they played and sang at various local concerts, events, churches. All my uncles, seven of them, were coal miners and played brass instruments in the colliery bands or sang in the miners' choruses which were typical of that area. So as a child, I was constantly surrounded by music played by these amateurs, who simply made music as a relief from the rather grim way of life they had to endure in Sheffield. Which at the time was one of the worst blots on any landscape in the world. At age nine, I had a piano teacher who, at the end of each weekly lesson, would write a simple melody in my manuscript book; and my task was to find a harmony for that melody. From that point onwards, writing music became as "natural" an act as playing it. And then I had organ lessons. This very same teacher who encouraged me to write also performed professional engagements; and at a certain moment he said, "Let's play a two-piano or a four-hands concert." We played repertoire from the four-hands and two-piano literature.

RD Did going to school in the English system permit you to continue playing?

BR Oh, yes. In school the option to study music was available, and I took that option. From the age of twelve I had regular music lessons in theory, harmony, counterpoint, and ear training and music history until I was eighteen and went to university. I entered the University of Wales; and my decision to do so was colored by a number of things. First of all, a strong feeling of affinity for the Celtic heritage; and second, I wanted to study Celtic literature. It is something that is at the core of my being, in a way that is hard to explain. Literature matters a great deal to me, especially that of the Celtic cultures, but literature in general. Later I became fascinated by the Welsh language itself.

**RD** And when did composition come along?

**BR** That was going on all the time. All through high school I did my technical exercises and composed small pieces. I had a number of friends who were instrumentalists . . . a clarinetist, a trombonist, a trumpeter, percussion, and piano. . . . I would make little pieces for each of us, in solo and in combinations . . . and then, obviously, when I went to university, new opportunities to compose and perform opened up.

**RD** I know you have done an awful lot of teaching of composition, which is a very difficult thing to describe. But was there some one person that you could say was your first composition teacher in a formal sense?

**BR** Yes . . . Reginald Smith-Brindle, a composer who had studied with Dallapiccola. He introduced the music of Petrassi and Dallapiccola to me, and also that of the newer generation of Maderna, Nono, and Berio. I was instantly and naturally drawn, especially to the music of Dallapiccola. It obviously had discipline and rigor in its technical concerns, at the same time as it had that Italianate quality which is hard to define. It is a music of intense passion—both lyrical and dramatic, a music of unfractured, *cantabile* lines.

**RD** What did you do with that new direction?

**BR** Immediately after completing my formal university studies, I went to Italy; and that's where I lived—first in Rome for a few months, and then in Florence.

**RD** So you had a very intimate time with Dallapiccola.

**BR** Yes, very much so, in 1959 and 1960. Whilst his music may be out of "vogue" at the moment, I think there is an underlying certainty in one's knowledge of his work, that here is a solidity, power, and magnitude of achievement that is rare.

**RD** I remember people being very snobbish about it, when the Schoenberg school was in its heyday. "Oh, Dallapiccola! He writes it as a tune!"

**BR** Thank God! Probably many of the same detractors couldn't appreciate Schoenberg's "tunes" either!

**RD** Were you working on pieces at the time and showing them to him?

**BR** Yes. He was a rather formal man—formal intellectually and formal personally. And he was formal in his use of language which, when he spoke Italian, had a slightly archaic quality of vocabulary and construct. He was also wonderfully kind and caring and most encouraging of me. The marvelous thing about those lessons was that they were not always preoccupied with musical concerns. He was an extremely well-read man, very cultured, with a vast knowledge of literature. So on occasion, the first forty-five or fifty minutes of an hour's lesson would be spent talking about Dante or Homer, Petrarch, Goethe, Thomas Mann, Joyce . . . and one wondered when or if there was going to be an opportunity to show him the music one had sweated over all week. One day when I arrived at his home, he was very excited when he met me at the door. Someone had sent him from Germany a tape recording which had been copied from a cylinder recording of Busoni playing *La Campanella.* He was so thrilled about this—the way that Busoni understood and heard this piece, and the way he played it. From these observations evolved what was always more than just a note-to-note lesson. However, I must say that, at the same time, he was a little pedantic about note order in serial procedures; and every pitch choice had to be justified according to the premises of the set. At that stage, he was drawn increasingly to the music of Webern, in terms of precision, manipulations, and the subdivisions of the set in his own music. I was moving away from that in very specific ways, and wanted to open out the sound world of my music. He was very good-natured about that. He recommended that I work with Bruno Maderna or with Luciano Berio . . . which at that time I was already planning to do.

RD Did you ever observe his work table?

BR I recall asking him, and he did show me his preparatory material, his sketches for several works. Everything was so neat! I couldn't believe it! You know, one's image of composers working in a tremendous passion and scribbling all over the place . . . not Dallapiccola! Everything was neatly drawn. The jar of sharpened pencils standing on the work table—and when he used an ink pen, he used a bag of sand as a "blotting paper." It was all idiosyncratic—very neat, everything under control.

RD Was this your first real confrontation with the serial method?

BR No. I had begun to adopt the principles while studying with Smith-Brindle. But of course, I was also studying the scores of Webern, Berg, and Schoenberg at the same time as I was drawn profoundly to the music of Debussy, Stravinsky, and Bartók. . . . I think that once I became aware of the historical growth of Schoenberg's ideas and of the musical opportunity these ideas afforded, there was no escape. In a sense, it was, as Schoenberg himself felt, a very natural outgrowth from one musical stake into another—and that the systematic ordering of pitch material, looking for an authentic way to control musical materials (as we had in the key system), was a perfectly natural thing to do. I was caught up in a natural way. I didn't adopt the practice in a theoretical, dogmatic manner. It was, in a sense, like repeating Schoenberg's experience (with the advantage of historical hindsight) some half-century later—discovering for one's self, step by step. In terms of my creative work, that is the only way I have ever been able to explore musical phenomena which engaged my attention and interest and then naturally assimilate and transform or reject and dispense with.

RD Moving to Maderna and Berio, was this an extension of the first experience, or was it a new experience?

BR It was an extension. Let me say, finally, that one has to remember Dallapiccola as being a very courageous man of great integrity, who did what he did during periods of his lifetime in a climate and context far from conducive to experimental, radical work.

RD You're speaking of his work ...

BR ... of his work during the Fascist years, a constant source of threat and, for him, an intolerable restriction on intellectual, creative, and social freedom. What he did was to make it possible for the generation of Nono and Berio and Donatoni and all of the names of the Italian postwar years to espouse a radical modernism previously unknown in modern Italy. To answer your question about Maderna and Berio, Berio too is a strict disciplinarian, but in a different way from Dallapiccola. Although he seems to have all the creative freedom in the world (and in a sense does, and stipulates and maintains the right to do so), it is a freedom informed by a profound musical intelligence which is highly selective and acutely judicious. It is also governed by a feeling for musical history, and not just a knowledge of the facts of music history. So that there are certain possibilities for Berio which are *musical* opportunities, and others which are *not* musical. I received strict training from Berio in that sense and was made acutely aware of how every tiny detail of a composition should contribute to the intention and effect of the entirety, rather than be a local sound event occurring and then gone. In many ways, I felt that he was a very good teacher, though usually in an informal context. I spent much time with him in rehearsals and very closely living in his home. It was not a formal relationship of teacher and pupil. He became a mentor and a friend; and he made it possible for me to do a lot of things. He has remained very faithful and supportive of and interested in my work. I found in Berio's music a basic sympathy with an attitude, with an aesthetic if you wish, with a musical technique, with a "language," with a whole approach to music as a phenomenon. The difficulty for me at that time was to not identify to the exclusion of myself. I was drawn so strongly to the man, his ideas, and his music. I think that time provides other opportunities; and I have explored in my own way the experience, and have assimilated and transformed what felt right to retain, and used that as a springboard for other and different musical explorations.

RD I recall in my article which introduced the 1985 Aspen Conference on

Contemporary Music that I mentioned a certain "of-a-kind" between your music and the music of Jacob Druckman, Luciano Berio, and Bruno Maderna. Yet you all maintained your individuality.

**BR** That's right. It would be hard to take note for note, chord for chord, progressions and syntax and all of that and say, "These four belong." But there is something about the ethos of the resulting music which relates us. I have no problem with that, anyway, Richard. You know, people ask, "Who are your predecessors? To whom do you relate most?" I find that easy. Debussy and Stravinsky, Mahler and Webern I would acknowledge as being *directly* influential in my own music—then Berio and Maderna. That sweep of music is the music which made me want to be a composer.

**RD** How did Bruno fit in?

**BR** Simply by being a God-sent enthusiast about everything musical.

**RD** I was privileged to know him, and I had the highest regard for him.

**BR** Whether it was Schubert or Mozart or Palestrina or Monteverdi or Stockhausen or Cage or whoever, he just loved the idea of making music; and he encouraged me, as he did so many others. He conducted the premiere of my *Actions for Six* at Darmstadt in 1963, and it was my first performance there in a big international festival. I had worked on that piece during the previous year studying with Berio. I felt excited about the piece and I knew well its relationship to Berio's *Circles* and also to *Differences.* It was on that mode of musical thinking that I was trying to model my own. Bruno conducted the very fine Kranischsteiner Ensemble. Mine was the last piece on the program; and there were, typical of Darmstadt, seven or eight long pieces, which in any case made people restless. But then, also typical of the festival at that time, the audience was split equally between those who loved the work and those who hated it. At the end of the performance, Bruno called me up, and after the first bow we left the stage. I wanted to stay off. I didn't want to go out again, because I could only hear the boos! He said, "No, come on. As long as they keep going, you keep taking bows." Of course, it became more and more provocative! Afterwards he said to me,". . . if you compose something, don't apologize for it. Otherwise, don't compose and don't offer your music to these people. When you compose music, you should be like Schubert and just say, 'That's the way it is.' " I have remembered that remark. And often, when the nerves have been no less jangled, I have gone out . . . and well, you know all about that! But he was wonderful in that way.

**RD** He had great courage. I knew him in the four or five years before he died, and we used to see each other in Europe quite a lot. What was it like to be in Darmstadt, and when were you there?

**BR** I was there first in 1960, and then in '63, '64, and '66. My recollections now are mixed as were my impressions and reactions at the time. First of all, it was tremendous to see musicians, mainly under the direction of Boulez and Maderna, mastering what seemed to be impossible technical or musico/aesthetic demands on their instruments. It was encouraging, because one felt this challenging new music *is* playable. As the years went by, the situation changed from one where these were the only people who could play this music, to one where students came from all over the world, in their early teens, with this dazzling ability to work on the most complex new music. All of this happened within a few years and made one feel that, yes, all of it is possible.

RD Was Darmstadt dominated by any one faction, or was it free? Earle Brown speaks about it not being dominated. . . .

BR No, it was not. Boulez had his theoretical base which he was developing and expounding, and Karlheinz Stockhausen had his. Messiaen, at the beginning, was influential and was something of a father figure. The "aggressive" theorist at the time was Henri Pousseur, on behalf of serial thinking derived from Webern. John Cage came along and infused a totally non-European flavor into the brew. So did Earle Brown. But being part of the "Italian group" there, I tended not to get overly involved with the polemics and the abstract theoretical concerns. We listened to the music, and we enjoyed the virtuosity of the performers and the general interchange of ideas and enthusiasms. I was happier being involved in rehearsals and performances and how to get at musical ideas involved in those specific works. It was balanced, as I've implied, by the visits of John Cage and Earle Brown, who, in the midst of all this evolving serial theory and practice came and said, in effect, "Yes, but there is another way." Earle is right that it was not dominated by one faction. There were allegiances; and often the slimmer the talent, the more allegiance to a dogmatic, inflexible way of doing things, as tends to be the case. But among the original creative minds and the inventive and imaginative musicians, there was an enormous amount of interchange. Milton Babbitt came there and brilliantly expounded on a different perspective of Darmstadt's prime concern—serial music. Bruno Maderna presided over this incredibly exciting spectrum with great interest, equanimity, charm, and unique musicianship.

RD Did you have any conflict in your mind about these two polarities, serialism and choice-or-chance procedure?

BR No. I think that I felt the internal crisis of serialism itself, both personally and in my observation of the larger context in which people were practicing it. However, let's not deny that there are remarkable musical achievements in the strictly serial mode by diverse composers. But I think that even the most extreme advocate in the late 1980s is aware of a certain exclusionary nature in its principles. At that point, it is a matter of individual choice as to how one wishes to make music. I certainly realized that this was too narrow for me in that its overwhelming concern is pitch. Serialism has a seductive beauty of its own as a principle, and those who espouse it unreservedly would assert that a rejection of it is tantamount to a confession of a lack of ability, imagination, inventiveness, creativity. Be that as it may. There are many powerfully creative minds in music who have, in no trivial sense, engaged this challenge only to reject serialism as a dominant principle in their musics at the same time as benefiting creatively from the experience. That has been my experience.

RD What disturbed you? Note order, or harmonic intention, or both?

BR I think it is a much more complex issue. However, let me simply say that I do not believe that there is only one way to make music; and so I choose to search for formal coherence in my work through other musical means. I believe very strongly in the role of musical intelligence in the creative process—a musical intelligence which is fed and nurtured by experience of a broad spectrum of musics, by visual and linguistic stimuli, but one that is deeply rooted in one's cultural heritage.

RD Have you come up with a new system for yourself?

BR No. It's not a system. I prefer the term "musical intelligence." Because I didn't accept and experience serialism as a dogma or credo in the first place, I didn't feel that a conversion to something entirely different or a revelation was necessary. There are many composers who today are suffering that problem: namely, that they have embraced in a very dogmatic way the technical procedures of serial composition at an early stage of their creative lives. Now, at a later stage of emotional maturity, there arises a conflict between technique and ideas which stem from other, deeper promptings. Often this results in a wild swing of the composer's aesthetic pendulum to an equally untenable reactionary position of seeking solace in the sentimental nostalgia of a previous century, or reducing all musical dimensions to a minimum and then placing the responsibility on contemporary society for having cradled them. So often, like converts from one sin to another, one takes on a missionary zeal in the new, which is totally contradictory of a previously held position.

RD And what about your pendulum? Where are you now?

BR As I first implied, it never swung violently or suddenly one way or the other. Of course, I experimented with this, that, or another musical concern—that is part of being a composer, putting oneself in new territories—I intend always to do that. But against the background of all kinds of music, and with a fairly catholic taste added to the kind of upbringing which I described to you (which can just as happily take an interest in a pub song, and somebody singing it in a drunken, out-of-tune manner as it can accommodate singing twelve tones scattered all over the range), I feel and believe in the power of music all across that spectrum. Therefore, as a composer, I am concerned to be inclusive rather than exclusive, maximalist rather than minimalist, with a propensity for the intuitive rather than the scientific, and a tendency to be plebeian rather than aristocratic, colloquial in idiom, not oratorical. I am concerned to be responsive to musical manifestations and promptings wherever they occur and in whatever mode they appear. And thus I detect a great strength, richness in the musical diversity and pluralism of our time.

**RD** Did coming to the United States change your perspective in any way?

**BR** Looking back over the years since I emigrated to the United States in 1975, certain changes in my work are clearly detectable. In 1977, in response to a commission from the National Symphony Orchestra, I wrote *Madrigali,* which is based on the Eighth Book of Madrigals of Monteverdi. This work you know well because you have conducted it. What drew me to that source, apart from a life-long love of the music of Monteverdi, was the experience of playing through the score of *Il Combatimento* at the piano. Whilst Tancredi is doing this and Clorinda is doing that, and Testo is giving all the details of complex events, the underlying harmony is relatively simple and the speed of its change, its harmonic rhythm, relatively slow. This wasn't a sudden realization; but it was a confirmation that the underpinnings of much great and complex music are essentially simple and *have* to be so. We are capable intellectually, in esoteric and abstract manipulations, to elaborate those fundamental elements to the *n*th degree. Whether or not there is a point at which that elaboration, that complexity, is counterproductive and thus severs itself from the essential and universal nature of music as it generally exists on this planet is a question worth pondering.

**RD** Are you implying that this underpinning is largely harmonic?

**BR** It's essentially harmonic. What we used to call harmonic rhythm has, in the last thirty or forty years, been required to assume a rate of change which may run counter to music's capacities. When we reach a point where harmony changes in every measure, in every beat, and sometimes within the beat, then I am not sure that the human receptive capacity for musical experience is acute and refined enough to cope with that rate. Of course, in a scientific age of discovery which "knows no bounds," cognitive scientists, especially in psycho-acoustics, prove otherwise. But I remain uneasy with the notion that one of music's functions is to live up to the demands of scientific proof. I think the speed, the flow of information in much music written between 1945 and 1965 was tremendous, stemming from this theoretical certainty that one could manipulate the dimensions of music this way. I don't think so anymore. I was never totally sure about it—never convinced.

**RD** Did you come to a gradual awareness of this, or can you point to one period of time?

**BR** It is cumulative, and I find it increasingly in the music of many composers of diverse stylistic concerns. As we approach the last decade of this extraordinary century, I find more and more composers grappling, in one way or another, with this question.

RD Did you come to terms with it?

BR In dealing with the question myself, *Madrigali* was an important undertaking. In addition to basing the work on the essential qualities of another composer's music, I deliberately slowed down the rate of harmonic rhythm in each movement, so that the complex juxtaposition of surface details (aspects of Monteverdi's music) could sit firmly on a sustained harmonic foundation, thus allowing these complex relationships to be more clearly effective and perceptible.

RD That piece also embraced certain aspects of repetition and aleatoire, as in the last movement. As I recall, a transcription of *Lammento della Ninfa* (which is synchronized) is surrounded by a ritornello made up entirely of non-synchronized Monteverdi fragments. These fragments are executed by the players, independent of the conducted *Lamento.* Do you apply that kind of approach to your pieces anymore?

BR Now now. Not in my present musical concerns, though I don't rule it out in the future. It is very appealing. It is very seductive to ask why should music have only one possible path in performance and only one possible formal conclusiveness. Why can't it, like much modern poetry, have an interchangeable format on the page? Why can't it have mobile qualities and capacities like much contemporary sculpture? Why should music not emulate the kinetic quality of certain visual arts? All of these non-musical correspondences we are aware of; and the application of these to musical concerns is both possible and exciting, but only if music's own essential qualities and capacities are not violated or weakened to the point of impotence. Maybe music doesn't have the capacity of language in that sense; maybe it doesn't have the capacity of the visual, the plastic. Any automatic and unthinking assumption that it does flies in the face of historical evidence.

RD Yet I can think of works such as Stockhausen's *Momente,* Berio's *Epiphany,* which use a kind of shuffling of the score, or the mobile quality of Earle Brown's music. . . .

BR I don't deny the impact of it; and I think that, along with many other ideas and experiments, it went into making this the richest century in terms of innovation in the history of art music as we know it in Western culture. The period between 1945 and 1970 was remarkable by the sheer weight of radical invention, not to talk about quality, which can be cited across the spectrum. As we approach the next century, we see it in retrospect as a much broader plane of activity which carries most of the mainstream essentials from the 18th and 19th centuries, as well as this century. Those "essentials" are particularly manifest for me in terms of harmony, because I've never believed that our Western art music really moved outside of harmonic constraints, and that even at the height of polyphonic periods, one is still very much aware of all the voices as sounding *together,* making also their vertical sense and impact. Maybe also the 200 years' conditioning of our listening makes it difficult for the 20th-century polyphonic period to register its impact as did polyphony in the Tudor church period, for example.

**RD** I share your view of the inventiveness of the postwar period until 1970. How do you feel about composers "stealing"?

**BR** Just before you arrived, I was listening to the "Trout" Quintet; and there's no doubt that the material for the variation movement is "stolen"! It would be difficult to argue that it is inferior or less original than the song from which the material derives. When composers reuse material from their own works, it can be for any motive, ranging from the highest and most noble one of discovering in it the capacity for transformation and development, to narcissistic parody . . . to the fact that the rent has to be paid at the end of the month and some pages must be delivered! It doesn't matter to me, if it is convincing. I am sure that all composers know the experience when, on completing a new work, they realize that the ideas and materials they have generated could with a shift of emphasis and intent become an entirely different piece of music. That, I suppose, is an impetus to move on to a new project. In a slightly different way Stravinsky was very honest and interesting on this subject. It came as a revelation to me that at the times when he was concentrating on a new piece, the only music he'd listen to was music in the same genre, meaning that if he was writing a Mass, then he listened to Masses by a specific number of composers who interested him. That came as a surprise to me. On the one hand, this genius through whom everything "original" passes. . . .

**RD** Is there a point where you feel you changed or were pivotal? And if so, could you maybe give one piece as an example?

**BR** There are more pivots than one. The piece I mentioned earlier, *Actions for Six,* 1963, the performance of it in Darmstadt and the composition of it the previous year, was pivotal. It was the first time that I had freed certain dimensions from serial control—aspects of notation, aspects of order, of pitch set—even though it was constantly informing the piece, in the end it was not a serial piece in any strict sense of the term.

**RD** It deviated.

**BR** Yes.

**RD** Intuitively or consciously?

**BR** Consciously, in response to intuitive promptings and "on site" discoveries!

**RD** Did cognizance of the past provoke any pivotal notions?

**BR** In the case of *Madrigali,* which was the next main pivotal point in my work, the idea was to use music by another composer from an earlier century. As a result, the overall stylistic nature of that piece is distinct from that with which I am normally identified. But what it did was to focus attention on the problem of harmonic rate and the nature and function of harmonic rhythm as an element of formal coherence. In music since then, I have been working more and more on slowing that down at a fundamental level, even though the surface detail can still be extremely complex in terms of instrumental articulation, ornament, timbre, rhythm, densities . . . and so on. Those two for me are the decisive ones, though I think of the composition of *Canti lunatici* and *Canti del sole* as another. Viewed as two parts of a large-scale work, I tackled this same question in a work whose overall time duration is close to one hour. I think that the extreme of harmonic "simplification" and slowness of harmonic rhythm was reached in *Canti del sole.* I felt no need to go further. But having cleared away a lot of harmonic clutter, so to speak, I felt free again to begin to build a richer, more complex harmonic base without returning to the condition I have worked so hard to alleviate. This I have done in subsequent works—*Suites I and II: Le Tambourin* and most recently in *Hiraeth* for solo cello and orchestra.

RD One of the pieces I programmed during my time at Juilliard was called *Quaderno* or *Quartet.* We later did it in Aspen in 1976 while you were there. That piece concerned itself with virtuosity, according to your program notes.

BR Collective virtuosity. . . .

RD Did you go beyond that?

BR There is a short hidden narrative; a short text by Lorca which talks about the spider spinning its particular web of mystery. It is from *Dreams (Sueño).* In fact, that's what the four players do. Now, remember that, historically, a string quartet strives to function as one unit. It is a string quartet in the sense that each one has to subjugate, some may say sacrifice, their idiosyncracies to the homogeneous, unified whole. Otherwise it is simply four string players. One of the aims of my piece is to identify and engage that collective in such a way that they sound like one player. In other words, to start with the historical nature of the string quartet (not to contradict it) and then to revivify that quality in meaningful, contemporary terms. The only time that individual players are heard as individuals is in the easiest and simplest task that they have to perform, which is a brief, three-note melody. Even that is very quickly covered up again by the very complex rhythmic synchronisms that make up the surface detail of the quartet. In this way, individual virtuosity is placed at the service of the collective virtuosity, resulting in a world that is intriguing.

RD You've worked with a number of poets who have passed away. What has your experience been with the living poets?

BR Or as Dylan Thomas used to call them, "more-or-less living poets." There are two only with whom I have worked closely. One is the American poet/novelist Gilbert Sorrentino, who has lived most of his life in Manhattan but now is professor of literature and poetry at Stanford University. He is of Italian descent, but very much an American poet, and was one of the contributing editors of *Kulture,* the literary magazine of the 1950s and '60s. The other is John Wain, the English poet and novelist, whom I've known for many years and is a friend from my years in Wales and also at Oxford. He is best known for his novels and his literary scholarship rather than by his poetry, which is a modest portion of his output. Over the years I have composed a series of works, now four in number, under the general title *Ballad*—each for voice or voices and ensemble. Each of these takes a Sorrentino poem as a starting point—all from a collection of fifty-two poems called *The Perfect Fiction.* A Wain poem called "Wildtrack" has been influential, in that I've used part of its text; but also its formal concerns as a long poem have served as a model to explore in purely instrumental works.

RD Can we speak about your Pulitzer Prize–winning piece, *Canti del sole,* which I did with Paul Sperry while you were in Aspen. . . .

BR Very, very fine performance, too, I must say. It stands very high in my experiences of that work . . . I thought that performance, of all the pieces of mine you performed, was especially gratifying.

RD Well, thank you. I did like that piece, too. But I wanted to ask you about the poetry for that.

BR As with *Canti lunatici* which you did with Carol Plantamura, the poems in *Canti del sole* are in English, French, German, Italian, and Spanish. In addition, the poets are so diverse in period, style, concern, and perspective, on the subject of moon and sun. Let me list them in order to make a point: Anon, Arp, Artaud, Baudelaire, Blake, Celan, Hüchel, Joyce, Lawrence, Lorca, Montale, Wilfrid Owen, Plath, Quasimodo, Rimbaud, Shelley, Sinisgalli, Dylan Thomas, Ungaretti, and Whitman. Two matters are very important in the preparation and performance of these pieces. First, those singing the two solo parts have, in addition to language fluency, to understand the nature of the work, the reasons for the juxtapositions of texts and languages, the sonic spectrum they represent, and the subtle and complex relationships and cross-references between text and music and language and vocal production. Second, the conductor of the ensemble/orchestra (they exist in both chamber and orchestral format) has to coach the players out of the subservient role of the accompanist into one which relates in every detail to the sonic and articulatory world of the languages. In other words, every instrumental sound and gesture also has a subtle and complex relationship and cross-reference to text, language, and vocal production. In short, the performers can only perform it well if they understand fully the *character* of the music.

RD In terms of performances, are there some performers who you find can grasp the character that you are after? And what do you mean by character?

BR Yes, and it has been my good fortune to work with some of the very best performers. The last part of that question is very hard! The other . . . it's easy to say, and it's true to say, I think, that those performers who have the best chance to understand a music's "character" are the ones who can immediately rule out all that is irrelevant, or almost all. In other words, they don't bring extra interpretive baggage and try to make it what *they* think music ought to be. They are able to see immediately from its characterization and its notation and if they start to explore it on their instrument, allowing its own terms to guide them, they are able to eliminate irrelevancies.

RD So elimination is step one.

BR Elimination of irrelevancies. But then to probe what the deeper aspects are, how best to understand the hierarchy the composer has given them, so that the weight is here for this and not for that, and this is prominent here and this is not. In the same way that if you are playing late Beethoven piano music, of course you may also bring the technique that you need for Scarlatti for some of the passages, but at a certain moment it won't serve you, it won't be adequate, you will need something else, because it goes somewhere else, and its concerns are not those of Scarlatti.

RD How significant is what you've just said in terms of audience recognition or audience perception? Or are they still too far down the ladder to even know that difference? Or does it occur to you that when your character is done properly, there's no other thing but the word "cultivation" left?

BR That I agree with. That last statement I think is right. And in a sense, one can only be concerned with that degree of cultivation. As I said earlier, music can and should be many things to many people. At the same time, when performers deeply understand the character of a music and present it accordingly, then listeners are caught up in the essential intent of the music and they in turn eliminate irrelevant thoughts and responses. I do think there is a point at which one can make very special contact with music when something significant happens emotionally, intellectually, spirtually, even physically. Physical sensation is probably the first overt sign of it. One can't quite ignore the person who says, "WOW! That music just went right down my spine!" Because if it did that, the odds are that it went up to the brain as well! Audiences are as complex as the music they are experiencing. Our task as performers/conductors is to have as authentic a view as we can; an authentic view meaning a deep understanding of the intention, the character of the music. I don't mean to imply that there is no place for interpretation. Quite the contrary in fact is true. I think the very best of performers are those who use a deep understanding of a music's character to bring an interpretative (I also use the term "creative") daring to performance, thus revealing even more of the music's hidden meanings. I shall never forget such a performance in the last months of his life by Bruno Maderna of Mahler's Ninth Symphony. It seemed that he took the last movement at tempi half the normal. As an idiosyncrasy, it would be an outrageous thing to do, but he did it with a deep knowing and love—and it was a revelation. I and everybody who enjoyed that experience said, "I never knew. . . ."

**RD** Certain music can withstand that treatment.

**BR** I'm perfectly happy that you and other musicians whom I've come to know and trust take my music and present it in a manner which allows it to establish its own character, and I don't need to be present to supervise it. Also, I'm always pleased, when I am present, to have revealed something about the work which I never knew. . . . That is always rewarding.

**RD** What are you currently working on?

**BR** There is an ongoing long-term project; namely, an opera, *Le Tambourin,* based on the life-work and letters of Van Gogh—part of which are the *Suites I and II: Le Tambourin,* which you have also conducted. In keeping with my general philosophical stance, which I've outlined in our conversation, I am using certain aspects of theater and the operatic tradition in their conventions; but at the same time, I am attempting to transform them into a contemporary statement in the theater. That project simmers away on a back burner as I address other commissioned projects—a large-scale work for chorus and orchestra for the BBC Symphony; a work for large orchestra for the new Suntory Concert Hall in Tokyo; a work for the Boston Symphony Orchestra, the New York Philharmonic, the Philadelphia Orchestra . . . indeed, several other commissions which seem, at the moment, to have become my preferred medium. As you know, I love orchestras, and I find in them the most interesting creative challenge—that of turning what otherwise might well become a cultural dinosaur into a living, vibrant, exciting contemporary medium by rethinking its capacities in relation to its accumulated history and repertoire.

# Narrative VII

Before resigning my position with the Buffalo Philharmonic in the spring of 1967, I had given serious thought to exploring conducting possibilities in Europe. The Buffalo days had been exciting, innovative, and rewarding; but I felt that I had done what there was to be done in my capacity as Associate Conductor. I also felt the need to expand my professional experiences and be totally on my own.

I had guest-conducted the Pittsburgh Symphony at the invitation of William Steinberg, and by the time of my final subscription concerts in Buffalo, I was convinced that I should move on. At these last concerts, I gave the premiere of Castiglioni's Concerto for Orchestra, performed the Mozart Clarinet Concerto, conducting from the clarinet, and presented scenes from *Boris Godunov* with the Bulgarian basso Niccola Ghiuselev.

This decision also brought to an end my close professional relationship with Lukas Foss, whom I had greatly admired and benefited from enormously.

The summer of 1967 in Europe was an adventurous time. Tracking down professional contacts and following up on a vast correspondence that began in the spring became a major project. It was a journey, mostly by car, from Italy, to Austria, to France, to Holland, that ended in the fall in London, where I lived until January 1968.

While in Salzburg, I was given special permission to observe a rehearsal of Maestro Herbert Von Karajan, who was conducting the Cleveland Orchestra in the Prokofiev Symphony No. 5. I was assigned a specific seat in the rear of the auditorium. This seemed peculiar since I was the only person in the audience. The Cleveland Orchestra performed the rehearsal brilliantly. I remember Karajan excitedly gesticulating to George Szell at the end of the rehearsal, expressing astonished disbelief at the perfection of the ensemble.

Later, I heard Karajan's production of *Boris,* and by sheer accident came across Bruno Maderna at the Mozarteum, where he was teaching. On that particular day, while examining posters in the lobby of the Mozarteum, a figure whizzed by, a figure that was unmistakably Maderna, one that Boulez affectionately described, on the occasion of Maderna's death in 1973, as resembling a "small pachyderm."

I introduced myself to Bruno and was immediately invited to dinner. At 7:00 p.m., in front of the Mozarteum, a small crowd arrived, and we all piled into a caravan of cars for a meal outside of Salzburg. Bruno began his meal with several orders of escargots, and everyone proceeded to have a hilarious time. I drove Maderna back to the city and spoke to him about my wishes to conduct in Europe, whereupon he immediately told me to contact his manager, Sylvio Samama, in Holland, who ultimately represented me in Europe.

This was the first of many such gatherings with Bruno after his concerts. I subsequently followed him around Europe and as a result joined the many musicians who held him in such high esteem, both as a musician and as a human being. He was loved as few individuals in our profession are. On a visit to the Donaueschingen Festival, I stayed at his home and we talked music into the night.

London seemed very active regarding 20th-century music and was still under the magnificent influence of William Glock of the BBC. I heard Boulez do some gigantic programs that included *The Rite of Spring,* Berg's *Three Pieces for Orchestra* and *Altenberglieder,* Stockhausen's *Gruppen,* Berlioz's *Lélio,* and others.

Aaron Copland payed a visit to London in September 1967, conducting the London Symphony in works by Carter, Ives, Tippett, and his own *Dance Symphony* and *Symphonic Ode.* He was most kind and generous to me about letters of introduction. During this time, I also paid a visit to Henri Pousseur (who had been Slee Professor in Buffalo), at his home in Malmédy, Belgium. Henri showed me his recent score of *Crossed Colors,* which used a text by a Native American chief, Seattle.

In December, I had managed to arrange at least one concert of British and American music at the American Embassy with the New Cantata orchestra. British works by Britten and Cardew, as well as London premieres of the Americans Schwartz, Babbitt, Subotnick, and Brown were presented. What was to have been a much lengthier stay in London was cut short for a number of reasons, including the invitation to act as Music Director–Conductor of a pair of chamber concerts as part of the 25th Anniversary Celebration of the Koussevitzky Music Foundation in Lincoln Center's Festival '68, in July.

Planning began in March, and it marked my first opportunity to direct a festival. After meeting with the president, Mme. Olga Koussevitzky, who brought me up to date on the commissioning program and enlightened me about the Foundation's past history, I met with Aaron Copland, then vice president of the Board of Directors, which guided the commissioning policy since the death of Dr. Koussevitzky in 1951. Members of the board were Leonard Bernstein, Richard Burgin, Lukas Foss, Howard Hanson, Peter Mennin, Gregor Piatigorsky, Gunther Schuller, and William Schuman.

Aaron asked me to join him at the Harvard Club in New York to discuss the possibilities. I had my notions and he had his. From over 150 commissioned works, about forty-three (including six string quartets) were appropriate for chamber concerts. Initially, considerations were obviously meant to reflect the time span of the commissioning and therefore the stylistic differences, as well as the variety of ensembles (instrumentation and size). A further consideration was to present a number of works yet unheard in New York. In addition to composers who received premieres of their works (Davies, Del Tredici, Nono, Takemitsu, Xenakis), the programs contained the music of Crumb, Fine, Henze, Martirano, Riegger, Shifrin, and Wolpe.

Among these commissioned works there were certain ones I was personally attracted to. *Revelation and Fall* (1966) by Peter Maxwell Davies was obviously a dramatic work, a work that appeared as a continuity of Schoenberg. The text of Trakl's prose-poem written just before his death in 1914 provided an immediate lugubrious quality that was not unlike the monodrama *Erwartung* of Schoenberg, text by Marie Pappenheim. This kind of Freudian self-indulgence was treated with expressionistic care. What struck me about the piece, however, was its brashness. Yes, it uses *Sprechgesang,* and yes, like bar 190 of *Erwartung* where the woman, after discovering her lover's dead body, screams for help—*"Hilfe";* in Davies we hear an even more shocking, sudden outcry over a bullhorn (loud-hailer) that remains hidden until the very last moment. But the brashness also lies in the abrupt juxtapositions of contrasting musical material and an orchestration of percussion instruments that range from an oil drum to a knife-grinder to a railway guard's whistle. The piece was written for Mary Thomas, soprano, and the Pierrot Players, later to be known as The Fires of London. Bethany Beardslee sang the New York premiere.

As a counterpart to this work, David Del Tredici's *Syzygy* (1966) struck me as also having certain influences from the Second Viennese School, such as his use of "mirroring" in "Ecce Puer," the first of two Joyce poems, where the music goes exactly backwards from its midpoint (but in a reorchestrated form). The second poem, "Nightpiece," underwent compressions and expansions that were also a form of "mirroring." Beyond this was a vocal part that was jaw-opening to look at. Such quick leaping around I had not seen before; and if we could find someone to sing it, I felt it must be done. That someone turned out to be Phyllis Bryn-Julson. It was Copland who called my attention to this work, and it was a world premiere.

By contrast, and much shorter, *Canciones a Guiomar* (1962) by Luigi Nono with a text by Antonio Machado was a lonely reflection by a lover, pensive and in the nature of a love letter. The vocal line is practically unaccompanied by the instruments, which, instead, furnish intermediary moments that sustain the subjectiveness of the confined lover as well as his implied physical surroundings. The piece reflects the spiritual and technical inheritance of Schoenberg and Webern and also establishes a new degree of expressivity in Nono's setting of poetry.

*Akrata* (1964) by Iannis Xenakis, for winds and brass, and *The Dorian Horizon* (1965–66) by Toru Takemitsu, for strings, provided instrumental contrast and contrast in compositional style. The Xenakis work is of extra-temporal architecture based on the theory of groups of transformations. Amid highly complex detail and dimension, an overt sense of opposites is initially established, for example, by irregularly linked sound-blocks of opposing dynamics that are constantly reshaped or—in the juxtaposing of silence with vertical sound-pillars of constantly varied heights, widths, and stances—that produce an inverse sense of proportion. Later, and in contrast, more varied and individual sound-profiles are grouped and combined. Perhaps the most minutely subtle effect is a kind of blurred echo brought forth by one note of a pitch combination played one quarter-tone higher or lower in order to release a measured acoustical "beating" or "out-of-tuneness."

At least one aspect of the Xenakis piece, the performance execution "absolutely without vibrato" relates directly to *The Dorian Horizon.* Three large sections are most clearly etched by the manner of performance (non-vibrato, vibrato, non-vibrato), and it is in these outer edges that the extra-temporal is also felt. Contrast occurs in the middle where a more expressive and less poised music is colored with vibrato and time-structured fragments (temporality). The sense of proportion is less complex than *Akrata;* rather, illusion and disguise and the much vaguer outlines these reveal are the amalgamating devices; that is, the echo, the shadow, the hint, the glimpse, etc. Asymmetric, atmospheric conditions prevail throughout the entire piece, along with fragmentary repetition, both literal and suggestive.

The Koussevitzky Celebration also contained orchestral performances by the New York Philharmonic, the Pittsburgh Symphony, and the Boston Symphony, with past commissions of Bernstein, Schuller, Copland, Schuman, Britten, Lopatnikoff, Piston, Stravinsky, Dallapiccola, Bartók, and Ginastera.

The whole project resulted in my conducting further performances at the Library of Congress in Washington, D.C., Princeton University, and a Columbia Masterworks[1] recording of Xenakis, Del Tredici, Takemitsu, and Nono.

1. *Music of Our Time,* MS7281.

# *Sir Peter Maxwell Davies*

BORN SEPTEMBER 8, 1934, MANCHESTER

**November 10, 1986** **London**

I arrived in London in the early morning from New York. Stayed awake all day. Saw Max Davies at 4:00 p.m. in his London apartment. He was leaving for Orkney the next day.

RD What is on your very active schedule of composition?

MD Right now, I've just finished an oboe concerto, which is for Robin Miller, who is the principal oboe of the Scottish Chamber Orchestra. And that is very relevant to what I will be doing and have been doing, as I've got a position which sounds rather grand and official . . . Composer/Conductor at the Scottish Chamber Orchestra. What that means is that I write pieces for them. I've already written two pieces for them, before the oboe concerto; and I plan to write pieces for the principals of the orchestra, and some orchestral pieces. And I'm conducting at least two clumps of concerts with them each year, in addition to recordings and touring. So it's a very nice connection.

RD When I spoke to you on the phone in the States recently, you said you were also writing an opera.

MD That's right. This is *Resurrection,* and this is a commission from the Darmstadt Opera. It is due to be performed in 1988, and I'm well into it now.

RD Is that the title, *Resurrection*?

MD Yes. Nothing to do with Tolstoy. It has to be called *Resurrection* even in German; because there are other operas called *Die Auferstehung,* or whatever. It's based on *the* Resurrection, many times removed, and it's probably going to be one of the big scandals! It's got for a start an electronic vocal quartet, and a pop group—which I've not used before. It's pretty close to the bone, a lot of it, even the text, and it's very political. And a lot of it is very funny and rude. Of course, it will be done in German, in translation. I've written the text . . . it will be modifed of course . . . in English.

RD Is it anti-Christian or pro-Christian?

MD It's not anti or pro. It's very anti certain types of puritanical, narrow

*Stephen Smilak*

Christianity—the kind that you sometimes get very fed up with in extreme Protestant Scotland, for instance.

RD You've spent a lot of time with these issues regarding Christianity.

MD Yes. Castigating it or setting texts which I found inspiring . . . I think my own religious feelings are quite simply summed up, basically, by just always being astonished that I'm alive, and that things *are;* and the wonder of that I think I find quite satisfactory. I've always wondered about it, every day! But when that wonder and that sense of awe begin to be, first of all, expressed through a series of symbols which then become ritualized, and which then become dogmatized; and then people start persecuting each other, even to the extent of taking other people's lives, because they can't get them to agree with the particular symbols by which they think they have somehow encapsulated ultimate truth; then I think you'll understand that I find the whole question of orthodox religions one which is, even from its inception, involved with problems. And one which sets up reactions in myself which are very complex. And I think that finds its expression in the work, too.

RD Well certainly Dostoyevsky felt that way, in "The Grand Inquisitor" of *The Brothers Karamazov.*

MD Oh yes. Certainly! But Dostoevsky does seem to have come down quite firmly on the side of the Orthodox Church.

RD And that fanatic quality . . .

MD Yes, that fanatic quality—it really is the *reductio ad absurdum* of Christianity. This is, of course, very much what *Resurrection* is about. I think a lot of people are going to be quite shocked; but I think while one has strength and one has any kind of feeling for religion as such, these "Christians," so-called, have to be shown for what they are—the total absurdity of it. And its anti-humanism on a very profound level—that it does reduce people to slogan-yelling zombies.

RD You say that your opera will be done in Darmstadt. Do you get a little quiver about that name?

MD Yes, I get a little quiver about that name, indeed. The commission has nothing to do with the Darmstadt International Summer School for New Music. It's from the opera house of the town, and it comes directly from the mayor. Who, by the way, has read the text and approved of it. So I'm at least relieved on that level!

RD Well, it's a very natural way to bring up that name and try to reflect on that name, Darmstadt. And I know that you had gone there . . .

MD Yes, in 1956.

RD Max, about Darmstadt . . . do you have any feeling about it having been an important idea, even though you may be disillusioned by it all now?

MD Certainly. I think for me it was very important. I'm sure that the Darmstadt aficionados would disown me completely these days, which one can understand. But I did go there in the '50s; and I learned a great deal. It was very liberating, indeed. Because, having been so-called educated at the University of Manchester, where I was thrown out of the composition class . . .

RD For what reason?

MD The professor didn't think I would make a composer. He said I was far too interested in these dangerous composers like Bartók and Schoenberg, and those terrible Renaissance and medieval composers. So I was thrown out. And going to Darmstadt, where people were actually talking about the technique of writing music, and putting a microscope on the procedures . . . I found that very stimulating. And there, taking serialism and Cage's ideas and Stockhausen's ideas about space and discussing things with Luigi Nono and Bruno Maderna (I was learning Italian then). . . . These things I found very stimulating.

RD Not necessarily accepting everything, but perhaps even reacting against . . .

MD Yes. I was very conscious then that I wanted to base my own work, right from the start, on techniques which would carry me through a whole lifetime, assuming that I was going to be living into and perhaps beyond middle age. I was aware that there were composers there who would peter out if they were not careful, within a few years . . . which they did. Although they had ideas which stimulated me a lot, they were pretty short-lived. And made a great deal of fuss of, but I thought that was rather a poor way. Although I didn't say anything to anybody, except a few close friends, I was very aware of the limitations of a great deal of what was being talked about. But for me, I think, its chief interest was that it became a huge catalyst to further investigation. And I did look particularly closely at the music of Boulez and eventually of Stravinsky and of Schoenberg . . . and of course, medieval and Renaissance music.

RD Your interest in the medieval seems to have dominated a lot of your thinking and it is still prevalent in one respect, and that's the old isorhythmic conception.

MD Yes, indeed, which I used as recently as the Third Symphony. I made it into a huge, huge architectural structure . . . right through the whole piece.

RD Can you speak a little further about that medieval pervasiveness?

MD Certainly. I think, in that early stage—that's in the '50s—when I was eventually studying with Petrassi, I was very keen on . . . not only the isorhythmic techniques, which were form building, but also on the ways in which they were applied, paraphrasing and indeed transforming *cantus firmus*. And in rhythmic procedures, in that a small cell would be used to fill out quite a long passage in the same rhythmic mode; and there would be a rhythmic modulation, if you like, and it would change for the next section, and you'd get that kind of contrast.

And also the way that, in the later music, there are modal modulations where one mode operates and then there is a shift to a degree, and then that operates . . . I found that very useful in my own work; because that is not a *tonal* modulation. It doesn't have all the associations. Even now, I find that I can use that kind of shift very, very constructively. And also, the way that they spaced the music, in that a tenor was in fact the thing that held it together.

RD What do you mean?

MD That you have a tenor in long notes and everything circulates around that. I realized that, thinking of what we call a bass, which is the thing that, in music after 1650 or so, everything circulates around . . . and you hear harmonies in relation to a bass . . . I realized that you need not hear harmonies in relation to a bass. And I always thought of a tenor. . . . The tenor could be in the top voice or the middle voices or the bottom voice, it doesn't matter. I certainly hear harmonies in relation to that principle, where usually longer note values are made pivotal in some way. In addition, I was quite interested in what they call "migrant" *cantus firmus,* going from voice to voice, too. All these things I found quite stimulating, and I applied them in my own way. But I don't want to lose sight, too, of the fact that I was devouring Heinrich Schenker at the same time and working on analyzing and learning to play by heart the Mozart and Beethoven piano sonatas and even the Beethoven symphonies. I know most of them note for note from that period when I was really working hard on trying to find out how they worked. Stimulated first of all when I was very, very young, by reading—I still think it's an extraordinary book, by Erwin Ratz—*Eine Einführung in die musikalishe Formenlehre,* and also Schenker's analysis of Beethoven's Fifth Symphony, which was the first of those that I read. The Ninth came later. And his big book, *Der freie Satz,* that came much later. Those things I found crucial, and also, going to the Dartington Summer School, where I remember particularly meeting Hans Keller, and spending many hours talking with him. Not so much talking to him as just listening to him talk, and going to his lectures about Haydn and Mozart string quartets particularly, and Mendelssohn quartets; where he was analyzing these things. I think that kind of stimulus, coming from the Dartington Summer School, and meeting there people like Elliott Carter, Luigi Nono, Aaron Copland, Stefan Wolpe, and Bruno Maderna . . . was life-saving.

RD I was going to ask a question, and I think you've begun or have finished answering it, I don't know. . . . If you were asked to explain or speak about your musical genetic code, what would you say? And I guess you're saying some of it.

MD It's a pretty elaborate one!

RD Yes! But let's pursue that, if you have further thoughts, because you've mentioned a lot of things that fit into that idea. Would you like to continue with that thought?

**MD** Yes. Right from even being a child at school when I was thirteen or fourteen, I was aware that composers like Byrd, Purcell, Haydn, Mozart had written many different kinds of music and had not been self-conscious about it. And at that stage, I enjoyed playing many different kinds of music. I played a lot of stuff by ear on the piano and thoroughly enjoyed it. I used to enjoy parodying popular styles for my own amusement. Then, it became evident, in the '50s, that this was not a very "in" thing to do, and that music was very "pure," and I never took kindly to that. I think that there have always been impure elements in many of my pieces.

**RD** So *that's* where it comes from! Your sense of irreverence!

**MD** *Yes.* This is something which is in the bloodstream, as a most important part of the "genetic code." After I'd studied with Petrassi and during the period in which I was going to Dartington and after Darmstadt, I started to teach at Cirencester Grammar School, as music master—a co-educational grammar school in the west of England, from 1959 until 1962. And being a composer, I wrote pieces for the school choir and the school orchestra, a junior orchestra, for school plays, whatever. But also encouraged the kids to compose. And I was very impressed by the lack of inhibition of their composition, particularly when dealing with theater pieces. Theater pieces which they composed in groups and made simple tunes and sound effects and dance numbers or whatever. The boys and girls experienced these things as a part of normal classwork, and I had them write their own music down properly, and rehearse it and perform it, and as individuals many made their contribution to the music for a school play or whatever. Their lack of inhibition I thought very, very interesting.

**RD** How does this fit into the code . . . ?

**MD** It fits into the code in this way: watching these kids compose, I realized that they were getting wonderful musical images which I was not getting; and that gave me . . . *envy,* if you like!

**RD** Now, I think I could skip to the first piece that brought us together; your early work, *Revelation and Fall,* which was a Koussevitsky Foundation commission. I did *Revelation and Fall* with Bethany Beardslee at the Koussevitzky Festival in Lincoln Center in 1968. I didn't know you, and I was very taken by that score. I found it not so easy to conduct.

**MD** It's a *hell* of a score to conduct!

**RD** You had flown in from London, and at one of the rehearsals you popped up onto the stage of Philharmonic Hall and I said, "Well, what do you think?" The first thing you said was, "Well, it's not quite crackers!" and that's how I met Peter Maxwell Davies. You then proceeded to tell me what was wrong. But I do remember it turned out, I think, rather well; because I think Bethany had a Schoenbergian kind of madness . . . shivering and shrieking. . . .

MD Oh yes, it was tremendous. . . .

RD It's not so important, but were you the first to use a bullhorn?

MD You know, Richard, I don't know!

RD I think you may have been the first that hid it before the singer used it, that was a shock!

MD It was a bit shocking, it still is when it happens. That's a wonderful thing to be remembered by—a hysterical shriek from a bullhorn!

RD Max, I have made a little list which I'd like to run down. I think I've conducted more of your music than any other living composer.

MD That's extraordinary.

RD *Revelation and Fall, Eight Songs for a Mad King* . . .

MD Which you did the first American performance of . . .

RD Yes, and *Vesalii Icones, The Martyrdom of St. Magnus, St. Thomas Wake,* were all first American performances. I also did *Stone Litany,* but I don't know whether I was first or not. . . .

MD That was a first American performance.

RD That's right, you were there, in 1976, that was with Jan De Gaetani. And then I did *L'Homme armé* subsequently with Adele Addison. God, she had a terrible time, when she ran out screaming at the end . . . she tripped over her priest's robe and fell smack on her face and was bleeding. She had to have stitches in her lip and leg. I also did *Mirror of Whitening Light, Ave maris stella* . . . those were North American premieres. I took your group, The Fires, on a European tour; and we did again the *Eight Songs, Miss Donnithorne's Maggot, Jongleur de Notre Dame, The Martyrdom.* You also wrote a fanfare for me called *Fanfare for Richard,* and then, of course, the great thrill for me was to do the world premiere of *The Lighthouse* in Edinburgh in 1980. So my God! I guess you could say I'm a fan of yours!

MD It looks rather like that, Richard!

RD But mentioning all these things only brings up this extraordinary output of yours. Let's talk about some of your Music Theater works and some of the aspects they contain, like the insanity versus the sanity qualities. Did you get that from Schoenberg? Did *Pierrot* do *that* much for you?

MD I think it was already there anyway. And I latched onto the Schoenberg thing, obviously. It's just something which I think has been in the air. And which probably was there in myself right from the start. Just for instance, coping with the kind of reality of the human predicament at the end of the war, when as a boy I remember finding it very difficult to come to terms with the fact that so many people had been killed in that thing that I had survived. Killed for all sorts of idiotic reasons, like being Jewish or something, in the most hideous way . . . and yet, the sun still came up. And if you were lucky, there was dried egg for breakfast. And I remember finding that particular *madness* quite difficult, as a boy, to cope with. I couldn't find words for it. There was no way that I could express that. I'm sure that kind of feeling somehow got through in the music later on. Because I couldn't, and still I don't think I can say anything about that. There's no way you can say those things in words.

RD Part of the code again.

MD Part of the code, yes. And something which I think again ties in with the period which interests me so much . . . the late medieval-Renaissance period.

RD The injustice, the torture . . . ?

MD Just that people coped with that kind of reality. And survived. And even produced works of art. I don't think they were insensitive.

RD Maybe that's why people, on the surface, could say some of your music sounds cruel.

MD Yes, maybe.

RD But you don't?

MD Not really. I hope it isn't.

RD But, I mean, to observe cruelty is not to believe it.

MD Not at all. Not at all.

RD Then, of course, you expanded that world of Music Theater like no one. Your observation of King George III is a little different. It's more human, very personal.

MD It's a very personal thing, yes. There you're dealing with somebody who knows that he's slightly addled, and suffers a great deal because of it. And is humiliated by the people round about. I suppose again one identifies very readily with someone like that. I think we all have that in us. It's something built in which you can readily identify with.

RD And the *Maggot?* Miss Donnithorne?

MD When I wrote that piece I quite literally became Miss Donnithorne for a period of about three weeks! I lived and breathed and felt Miss Donnithorne! Her tragedy! I really identified with that one!

RD What a thing . . . to be stood up and to spend the rest of your life in that position.

MD A prisoner in your own house, with a rotting wedding feast on the table. Awful, awful! Yes, I remember that one. I thoroughly enjoyed writing that. And musically I think it's a far stronger piece than *Eight Songs for a Mad King.* It has a much more subtle kind of tragedy even though it's not quite so loud.

RD Let's move now toward your Orkney experience, because the piece that I felt typified it is *The Martyrdom of St. Magnus,* and the George

Mackay Brown text, and I suppose the historic input that you began to realize up there. You didn't go there for that reason, did you?

MD No, no. I went there simply because I fell in love with the place and met a few nice people, including George, and that was that.

RD It made a profound mark on your whole attitude, didn't it?

MD It did, yes.

RD That's something that now is talked about, but I remember seeing you go through it. Because "Mr. Spiky" leveled out a little bit, I mean, if you compare the Orkney works to *Revelation and Fall.* You were all "nails and spikes" and all kinds of angular things! You once gave a very vivid description, Max, that applied to *Ave maris stella.* We were talking in the evening somewhere, we were just relaxing, you were speaking of an image of a cliff falling and the sun and time . . . what was that?

MD I can't remember exactly what the image was, but I think it probably had to do with time, in relation to *Ave maris stella.* I remember while I was writing that, I think I made the analogy with the timelessness that seeped into the last section of that; the feeling of time having stood still, or not functioning in its normal way, which was very much a feeling that I wanted to convey in that last section. When I looked at the cliff and saw these layers and layers and layers of geological ages in the exposed face, and realized that, if our time scale were different, and you actually looked at the place where the sun rises and sets in the sea; if you were living on a time scale that was very speeded up, like one of those cameras that takes a photograph every now and then, you wouldn't see the sun rise and set and pass across the sky at all. What you would see would be a light which was there permanently, connecting those two places. And sometimes I wondered about time and about our perception of time, and, were they actually able to do so, how those cliffs would perceive time. Because actually looking at them, you feel that they are falling, that they are in motion. I have this illusion, sometimes. And eventually, of course, they will fall.

RD That's a fascinating observation. You must have experienced a great expansion of thought there!

MD Yes, it was quite extraordinary. I remember feeling that it was as if I were outside time.

RD And you think you've applied that then to the last section of *Ave maris stella?*

MD Yes, working out ways that one could perhaps recapture something of that particular magic moment, which seemed to stand outside time, through the technique that I was using in the piece. And, of course, to achieve that, one had to create the impression of enormous activity before that moment, with everything speeding up, speeding up, and the pace of the piece going very fast. And then cutting, and gradually spiraling in . . . and that's the right term, because I was very conscious of using numerical schemes in the work, which were at that instant spirals, rhythmic spirals . . . I think you know what I mean . . . which spiraled into the very last section, where the pulsations are very slow, and not regular. So it's as if your heart just skips and slows further and further and further. Also, I remember, to create that illusion too, of a great burst of light at the end of the piece, centering attention on an ever-diminishing volume in such a way that the ear must, as it were, grow bigger and bigger, out of the head, in order to be able to take in the very small sound.

RD That's lovely.

MD And at that point when your ears are very, very big out of your head, then to have the whole ensemble play something very, very loud, as loud as they can. It's only a tiny ensemble . . . half a dozen people . . . but under those circumstances, it sounds so huge that, with your ears in that expanded condition, it's very shocking; and it's like some huge blaze of light right at the end of the piece. That kind of thing has always interested me anyway, to condition, purely through musical means, with the right notes, your (ideal) listener, through the means of a small ensemble, so you can make him listen in certain ways and almost stop his heart beating. I think that happens in Balinese music, also. I find that they do extraordinary things with the way time is perceived. Those things, I remember, were very much encouraged by being in that situation in Orkney. Just being able, in total silence, in total solitude, to concentrate. And concentrate very, very intensely on those aspects of the musical language.

RD That particular piece of course is without text, because that's another very strong aspect of your music; you're frequently involved with texts.

MD Yes! I like writing my own text. I did *The Lighthouse* text and the text for the opera *Taverner,* and I've done the texts for various pieces I've written for children, the opera texts of *Cinderella* and *The Two Fiddlers,* and the smaller pieces like *Kirkwall Shopping Songs* and *The Rainbow.* I think that's a very essential part of my . . . dare I say? imaginative output! People are always saying, "Oh, you can't possibly write your own text! You don't know anything about it. You're not an expert." Well, probably I don't know anything about it, but I can at least hear the music that I want to go with those texts; and so the text and the music are born at the same time. They're part of the same conception. I think, *if* you can manage it, it's a very good thing to do. A lot of composers have done that in the past.

RD So we have your texts, Mackay Brown, and who else?

MD Recently, it's been all George Mackay Brown or me. In the past it was

Latin or Greek, and I did use Randolph Stow for *Donnithorne* and *Eight Songs.*

RD And George Trakl for *Revelation and Fall.*

MD Yes.

RD I've heard two of your symphonies, but I don't know the third. And I understand you might write a fourth.

MD That's right. I'll do another one for the Scottish Chamber Orchestra more concentrated, but probably as long.

RD What are you doing to that thing called the "symphony"? Why do you call it a "symphony"? What's the formal attraction there?

MD The possibility of a very large-scale, purely abstract work for orchestra, and I think "symphony" is as good a name as any.

RD And the movement aspect?

MD Yes, that appeals to me.

RD And what's this connection with Sibelius? I see people comparing your work to Sibelius.

MD That's my fault, because I have drawn on some ideas from Sibelius.

RD Can you speak about it?

MD Yes, of course! Particularly his way of transforming material. In the Seventh Symphony, I think it's very strong; in *Tapiola,* I think it's very strong . . . the way that he will slowly transform one bit of material into another.

RD That is fascinating. Sometimes that material starts out like a horn growing on your head . . .

MD That's right, and then the horn just takes over . . . the rest drops off. It's a very, very individual thing. And also the way he can articulate time, with almost nothing happening. A very sparse texture, with perhaps a few little pulsations going, but you feel that it's multilayered and that it's not just a simple rhythm expressed through the surface note values on the page. There are pulsations going on inside that. Bigger articulations which, I'm fairly sure, have to do with his perceptions of the way that landscape was working, and the seasons working on that landscape. Although he probably wouldn't have articulated it in words, it gets through, that kind of perception in his music.

RD And what about the ending of the Fifth Symphony . . . a most astonishing ending.

MD It is indeed. I pinched that directly at the end of my First.

RD I mean, that's really a frozen moment.

MD Quite astonishing. And another thing that I remember fascinated me about that Fifth Symphony is the way it starts . . . as though it's been going on for twenty minutes. What a lovely illusion to set up. I was very interested in the way he did that—those are relatively small things that influenced me tremendously, I think. Also, when I was a boy in

Manchester, and going to the Halle concerts, John Barbirolli was a great Sibelius fan and he did Sibelius symphonies at every opportunity; so I got to know them well. I remember the first time I heard the Seventh Symphony at a Halle concert, my hair stood on end! . . . Sheer awe of this extraordinary piece. He's had a lot of misunderstanding. On the surface you might think that he's very traditional. What I find interesting is the way that he articulates his time and the way he transforms his material. And in some pieces, there's almost an absence of material.

RD Max, I think we've touched on many aspects of your music which I somehow knew from doing. Is there anything that I've overlooked?

MD I don't think so. We've touched on every aspect, the symphonies, theater pieces, the kid's music, chamber music. . . . I can't think of anything else.

RD The Violin Concerto!

MD That was a completely new thing; and again, I hope I responded to the situation of having Isaac Stern to write for, who I think is a wonderful musician, and I loved working with him. He worked very hard.

RD You were happy with that?

MD Yes. He really did work. My goodness! This is purely anecdotal, but I remember after the first performance in Orkney when he'd done this and it was live on the television and with tremendous exposure. Radio and television took it direct which is quite something. . . . I don't know how many fiddle concertos have had that treatment. But poor Isaac, I felt sorry for him, with this very difficult new piece. There was a reception afterwards, and about two in the morning, he said, "Well, I want to listen to a tape of that." And for a couple of hours just ran through it and said, "Well, what do you think of this bit here?" and "What should I do with this bar, here? . . . I don't like the way I played that. . . . What about this? I didn't phrase it right." Such a workaholic! Not bed, not sleep . . . *work.* And he had a recital the next afternoon. I really appreciated that, because he was terrific!

RD And then, lastly . . . two farewells! One to the Magnus Festival, and now you tell me you're saying farewell to The Fires of London.

MD Yes. The Magnus Festival because it has been ten years that I've done it, and I'm now president of it. I'm not involved in the day-to-day workings. So that, if it has taken root, which I think it has, it will continue and flourish. But I think ten years is enough for me to be artistic director. Of course I will help. I have contacts and that is always useful. And I will write pieces for the Festival, obviously, if I'm asked. But I think I must be seen to withdraw. Otherwise, it would look like a one-man show.

RD And The Fires?

MD The Fires of London are now twenty years old, and of course have been and still are very successful. And I think that's the time to call it quits.

Recently, I haven't written many pieces for them. In the last couple of years, none. . . . I've just written a new one for their final concert. And whether I like it or not, it's been my work which has been the main feature of many of the concerts, although we've done so many composers apart from myself . . . commissioned an enormous number of pieces.

RD You mean the group will no longer exist?

MD That's right. There will be something called Fires of London Productions, Limited, which will be got together when there is an opportunity to do one of my music theater pieces, like the opera on St. Francis which I'm going to write. It seems that will be done in this country, and I would like to supervise that myself.

RD So that is the end of a very active epoch.

MD It is indeed. I'm very sorry that it is coming to an end; but circumstances change.

*Charles Abbott*

# *David Del Tredici*

BORN MARCH 16, 1937, CLOVERDALE, CALIFORNIA

**October 21, 1987** **New York**

RD What got you started in music?

DT Well, I think my beginning was unusual, in that my family was completely unmusical. I was the oldest of five children; and not until the age of twelve did I start piano lessons. I had absolutely no music in my past. I don't remember going to a concert, hearing a record, anything. But it was love at first sound the moment I began to play the piano at age twelve. By age sixteen, I had made my piano recital debut and played a number of times with the San Francisco Symphony, and had a beginning career as a concert pianist. I learned very fast, but rather late. You might say I was an old child prodigy. I loved to practice . . . three to four hours every day. My parents were amazed, and though they knew nothing about music, always supported me and approved. I felt a little like the first astronaut, so remote seemed this relation of my practicing to the activities of the rest of my family.

RD And what were your piano teachers like?

DT I began studying at parochial school with a nun of the Holy Name Order, Sister Mary Engracia. But very quickly she suggested I go to a man named Bernhard Abramowitsch, whom, I realized more and more, I was very fortunate indeed to have found. He was an immigrant from Germany and was a pianist in the great tradition of Schnabel. He got me interested in playing lengthy works of the Romantic period and in trying to project the form of the whole piece, making it as coherent an aural reality as possible. My whole training was geared to looking at music from the largest perspective. I was very fortunate, because what he was really doing was giving me composition lessons.

RD And you didn't realize it at the time?

DT I had no idea. Curious, too, that during this whole piano playing time, it never once occurred to me to compose. Abramowitsch also played a lot of contemporary music, on the Composers' Forum in San Francisco. Whenever he became too busy to learn some of these pieces, he would palm them off onto his young prodigy student. And I, without realizing it was something unusual, learned almost all of the Schoenberg piano literature by the time I was seventeen, as well as a whole host of other contemporary compositions. The piece which perhaps made the greatest impression on me was the *Three Études* by Robert Helps. They are dedicated to Abramowitsch, but being practically impossible to play, were quickly passed on to me. I learned them, played them in a recital, made a tape and sent it to Helps. He was stunned that anyone could really play them, let alone a kid! A lively correspondence and friendship quickly developed. The association stirred thoughts of coming east. Bob was at that time connected with Princeton, as was Roger Sessions, whose Second Sonata and *From My Diary* pieces I had played and enjoyed. Helps, Abramowitsch, and Sessions were all great friends. So there was a Berkeley-Princeton connection. When I graduated from the university, having gotten interested in composition by this time, it was natural for me to go to Princeton to study with Sessions and Helps. Bob was an influence on me as well, because he was the first person I met who was both a terrific pianist and composer, and had to juggle the two. I was beginning to face that problem myself. I had started having this horrible feeling that all those years of practicing—all that skill—I was giving up . . . and for God knows what. I had written practically nothing—a few modest settings of some Joyce texts, the *Fantasy Pieces* for piano . . . I really had nothing much yet. But at the same time, I was increasingly interested in composing. Playing the piano was becoming a vanishing enthusiasm.

RD You spoke from time to time about your Aspen experience with Darius Milhaud. Did you go from San Francisco to Aspen?

DT It was between my junior and senior year at Berkeley. I went to Aspen as a pianist to study with Leonard Shure. For the first time in my life I was unhappy playing the piano. I thought, what else do you do if you don't play the piano? Sing? Compose? For me there was no choice. I wrote, in what should have been practice time, a piano piece called *Soliloquy*. And a composer friend of mine, Robert Morgan, suggested that I play it at Darius Milhaud's composition seminar. After Milhaud heard the piece he simply said, "My boy, you are a composer." And that was that. I was thrilled and went back to Berkeley, played the same piece for Seymour Schifrin and was admitted to his graduate composition seminar.

RD And as a budding composer, were you under the influence of the Schoenbergian organization?

DT At that time I wasn't aware of such a thing as Schoenbergian organization. It was more the feeling that dissonance was exciting, a dazzling new phenomenon. Schoenberg was wild. And I had played his music. When it comes to composing, you are what you eat, I suppose; so my music was a blend of everything I did, of both the romantic and the modern. In fact, that first piece—*Soliloquy*—is the most dissonant piece I've ever written. But it was written entirely "by ear," without any thought of a row. I certainly think that my return to, or embracing of, the tonal language as a living compositional tool was a direct result of my having had a very involved, passionate career as a pianist, playing mostly Romantic music.

RD So there's a connection.

DT Absolutely. Having had a teacher who encouraged me to play the piano in a creative way got me thinking about music from a composer's point of view, and set a very special tone from the beginning.

RD But . . . you and I met over a piece that had quite different qualities . . . and that was *Syzygy* in 1968.

DT You did the world premiere at Avery Fisher Hall, then called Philharmonic Hall. How could I ever forget that event! It was by far the biggest concert I'd had to that point.

RD How did you come to *Syzygy*? What is the connection?

DT You mean the aspect of *Syzygy* as a highly organized, tightly controlled piece, full of mechanistic . . .

RD . . . machinery . . .

DT . . . absolutely *un*Romantic machinery!

RD How did that occur?

DT Well, I suppose it's the other side of my musical coin. The earliest pieces I composed, being a pianist, were completely improvisational. I just did them. Improvised at the piano, found what I liked, and wrote it down. At a certain point, that process just stopped. I then spent a terrible two years trying to compose, but with little result. I worked on a string quartet, and never worked harder, but none of it made any sense. So I put it aside. It was a crisis. I could not find a way to compose any more. I had lost my innocence. Somehow just improvising didn't work. I had to learn to compose over again; so I made a kind of rule for myself: composing must always be a pleasure for me, or I wouldn't do it. Compositional time had to be play time and had to remain simple. Strangely enough, thinking that way, I suddenly got interested in the idea of using musical devices: a sequence, a backwards or forwards treatment of a theme, a rhythm. The twelve-tone row is of course a device, but I never really used that much. I got passionate about these mechanistic means of musical motion.

RD Well, the first section of *Syzygy* is a mirror.

DT In fact, that's the most extreme example of my newfound predilection.

RD . . . which is a kind of Bergian idea, isn't it?

DT I didn't really know anything about Berg then. I was just involved in devices that seemed like *my* discovery. The piece you mentioned is a setting of "Ecce Puer," a four-verse poem of James Joyce. I had set the first two of these verses—and each was full of little mirror inversion devices, suggesting already a backwards and forwards motion. But I had *no* ideas for setting the last two verses, nor even any sketches. I remember suddenly waking up in the middle of the night thinking, "My God! I could just run the whole thing backwards." All the verses were strophic, so all the words would fit. But I felt so guilty—I was getting so many "free" notes with but a single thought—that I couldn't accept this idea for a long time and had to talk myself into it. "David, it's all right," I'd say over and over. That's a very interesting thing about using devices . . . for one thought, you can get many, many notes. I mean, if you decide to play a section backwards and it sounds good, you've instantly created a new section.

RD Did you find it was good for you in the end?

DT Yes. Besides getting me to compose again, I started for the first time to write pieces with some kind of breadth. I mean, when you look at all the music of the past, you see they are full of devices. Mozart, Schubert, Beethoven would not exist without the sequence. It's interesting. When my students are writing tonal music and I try to suggest the use of sequences, they usually resist, because they feel they haven't "thought up" all of the notes. The notion of trying out a musical idea at various sequential degrees, approving the result and thus suddenly having a lot of music you didn't think up "note by note" somehow seems wrong to them. It's almost a moralistic thing, like the Protestant work ethic. If you didn't "work" for each note, you don't deserve it. Surely Mozart and especially Schubert didn't "work" for each note. There wasn't time. Devices were a part of their natural musical breathing.

RD Was there something beyond this mechanistic approach that concerned you? The poetry? The words?

DT Well, if I might just backtrack slightly, in tandem with my becoming interested in using musical devices was my interest in the poems of James Joyce. Early on I set a number of Joyce poems for piano and voice; but *I Hear an Army* (1964) is the first piece I wrote without piano. It is scored for string quartet and soprano and is the first piece in which I used devices in a serious way. After that came another Joyce piece, *Night Conjure-Verse* (1964), and finally *your* piece, *Syzygy* (1966). Each uses a soprano, each gets longer, and each has more instruments than the preceding piece. With each, I experimented more and more with the distance I could go using devices and still have it sound like "real" music. On top of that, I wanted it to sound expressive. I *am* setting a text, after all—what is the point if the music doesn't somehow reflect the text? It was great fun trying to reconcile those two seemingly irreconcilables . . . the mechanistic and the expressive. This reached its extreme point with *Syzygy*. After that, I felt I'd gone about as far as I could with devices and still maintain a flexible musical result.

RD What was your next step?

DT After *Syzygy*, I wrote a piece called *Pop-Pourri*, which you premiered in Aspen. And that was a sort of a turning-point piece; I used my first

Lewis Carroll *Alice* texts, a Bach chorale, which of course is completely tonal, and a Litany of the Blessed Virgin Mary, a reference to my Catholic background. The piece had everything mixed together—hence the title, *Pop-Pourri.* It was not really tonal, except that I used a Bach chorale as a kind of "found" tonal object. But in hindsight, I would say that that was really the beginning of a whole new direction. The Litany of the Blessed Virgin Mary is a kind of unsettable text—unsettable in the normal sense. I mean, all those repetitive phrases—"Blessed be the holy name of Mary," blessed be this, blessed be that—except *that* kind of unsettableness appealed to me and was appropriate to my then-mechanistic *Syzygy* style. I thought of the Litany of the Blessed Virgin Mary as a kind of non-sense. On the other hand, I set Lewis Carroll's "Jabberwocky," too, in *Pop-pourri*; and that is truly a nonsense poem. With nonsense I felt somehow I could do whatever I wanted to the words. When I began to set the *Alice in Wonderland* texts, I took the craziest ones first, like "Jabberwocky," because they were, in a sense, the most adaptable to my atonal language. The Jabberwocky is a monster; the words were cracked, I could do as I pleased. There is no expectation or tradition for setting nonsense verse. It was very liberating. As I went on with the Alice pieces, *An Alice Symphony, Adventures Underground,* I think the most seriously tonal might have been *Vintage Alice.* It happens to use a poem which parodies "Twinkle, twinkle little star, how I wonder where you are"; and I used, in the piece, the famous tune always associated with that text. Because of the story, I needed, as well, a musical reference signifying the presence of the Queen of Hearts; so for that I used "God Save the Queen." I had then two "found" tonal elements. This time, rather than just present them, as I had the Bach chorale in *Pop-Pourri,* I really worked "Twinkle, Twinkle Little Star" and "God Save the Queen" into the texture. So in the service of the text, a kind of tonality was forced on me. But it was an odd usage. For example, I would use the one theme very, very slowly, then the other four times as fast. I would put them in three different keys at once; I was using tonality, but the effect was rather fractured. I certainly wasn't tonal in the Bach-Beethoven-Brahms sense. But I was beginning to get a feel for tonal materials. I was subjecting tonality to all of my *Syzygy* machines.

The overriding image I remember for *Vintage Alice* was what the Queen says to the Hatter, "He's murdering the time. Off with his head!" "Murdering the time" fit in very nicely with my earlier techniques of rhythmic distortion—simultaneous fast and slow, acceleration and *ritard.* I mixed up these rhythmic ideas with the newfound tonal material. It was my own brand of tonality. Crazy, I hoped, like its subject.

RD You seem to respond to literary suggestion. How important is that in your composition?

DT Well, it seems to have been of the utmost importance, since all the highly organized mechanistic pieces concern themselves with James Joyce; and then the tonal music was all connected to Lewis Carroll. I've always needed a text to get me going.

RD The text obviously would give you a shape, no?

DT No! That's exactly what a text *never* gave me.

RD If it didn't give you a shape, what did it give you?

DT An idea. With *Vintage Alice,* the image "He's murdering time" permeated everything, colored all of the text. In *Jabberwocky,* the idea was craziness, insanity, the monster. There is always an overriding image from any one text, and that image would eventually suggest a musical means. That is what attracts me to a text. I like to make some aspect of the poem *become* the musical means. For example, in *Adventures Underground,* I set the poem "The Mouse's Tail." What's remarkable about that poem is that it appears on the page in the shape of a tail. It starts at the top of the page with very thick letters and gradually winds down to the bottom with the tiniest print available. Very original. I thought, well, to set such a special-looking poem, I should make each page of music also look special—like a tail. I spent a lot of time figuring out how it would sound. An audience after all can't *hear* a tail! So it must look like a tail and yet sound like music. Again, it was the kind of restriction which excites my imagination. And it yielded a whole brave new world of musical sound that I would not otherwise have gotten.

RD Is there a slight transformation between the text to a visual image, to an aural image, or does it go straight from the text to an aural image?

DT It depends. Of course, in the *Mouse's Tail* the special aspect *was* the visual shape of the poem. It didn't matter particularly what it said, because the transformation of the visual idea was what gripped me, as did the "murdering the time" image in *Vintage Alice.* As monstrous nonsense did in *Jabberwocky.* In fact with *Jabberwocky,* the instrumental choices were poetic. I used a rock group—two saxophones and two electric guitars—just because I needed a three-dimensional, wild personification of the monster, Jabberwock, when it first appears. It's the most dramatic moment in the poem. I built the whole orchestration around the fact that this rock group would enter just then, and only be heard there. As the monster is slain, so did I, as graphically as possible, depict its dying in musical terms. I let myself be hypnotized, seduced by the poem, or by some overriding image the poem projects. Eventually that preoccupation leads to the musical means to turn that image into sound.

RD It seems incredible that after Joyce, you have stuck with one writer—Lewis Carroll—for so long, about twenty years.

DT I didn't think of it as one writer. These different poems suggest wildly different possibilities. Perhaps most interesting was trying to integrate into the Alice tales the true story of Lewis Carroll, the man, and his infatuation with Alice Pleasance Liddell. She was the real-life Alice for whom he wrote the books. I was thrilled when I discovered in Martin Gardner's work *The Annotated Alice* that many of the nonsense texts were actually disguised versions or parodies of poems by other authors. And these other poems always concerned unrequited love—even, in one, the unrequited love of a man for a girl named Alice. I wanted somehow to put into the Alice stories the real feelings of Carroll the man. I'd never thought before about that aspect of Alice. To intermingle the Alice poetry with the revelatory love poetry I'd discovered, because that was the overriding idea/image for *Final Alice.* Now it was not so much the Carrollian nonsense—a camouflage—which fascinated me, but rather what he had *not* said, the poems Carroll had in effect suppressed, the ones which spoke about love.

So for *Final Alice,* I made a libretto of many texts, weaving the love poems in and out of the actual Wonderland tale, in this case, the trial scene which ends the book (hence the title *Final Alice*). I used these original love poems as depositions of evidence which the White Rabbit presented before the court. It fits in as a dramatic device because that was already happening in the court scene. I simply augmented the amount of evidence so as to capture the touching sentiment of these love lyrics. My music became more and more tonal. It was exciting. I had been recreating monsters and mouses' tails and fooling around with "Twinkle, twinkle little star" . . . but had never actually tried to conjure up love. It was a new musical emotion. I wanted to make his infatuation, his passion live. So I just grasped at these chords which, in the past, have always dealt with such things. And that is called tonal harmony.

RD And that harmony suited the situation more than anything else.

DT I remember halfway through writing *Final Alice* I thought something would happen, as it had in *Vintage Alice,* to obscure somehow such blatant tonality—that sections would be refracted and splintered, that I'd add something contrapuntal in another key, or the rhythm would be jerked about. But nothing like that ever came to mind. I was left with these pure tonal "things." The situation gave me a kind of musical nervous breakdown. I thought, "My colleagues will think I'm nuts! I can't be so tonal in 1976. It's crazy. It's not legitimate." On the other hand, I had to look deeper to that part of my personality which had always done the composing, and it was as excited about the tonic and the dominant as it had been about retrogrades and inversions. So I went with the excitement factor. I really had no choice. By tonality, I mean use of the tonic, the dominant, the chords of tonality—functional tonality. I don't use the word in the generalized sense. In my usage, the music lives or dies by the tonal progression of the chords, as in the music of Beethoven, Schumann, Mahler—the tonal composers of the past.

RD The other thing that strikes me now is—you sound like a composer who should write a piece or many pieces for the stage.... Why did you stop short?

DT All the pieces are long. *Child Alice* is two and a half hours long, as long as some operas. God knows, I've been asked. Perhaps it's that I never found an appealing libretto. Perhaps it's the idea of collaboration. I find it frightening. I don't know. It makes me think of Berlioz. The works of his which people really love are not the operas, but pieces he wrote for the concert stage, which are opera-like. *The Damnation of Faust* or *Romeo et Juliette,* for example. His concert pieces sound as if they should have been operas, yet they aren't. I feel a little in that category. I love making the concert stage, the concert hall, seem like an opera house. It is so unexpected, disorienting—the audience doesn't know what to expect. It's a more interesting world for me than opera, though I *do* love the opera. It is also surprising to me why Mahler didn't write an opera. God knows he would have been the perfect man for the job!

RD If we can continue in the Alice domain, I remember also doing *All in the Golden Afternoon ...*

DT That's perhaps my most decadent piece.... And you, Richard, were the first conductor who really *loved* it.

RD Well, I did.

DT That thrilled me, finding someone who felt as strongly about the work as I did. It was wonderful. You made the piece come alive. That meant a lot to me.

RD I'm glad, because we did the piece not only in Aspen but in Rotterdam and The Hague. But, despite my affection for it, came criticism like, "Well, it sounds like Strauss, it sounds like maybe Mahler ... it's all German-sounding." And people used that as a form of criticism. Were you aware of this Straussian aspect that they talk about? Or this Mahlerian aspect?

DT Oh, sure.

RD You were doing it consciously?

DT No, not consciously, but there is a curious thing about tonality ... there is really no precedent for using it in the late 20th century, so there is no way you "ought" to hear it. If you use tonal chords and a Romantic harmonic vocabulary, to some degree your music is going to sound like someone else. Anyone who uses atonality probably is going to sound something like Schoenberg. Does that mean their music is not original ... or good, or bad? I was aware of the fact that an aspect of the music would sound like other composers, since it was their tonal language; but because I was so excited about the composing, I felt that there must be something of me in it, too. I just went with that simple visceral feeling. If at any moment when I'm composing one of my tonal pieces, I feel dead, or it sounds to me borrowed or unfresh, I throw it out. I am very sensitive, particularly because I am using a language which has been well worked. It's a double jeopardy, in a way. Because I'm using a language in which so many great pieces have been written, incompetence is all the more evident. This is controversial to say, but if you write something atonal, dissonant, chance-filled, you can often get away with murder. It's more difficult to separate the good from the bad; standards are not quite so clear. But if it's tonal, there's just too much good music people know and love. You put yourself on a certain line when you use tonality seriously. Also, I think listeners, especially critics, need to retrain their ears. It took a long time for them to develop a positive response to atonal music. It sounded like nothing, or *noise.* Gradually though, as they heard it over and over, their ears began to differentiate the good atonal pieces from the less good. Schoenberg from his imitators. I think the idea that now those retrained "atonal" ears are suddenly asked to listen to a completely tonal piece in the 1980s is a shock. I mean, all that time spent trying to understand atonal music, to like it, to think of atonality as the only legitimate direction for musical "progress." There's no given appropriate critical response to present-day tonal composition. Every critic is on his own, must take a chance with his instinctive reaction. Often, I think, they are scared if they like it, afraid they'll sound naïve, reactionary . . . I don't know. I can't read critics' minds. Tonality is, strangely, a new thing, and there's no "proper" way yet to respond to it.

RD Do you have any thoughts about the word "modernism"?

DT I have no thoughts about the word "modernism." I have more thoughts about the word "post-modernism," of which I'm said to be an example.

RD What does post-modernism mean?

DT In music, modernism is atonality. Post-modernism is what came after; and that, in my case, is tonality. Similarly, in art, it is Realism after Abstraction. Post-modernism re-embraces an artistic friend from the past and gives it new life.

RD So there's a return aspect.

DT A return, and a kind of renegade conservatism.

RD Well, there was a period in those hyper-modern times, and in painting also, that said something like, "If it isn't new, it isn't art."

DT The whole idea of being new has changed. Had to change. In 20th-century music, "new" meant more dissonant, more chance-like, or somehow more rigorous. Schoenberg pushed tonality out the window and embraced the twelve-tone system. Then little by little, the idea of chance took hold. At a certain point, the most rigorously organized atonality sounded just about the same as chance-oriented music. When that point was reached, I think people realized that something was wrong. There had to be something more to music than this! Also, new music was not touching its audience. Though I've often been asked the questions, "Were you aware of the lag between the composer and the music and the fact that modern music was dying out? Did that have anything to do with your return to tonality?," my answer is always no. The impulse to use tonality seemed to come from a deeper well.

RD You were more interested in touching yourself, then?

DT ... to pardon the expression! Aren't we all? When times get tough? Would you like to rephrase that one?

RD No, I don't think I'll touch it!

DT I liked your question. I remember clearly thinking then that my music, being tonal, would be rejected. People would laugh at it.

RD Intellectually not acceptable, is that it? But did the audience like it?

DT I hadn't thought about the audience. I grew up in the tough '60s and '70s, when composers did not have real audiences. It was mostly colleagues in sparsely filled concert halls. That was my point of reference. So I was scared to death that becoming tonal would alienate what few composer colleagues I had. The fact that my music was warmly embraced by the audience was a shock. I remember the premiere of *Final Alice,* and the audience at the end just stood up and cheered. It never occurred to me that such an extraordinary reaction might occur.

RD There's another thread in your composition that I think is evident, and that has to do with virtuosity. It seems that most, if not all of your pieces demand considerable performance capability ... from voice, for instance, to instruments. Can you trace that in your development?

DT Well, I think it has something to do with the fact that I was a piano virtuoso. I knew what it was like to play difficult music, so it was natural of me to expect the same of others. Why it transferred to the voice, with which I had very little experience, is due perhaps to the fortunate circumstance of having met Phyllis Bryn-Julson so early on in my career. In fact, she sang my first difficult vocal piece, *I Hear an Army* (1964). And she did it so easily that I wrote another piece *Night Conjure-Verse* (1965) that was even more difficult to sing.

RD And then she did *Syzygy* (1966).

DT Yes. Finally she did *Syzygy,* which is most difficult of all. After she had sung all three pieces, I realized suddenly just how lucky I had been to have met Phyllis, who could sing anything. On the other hand, I feared that I might have written three white elephants which no one else in the world would ever be able to perform! People talk about the interaction of composer and performer. If I had not met Phyllis, would I have written such wildly difficult music? If the first singer had said, "This is impossible," I probably would have believed her. I didn't know. I wasn't a singer. Really, then, Phyllis Bryn-Julson shaped my vocal style, and I went right on with this virtuosic writing into the *Alice* pieces.

I think I've always been interested in pieces which are long, that contain the extremes of emotional expression. I love having hysteria and violence mixed together with tenderness. Naturally, virtuosity fits with this idea.

RD Your virtuosity in terms of the voice borders on punishment, practically.

DT But for the singer who can do it, it's heaven. For the singer who can't it's pure hell. If you have a soft high A, I'm your best friend. I think Barbara Hendricks loved me for the "5000" high As I put in *Final Alice.* But I think many other singers cursed me for the same.

RD This is a little different direction. But let's consider strains in music; and let's consider your Catholic background. Do you feel there is a possible Catholicism in your music? Or is it even *possible* to have a Catholicism in contemporary music?

DT Well, when I think of a composer like Messiaen, I think there is something very Catholic about his music.

RD What is it?

DT I've always felt close to Messiaen. Even in the beginning, when I really didn't like his music, I still was on his wavelength. I knew it was a Catholic wavelength. The idea of combining impossible things: ondes martenot, bird calls, tight little mathematical units that keep repeating, excessive length, romantic harmony—it doesn't make any sense. It shouldn't work, fit together, but it does. It's very Catholic. Although our music is quite different, we are both attracted to excess and the heterogeneous mix. I love the rigorous musical constructs. I love tonality. I put them together and enjoy the electricity from this "impossible" combination.

RD Well, his music has a certain kind of reverence because of the text. I don't think your music is exactly reverent, is it?

DT He's certainly reverent in his *titles,* but the music itself doesn't necessarily sound so reverent to me. I was the last generation to have an old-fashioned upbringing. That is to say, I went through twelve years of schools with nuns and priests as teachers and as role models. I know what James Joyce meant when he spoke in the *Portrait of the Artist* of the problems of being an adolescent Catholic. I lived it. In fact, *that* is why I was so attracted to Joyce. It was not just the poerty, but Joyce the man, writing about *my* problems in *his* books. I remember having to say rosaries and penances, counting the days off from Purgatory for saying this or that prayer. The Church tried, then, for a very exact organization of every aspect of emotional life. I am an old-time Catholic, and that fact, I think, has something to do with my constant striving for a tight musical organization, while at the same time trying to project a wild, out-of-control quality.

RD It sounds as though you used that experience, and not necessarily rebelled against it.

DT I didn't think of it as rebelling. It simply was, I suppose, in my genetic code. That was me. I suppose one is fortunate, *if you survive,* to have strongly opposing forces always at work.

RD If that is one possible strain in your music, could we speak about other composers that might have influenced your aesthetic thinking, development, etc?

DT I can think of two who influenced me enormously . . . one was Aaron Copland and the other was Leon Kirchner. I met Aaron very early on in my composing career. I had written just a few Joyce settings; and a friend of mine, the composer Stanley Silverman, said, "Why don't you send them to Aaron?" Of course, I had never met Aaron, this famous man, but I did send him my cassette. Several months later, mysteriously, I got my first commission from the Fromm Foundation, and an invitation to come to Tanglewood to study. So I went to Tanglewood, met Aaron Copland, and heard my newly commissioned piece, *I Hear an Army.* That was really the beginning of my career. Though Aaron never admitted it, I'm sure it was he who made those two things happen.

RD Did you see Copland after Tanglewood?

DT Yes. I would visit him often and played him every piece I wrote. Although I never really studied with him . . . he was an enormous influence. He was someone who *really* could compose. Though he wrote wonderful music, his attitude about composing was still simple and straightforward. He didn't make the process seem complicated. These were, you'll remember, the complicated times—the '60s and '70s. I was at the same time an assistant professor at Harvard University. When I was talking with colleagues, composing music seemed so cerebral, so difficult. Then when I would visit Aaron and play him something, it would all seem so simple again. I would ask him, "Aaron, how did you write that theme from *Appalachian Spring?* How did you get the idea for those *Variations?*" And he'd say, "I don't really know. It just kind of came to mind." Writing music was for him a natural thing, he trusted his musical instincts. It was not a torturous, intellectual exercise. His attitude gave me courage. "Well, if it's good enough for Aaron, then it's good enough for me. And perhaps those colleagues who talk a wonderful line but can't write such wonderful music, perhaps there's something wrong there. . . ." I began to realize that musicality, such as someone like Aaron had, was the main thing. Perhaps even the whole show. Aaron was not someone who theorized about music—he simply did it—a real professional.

RD Was it an extraordinary experience to play your music for him?

DT Yes. He wasn't a teacher in the didactic "composition seminar" way I'd experienced at the University. I would play him a piece, and at first he might not say anything at all. At a certain point, however, Aaron might casually say, "Are you sure that ending isn't a little too long?" He'd often make his comments offhand, but they'd be right on the mark. Aaron had a wonderful way of getting it all, right away. But he'd let you know his opinion, or give a suggestion very gently—in a way that wouldn't offend, be resented, or might even go unnoticed. He never imposed his personality. I got so I was very sensitive to any comment Aaron might offer and valued it, while trying to guess all its possible implications. Aaron was a master of the light touch.

RD He certainly was. I have had my share of associations with him, and of course the event that ties the three of us together is the 1968 Koussevitzky Festival, where *Syzygy* was first played. It was Aaron who said, "I want you to look at this piece by David Del Tredici." And I didn't know who you were. It was in those days that I got to know your music.

DT Aaron was wonderfully supportive of my music. I'm sure many of my early performances were a result of Aaron's having said a nice word about me. But Aaron would never let me know he'd done such a thing. He would deny any connection, but things *did* seem to happen ... Aaron is unique in contemporary music for his generosity towards composers. Everything people have said about him is true.

RD What about Kirchner?

DT I met Kirchner while I was a student at the University of California in the '50s. He taught then at Mills College and heard me give a recital. So ours was an early association. Years later when I was living in New York City and rather desperate, economically, he suddenly appeared at my apartment door and said, "I want you to come to Harvard and teach." He said he remembered me playing my *Fantasy Pieces*, and he liked them. It was like a light at the top of the stairs, and Harvard was my first teaching job. Kirchner was marvelous. Though his manner was different from Aaron's, they were alike in the most important thing—music was for both an emotional, vital reality. It's interesting that the composers who influenced me the most were not the people with whom I "officially" studied.

RD One last question, David. What are you composing now?

DT Well, Richard, I think I've finally left *Alice in Wonderland* behind. Or rather, forsaken one children's text for another. I'm setting the poem *The Spider and the Fly* for soprano, baritone, and orchestra. It's a piece I began about three years ago, but put aside to write three other pieces: *March to Tonality*, *Haddock's Eyes*, and *Tattoo*. I've recently come back to it. And though the poem, like many of the *Alice* texts, is short, the amount of music I have composed for it is immense. I'm in that terrible spot I've been in before, not knowing how to wed one means to another. . . .

I often write the music before I know just how or where the words will fit. Who's going to sing what. It's a very nervous-making time. Though most of my music is for voice, nobody, interestingly, has ever called me a vocal composer. Perhaps it's because I write so extravagantly and virtuosically, or perhaps so much of each piece is purely orchestral. I like to create an environment that's appropriate to the song, and then let the song enter that environment. To let the text "pass through" the piece, as it were. Or even, as in *Child Alice,* let the poem be heard in two different versions. In that later piece, one setting was done from the child's point of view, as Alice Pleasance Liddell might have understood the poem; the other was presented as the adult sensibility, as Lewis Carroll might have felt as he penned those verses.

**RD** Do you see this piece as a lengthy piece?

**DT** Yes, it's my standard extravagance—forty-five to sixty minutes.

**RD** Do you see any cycle coming on?

**DT** Please God, no! But what has often happened with works of mine—certainly *Child Alice* and *Final Alice*—at the start I always think they will be little five-minute pieces. I really do. And then they grow and grow of their own accord; and it's a shock for me when I finally have to accept the fact that this tiny germ of an idea has become a huge, sprawling thing.

**RD** And how does *The Spider and the Fly* seem right now?

**DT** It's definitely huge and sprawling, but needs still a continuity. I like my works, however large, to fit together neatly, like a crossword puzzle. I am reticent about talking of what I hope yet to discover.

# *Iannis Xenakis*

BORN MAY 29, 1922, BRĂILA, RUMANIA

**November 27, 1986** **Paris**

Today, about mid-day, near my hotel, I saw the student demonstrations begin at Montparnasse and continue down rue de Rennes. At first, just a few students appeared, then little by little banners appeared, then more students and very soon hundreds of students began passing by. I watched for about an hour and the numbers grew into the thousands. The basic issues of protest apparently have to do with a unversity bill proposed by the government, which would raise tuition costs and change entrance regulations, which they feel are elitist. However, what I found extremely fascinating was the aural experience. From these thousands of non-violent students came singing, unison-chanting, and whistling, that made unified sound waves ripple down the rue de Rennes for blocks! The aural effect was absolutely thrilling! The only interruption or "punctuation" occurred when someone threw a bag of garbage from a window, which landed on the pavement with a dull sounding "splat." This produced an exchange of fist-shaking from the window to the crowd and back, accompanied by chaotic verbal declamations. A "rare event," if you want. It was fantastic!

My appointment with Xenakis was for 6:00 p.m. at his studio near place Pigalle. Because of the demonstrations, Paris traffic was a mess and taxis at a premium. I arrived late. His one-room studio was stuffed with books and scores. There was a stand-up desk and a crowded seating area in the middle of the room.

RD Some time ago you were developing a system which you said would provide new possibilities for composition, teaching, and research. As I remember, it also had to do with exploring the creativity of children. What has happened to that idea?

IX Yes. This is a system that we have developed over the last ten years here in Paris in my laboratory, CEMAMu (Centre d'Études de Mathématique et Automatique Musicales). The system is called UPIC (Unité Polyagogique Informatique du CEMAMu), and it's based on using a large, special drawing board. Anybody, even myself or you, or children, can draw lines or graphics with an electromagnetic ballpoint, and they are transformed by computer directly into sound, based on acoustics. Thus

bypassing the traditional steps of computer-assisted programming, perforated cards, etc. You can compose or do any training or pedagogy for the ear or for writing; because the writing is not the usual musical writing. It's a much more universal one, because it is with lines. For instance, a note that is held is just a horizontal line.

RD That's the linear aspect. How do you achieve the vertical?

IX The vertical aspect is for the pitches, exactly like when you write for an instrument: when you go up it's higher; when you go down, it's a lower sound.

RD And polyphony?

IX Polyphony, yes of course. Hundreds of lines at the same time. And you have to design also the elementary wave form which is first, approximation to the timbre. You can write sine waves or whatever curve you want. You can obtain a difference in the timbre and also the dynamics by designing the dynamic envelope. The interesting thing is that you can design, listen, and then start again . . . correct or throw away what you don't want.

RD Have you used it yourself?

IX Yes. I've used it for two pieces. One was written about ten years ago. That was because nobody else understood the method at the beginning of the UPIC system. So I wrote a piece which is called *Mycenes-A*. But, there is no synthesizer, or other source. It is only by drawing that you obtain the sound, directly.

RD And did you develop this by yourself?

IX Well, yes. I developed it and I have people working with me, programmers and electronic engineers. It is supported by the French Ministry of Culture. Inside the Ministry there is a special organization called Direction de la Musique, and they pay for these salaries and whatever.

RD Once the so-called piece is done, is there then a "memory" that can recall it?

IX You can store it on a digital tape or on an analog tape. Then you can play it back in concerts on speakers.

RD But the sound source is the computer?

IX The sound source is your drawing which is then translated by the computer into sound.

RD The sounds themselves are electronic sounds?

IX No. The sounds come from the design that you have done.

RD What kind of sounds?

IX That depends.

RD Artificial sounding sounds?

IX Yes, rather. But you can also imitate instrumental sounds or voice.

RD Beyond your use of it for your own composition, you say that it can be used as a pedagogical aid for teaching students.

IX Students or children. We went to many places in France, in centers, in conservatories, and abroad also, in Holland, Germany, Portugal, Greece, and in Japan.

RD And was it stimulating for the children?

IX I think so. They were very . . . how do you say . . . enraptured? Because it's like a game. I noticed that with children of five or six, they have imagination, but they have no milestones, structures in their mind, in order to organize things. They have to be helped somehow by monitors or by teachers. When they reach nine or maybe twelve years old, they are much freer and they are not yet under the influence of what they have heard from the mass media, radio, disks, TV, etc. They can adapt very quickly and show their inventiveness. Whereas, when they are older, let's say fifteen or sixteen, they are already "made." That is, they are distorted. Even so, students who are training to be physicists and in technology have also used the system. They made some music with a certain originality, but usually tried to do what they knew. Like conservatory students. Instead of reacting in a very new way to a different tool.

RD That would be an interesting new generation of young people if it were universally accepted. Is that what your aim is?

IX It is a basis to learn music, to teach music, and also to make music which is closer to the human mind; because you have only your hand to use. You don't need to be a computer scientist to do programming, you don't need that. You don't even need to be a trained musician. What you need is to have some ear, some mind, and fingers.

RD Is there a visual image they try for, or are they just reacting to the sound they hear?

IX In the beginning they try to do designs, nice designs. If they are children they use the imagery of houses or cats or suns. But then they have this immediate response from the machine and they start listening more carefully to what they design. That is, they adapt the design to the aural effect. It depends on the intelligence.

RD Do you think that people would begin to understand your music better if they would be trained in this way?

IX They could understand any music better, I think. For instance, you can design a chorale of Bach; and if it's properly designed, you can play it back. You have a Bach chorale with whatever timbre you want. Because you have designed it.

RD So, in other words, it doesn't rule out using the past.

IX The past is to be performed by performers who have devoted lots of study over the years. But in this case, you have a direct transformation from the score into sound without going through any intermediate machine, like the human machine . . . the orchestra. I don't say it's better. There is much work to be done, because the refinement of instrumental color is not yet achieved or realized by the technology. But it will come in the near future.

RD Are there more mature students or young composers that are fascinated by this process?

IX Oh, yes. There are several composers who have done compositions using this system, like François Bernard Mâche, Jean Claude Eloy, like Julio Estrada from Mexico, like Wilfried Lentsch from Germany, Jim Harley from Canada.

RD How many machines like this exist? Are they only in France?

IX Today there are about seven or eight machines. Most of them are in France. Two are in my laboratory. Two others are in La Villette, and another will be installed there in January, which is where the Scientific Museum is located and also where a conservatory will be built. So there will be three there, belonging to an organization that is inspired by my organization, and which also organizes workshops everywhere. Then there is another one that is being installed at the University of Strasbourg in the Musicology Department. There is another one at the Paris University and then there is one in Athens.

RD That's an extraordinary development. You say that this project is about ten years old. Is there one pupil that is now ten years older and has stuck with it, like a violinist practicing?

IX I don't know. There are some American students, composers, who are working on that aspect. We are not a conservatory or a training organization. No, we are doing this, and composers can use it if they wish. Now maybe at La Villette they will start doing systematic compositional and pedagogical research.

RD It's an incredibly new branch of technology in the world.

IX Technology of today, that is, computer science, has put into the hands of the composer or musician a kind of writing, much like when writing was invented 4000 or 5000 years ago. It was a completely new era, because people could transmit their thought, but also themselves. They could, by writing and erasing, consolidate and think better. This is now happening because of the technology of today.

RD Is it also a hope that it will go beyond the sort of library or collection aspect of computers? I have two young boys that have gone through computer classes, and they were bored stiff. Because once they learned this present method of using the computer, it wasn't so interesting for them. It's not easy to be inventive or truly creative with it. Might this be a way of stimulating the artistic side?

IX Absolutely, yes. In other centers like in France or the States, where they do music on computers, they are stymied because of that interface problem between the user, that is, the musician or the child, and the difficulty in programming. Even the gadgets that they use produce the same difficulty. So they are stuck or they produce very uninteresting results, because the quality level and the possibilities are narrow. I don't say that that will not improve in the future, but for the time being this is the case.

**RD** Besides yourself and UPIC, are there any other leaders in this domain?

**IX** Not yet, because the idea to base composition on design is the original thought. And most of the other people have not conceived that. They do some drawing, but in a very narrow and limited way. They don't use the whole capacity of the human brain to have an interaction between the design and the sound. Which you do when you write music on a score in the traditional way. But in that case, you have to train yourself for years to know the writing and also to understand what you are reading. In this case, the same thing exists . . . that is, you are designing, but the interaction is almost immediate, that is the difference.

**RD** I imagine that the designing aspect has great fascination for you, from your background as an architect. Would there be any use for the word "poetic" in this music?

**IX** Well, poetic . . . when there is an artistic standard. Then it can be poetic, in that sense.

**RD** So that part doesn't change.

**IX** No. It depends on the skill and the imagination of the user, always. But it makes things easier to express one's imagination.

**RD** What about the cost of this machine or system?

**IX** The price is relatively high, because it's like the cost of three professional stereo tape recorders.

**RD** Who builds it? Is there a company?

**IX** No, we build it. From here and there, taking parts that exist on the market. For instance, the board exists on the market. The only thing is that it's very expensive. People who use such a board are in industry, designing, for example, cars or architectural things. It's a limited market, so it costs a lot. Other devices that are bought, like "bugs" or components, are much cheaper.

**RD** "Bugs"?

**IX** When I say "bugs," those are transistors, and all those kinds of logic circuitry, which are very cheap today. They are astonishingly cheap, although they are very complex, because they are used by millions of people now in the small personal computers.

**RD** I could imagine that Japan would be fascinated by this whole project.

**IX** Well, yes. They invited us to come. Maybe they would like to have such a system in Japan. They are building a center in Fukushima, where every year there is a kind of international symposium about art and technology and science. I have been there twice. I think that the relationship between the visual and the hand that makes the design, and the brain, is a fundamental wave of the future, for anything . . . instead of having a keyboard and a program or whatever, like the MIDI system that puts together all sorts of synthesizers and other devices. That is very easy to do, but you have to remember all sorts of things with the keys and the meaning of the keys, what they activate; whereas when you design and see what you have done, that is the closest to the human mind.

RD It's interesting that you make the parallel between that and writing. Have you applied some of your other theories of probability, etc., to this system? Can you do that, too?

IX In that machine? No . . . not for the time being. I hope to add some analogic or stochastic functions or transformations to what is done by hand, so that anybody could use them just by pointing out or defining some very elementary parameters.

RD In the meantime, you have to battle with traditional instruments, and you are now battling with saxophones. You are writing a piece for the Rascher Quartet.

IX Yes, correct. This year I composed six compositions.

RD Can you speak about those?

IX The first one, *Keqrops,* is for piano and orchestra and was premiered in New York about ten days ago by the New York Philharmonic, Zubin Mehta conducting, with a wonderful pianist, Roger Woodward. I wrote another piece for orchestra called *Horos,* which was premiered in Tokyo, commissioned by the Suntory Foundation. Suntory is an industry of Japanese whiskey, a whiskey which is quite good, I must say! They also built an important auditorium in Tokyo, with about 2000 seats, with good acoustics, but less interesting architecturally. Then I did a short piece for trombone solo, for my friend the trombonist Benny Sluchin. It will be premiered in the States at the International Association for Trombone.

RD It is incredible how these organizations develop! There is one for tubas also, and for clarinets. . . .

IX I wrote another piece for the Xenakis ensemble, from the Middleburg Festival in the south of Holland, that was founded by Ad Van't Veer. It was for him and the harpsichordist Elizabeth Chojnakca. I wrote another piece for the Arditti String Quartet, with piano . . . they are the best in Europe. They play a lot of contemporary music and are very dedicated people. Claude Helffer will be the pianist in this piece, and it will be premiered next year in December. The last piece I can speak of is for the Ensemble Intercontemporain which Boulez will premiere next January in Paris. It is for the tenth anniversary of the Ensemble Intercontemporain. That makes six, doesn't it?

RD Your direction is unique. Do you feel that your music has evolved?

IX Yes, I think so. From what I hear from the people who listen to it, and perhaps from myself, it is changing.

RD You seem to defy evolution because of the many primary principles that you are always applying.

IX Ah! The principles might be everlasting, but how you address them, that is different. That is the music, the final result. Now when I am asked for program notes, I say I don't have any. This is because I don't want to influence people. I say, music is made to be listened to and not to be described. Because any verbal description fails. When I was young, I used to write program notes which were rather scientific, concerning some of the principles that I applied to pieces. But the critics and the public were misled. They thought *that* was the music. They thought it was just a kind of mathematical music or a cold music, instead of carefully listening to it. Except for a few compositions, I never used only mathematical reasoning or techniques to compose. For instance, in 1954 I introduced the "mass" concept of sounds, with many glissandi, in *Metastasis,* long before other composers were taken by it. That piece produced a scandal when Hans Rosbaud conducted it in Donaueschingen. And it was the same two years later with *Pithoprakta,* conducted by Hermann Scherchen in Munich.

RD Just a footnote about the glissandi issue. I talked to Ligeti recently and I asked him about this issue, since Stockhausen had claimed to be the first to have used it. Ligeti said, "I can assure you that it was Xenakis who did the first piece with glissandi in this manner in 1954." In any case, concerning composition, often it has been said that you are basically in the absolute. You say that you don't *only* use theory. When you don't use theory, Iannis, where does that "other thing"come from?

IX I think that, after a while, when you have thought a lot and gotten used to the theoretical means you have developed, it becomes second nature to you. You don't even need to calculate. It's almost automatic. Which is both a bad and a good thing at the same time. Bad because you are like being caught in a kind of web that you have created for yourself. On the other hand, it gives you a much wider sense of things that way. For instance, for many years I have worked to produce some kind of structure, stochastic or whatever. Now, I know what it means, musically. I am further away. I can see it, I mean I can hear it. I can understand it. . . . Therefore, I can put that together with the other structures, and have on a higher level a kind of vision of things, and therefore a vision of the architecture of a piece.

RD Do you feel that many composers have used aspects of your music as an aid for their composition?

IX Well, yes, mentally, and the sonic aspect, as you mentioned, the glissandi. Many people, the Polish School and others, have taken, let's say, the superficial aspects, not how it was built, and used them. Picasso used to say that you take so much time to create something, you create that with your guts. Then it's there and anybody can use it, because it is finished, it's "ready-made." That is what happens. I don't teach personally, because what does teaching mean? It means to push the pupil to do things. That I can do. But when he shows what he is doing, and you have to correct and try to influence and to impose on his personality, I don't like that. I prefer that he would react in his own proper way. And this is also my experience. I think that there is no difference between having a teacher and listening to music . . . to good music. Good music also teaches. That is the best teacher, and if you don't have the perspicacity or the intelligence or the ear to understand what has been done or what is being done, then all the worse for you.

RD This systematic thinking that came out of Darmstadt produced a lot of bad music.

IX Yes, of course. Because it was a kind of Fascism. I remember some people saying back in the '50s that even jazz would be serial music in the future. Which was absolutely wrong. It was exactly at the height of the imposition made by that music in Darmstadt and other places in Europe when it collapsed. Suddenly it collapsed. Serial thought as Schoenberg developed or devised it, I think is very interesting. It somehow linked all the tendencies of other sciences and human thought to try to purify their science from language and what they were doing. This had started already in the 19th century in mathematics and logic. With Boole in logic and the revolution of Hilbert and the German School in mathematics . . . and other people, Cantor in the theory of "set theory," and so on and so forth. That was a very important trend. In painting and architecture also, you find at the turn of the century a kind of reaction against all "decoration" . . . in architecture with Gropius and Le Corbusier, with the Dutch School, and that was a purification from an overwhelming tradition. The same in painting, with Abstract painting that came also from that kind of direction, to start something different; and Schoenberg was deeply involved, at least sentimentally, with this trend. Although the serial system didn't start exactly with him, it started before him, with the dodecaphony of Hauer and even other people. For instance, there is an absolutely unknown composer, a Professor Loquin, who was at the Conservatory of Bordeaux in 1895, who said there was no reason why we shouldn't have a specific hierarchy of the tones of the scale. Therefore, the twelve notes should be on the same, equal level. That was the kind of thing that was starting at that time. But to declare that it was the *only* way is incompatible with another, more general thought; which is the, let's say, "deterministic" vs. "indeterministic." And in indeterminism, you find "probabilities," probabilistic logic, which deals with "masses" that I used, or with "rare events." As a result of the impasse in serial music, as well as other causes, I originated, in 1954, a music constructed from the principle of indeterminism; two years later I named it "Stochastic Music."

RD You have spoken about how this concept utilizes the laws of rare events, the law of large numbers, and the different aleatoric procedures . . .

IX And therefore deals with symmetry and asymmetry, which is the main core of music and perhaps of most of the things that happen in this universe. And also periodicity, which is for the in-time repetition of things. As biological beings, we are bound to this kind of thing. We are at the same time bound by heredity . . . what does it do? It tries to retain, to repeat. Because we have to die. But at each birth, most of the past is in yourself, from your parents. And so on and so forth, for generations . . . the species. But at the same time, there is no possibility for an absolute duplication, one that is absolutely the same accurate repetitition. So there is a diversity that builds up, and this is why we have so many species on this earth. This is why we change, ourselves, not much perhaps during the last thirty-five years; but it takes time, of course, geologically and biologically speaking. We are different from what we were, say a million years ago, or three million or thirty million years ago. So, in spite of this necessity to hold what you have, that is, the permanence or the renewal of the same thing, there is a diversity which is unforeseeable. And there is where this "indeterminacy" enters, which is also rooted in quantum mechanics and the working of the universe. Although there are rules, laws, these laws are perhaps only one aspect of today's knowledge, and we don't know what happens, really, in fact.

RD Often the critics and some composers have complained that the serialists not only ran into a dead end, in terms of being the *only way,* but that they also "dehumanized" music.

IX I think that they are partially wrong. Because serialism tried to produce new rules, and rules of some kind always existed in music. For instance, with tonal music, there are rules about how to do that well. Otherwise, you are mistaken. So there were rules and there were mistakes. And the composers of any time were involved with those rules automatically, or they tried to learn them, or they trespassed them. It's no wonder that serial music also has rules. But the mistake was to consider that these rules were absolute and forever, the only ones. That was the mistake. There is no such thing as the "Absolute Thing."

RD Considering society in general, how can it influence a composer?

IX You mean the pressure of society? There is a way that it could influence the composer. When he doesn't earn enough money and says, "Why am I not played, performed? Why are people not interested and why don't I get enough money?"—And he comes to the conclusion that music should primarily exist in order to attract a public. If he does that, then he goes into the path of the public. That is, he starts becoming less original, less different, than he could be. In that sense, commercially speaking, there is a pressure on the composer, on the artist, in general. So he follows fashions. Fashions that grow here and there because of that commercial pressure. I think that the duty of an artist, his endeavor, is to be independent of those kinds of pressures. Otherwise, why does he write?

RD Why do you compose?

IX I compose because . . . well, I could have done other things in my life, too. I was interested in many things. But then finally I decided to do music, essentially. And I decided to do that because I thought I would be less unhappy doing music than doing other things. That's it. There is no other reason.

RD That's a good answer! I've done quite a number of your works. You were present when I did the world premiere of *Anemoessa* as well as the North American premieres of *Syrmos* and *Anaktoria* and you heard my performance of *Erikhthon.* I also did *Empreintes, Pithoprakta, Polla ta Dhina,* the music you wrote for the Aeschylus *Oresteia* and I recorded *Akrata* for Columbia Records.

IX *Anemoessa* was not performed again. That was the only performance. Maybe it is bad music.

RD I don't think so myself. . . . What are the similarities or the dissimilarities in these works that I mentioned, or is it too vast a subject?

IX I can point out some dissimilarities, because they stem from a theoretical basis which is different. For instance, *Syrmos* was written in 1959. I wrote it when I was struggling with "quanta" of sounds and Markov "chains," and the meaning of all that. *Akrata* was the result of another theoretical study that I did which was based on "group theory." I wrote other pieces similarly based, like *Nomos Alpha,* for solo cello, which I wrote for Siegfried Palm; and *Nomos Gamma,* for orchestra scattered throughout the audience. *Oresteia* was a completely different thing.

RD Well, you were dealing with a staged production of the Aeschylus play.

IX That was one thing. The reason why I wrote it that way was because I was trying to reinvent the music of that time, with my poor means. That is, from the fifth century B.C. we have had absolutely no trace of this music anywhere, except for some theories which came later.

RD But if anybody had it in his genes, it would be someone like you.

IX Maybe. I don't know. In any case, I put my genes on the table and I tried to write the music!

RD And what about *Erikhthon,* which I remember doing with the pianist, Geoffrey Madge.

IX *Erikhthon* was based on another idea of "bushes" or lines of sound. Traditional polyphony is based on one melodic pattern which is repeated or imitated; or, like in a fugue, there is a subject and a countersubject which is different, and then you plot with them. Instead of just having one line, I used a variety of lines at the same time. That's why I call it a "bush" or arborescence. A "bush" is what you find in nature, like a tree, and it is made out of many branches that overlap, or there are diverging or converging branches, etc. If you take that as a new object, then you can rotate and impose on it various dramatic transformations, that is, in the pitch versus time domain. So that was the basis for *Erikhthon,* for piano and orchestra.

RD When I did it, my imagery for that was "vein-like." It was like veins.

IX A vein is a tree-like structure, absolutely.

RD I remember having the idea I was inside the blood vessels!

IX That's very good! Very good image. It's the same.

RD I even told the musicians to try to believe that they were inside their blood vessels.

IX They didn't agree, probably.

RD Well, they thought it was funny. But what about *Polla ta Dhina* where you use children? The children's voices sound so pure.

IX Ah, yes. Scherchen asked me to write a piece for the Stuttgart Festival of Popular Classic Music, in which he conducted Ravel's *La Valse.* I told him, "I cannot write 'light' music. And he said, 'Never mind. Write whatever you want.' " So, I took a hymn from Sophocles about Man—how wonderful and terrible Man is, because he can do fantastic things, but he can also destroy and produce destruction. I took that and made the orchestra part; and in order to bring out the meaning of this text, not in a solemn way but in an absolute way. I chose children, because children are fresh, they are innocent. So, I put into their mouths the text of Sophocles sung on one note.

RD Could we speak about your use of primarily Greek titles for most of your pieces? Is that because of your background or do you find them more descriptive? They don't *mean* anything in terms of the music, do they?

IX Sometimes they are like a simile. For instance, *Syrmos* means a continuation of something, like a train, which is the case with Markov "chains," it's a train of events. *Akrata* means pure. Why pure? Because it's based on some basic structures of mine, the group structures, which have been worked out and investigated by psychologists like Piaget, for instance, who did that with his own children. He discovered the evolution of this mind structure in children from six or seven years old up to twelve years old. After that we stop. We have the same structure, it doesn't change anymore.

RD And *Anemoessa,* for choir and orchestra? You said the title translates, "full of wind."

RD *Anemoessa* was a much more distant kind of impression. I don't know . . . it's also a matter of phonetics.

RD Well, the phonetics were vowels, as I remember. I can't say it sounded like the wind howling or some other obvious image, no, but I suppose you do conjure some aspect of what wind is. However, if I was not told that the title means "full of wind," I don't know that it *means* anything.

IX It shouldn't *mean* anything, it's just the shape maybe, the way that the wind moves or behaves or exists in continuity.

RD I remember one tape piece of yours, *Bohor*, in New York at the Whitney Museum, and it literally blew out the place.

IX It was too loud.

RD Well, the ceiling was terribly low. You say *Bohor* can be played on a beach?

IX I played it on a beach in Royan at night, and we had some fireworks afterwards.

RD It has been said that in your work there is no cleavage between musical material and musical organization. Can you speak about your sense of architecture in music and your sense of reasoning?

IX Perhaps what I could say is, that the various directions, orientations I had in my research did not stem from each other; but they were parallel. They converge finally because of how you put them together, much the same as how we usually live. Sometimes you think in a very logical pattern, trying to demonstrate things; sometimes you can take the same thing as a revelation, saying, "This is it!" There is no explanation of that. And you live with this. You suddenly have ideas and you act according to those ideas. Sometimes you are reasonable and say, "I have to do that because . . ." And you act that way. So I can say, therefore, the problem of architecture in music is to bind together these things so that the result is self-sufficient on every level, and not made out of bits and pieces.

RD Just one last thing. Do you consciously avoid our past musical tradition?

IX Consciously . . . not in a systematic way. Sometimes the musical translation of what I do theoretically looks like something that has been done in the past. But that is fortuitous, and my personal determination is not to use, as many other composers have, citations and things like that. I don't like that. I think that my involvement in music, my privilege and also my duty, is to try and do something different. Otherwise, what the hell! The others have done much better than me in the past.

RD You are interested in something beyond the reach of what we know now—like making a leap forward?

IX Yes, I like to do that. I don't say that I achieve that, but it is what I try to do.

# Narrative VIII

My first official concert in Europe took place in Madrid, Spain, on May 9, 1969, with the Orquesta Nacional at the invitation of the Spanish composer Luis De Pablo. The program included the Spanish premieres of Ives's *Fourth of July,* Stockhausen's *Punkte,* and Berg's *Three Pieces for Orchestra,* Op. 6. The day before my first rehearsal, I heard a program in the Teatro Real conducted by Cristobal Halffter in which he presented his cantata *Yes Speak Out* commissioned by the United Nations. It caused a complete sensation, for it was perhaps one of the first politically oriented works heard in Spain in recent years. The audience demonstrated for nearly half an hour, and I could only assume that it was anti-Franco.

My work began the next day. The first rehearsal was a disaster! In attempting to rehearse Stockhausen's *Punkte,* I found that the material sent by Universal Edition did not correspond to my score, which was a later version. Not only that, but the musicians began shouting (in Spanish) that their parts were not properly prepared. The pagination was completely foul. Pandemonium reigned. I went on to Berg, Opus 6, and soon realized that it would be a grizzly job ahead trying to make *anything* work.

At the next rehearsal, I was to begin work on *Fourth of July,* but I was told that the material was "somewhere in Spain." It finally showed up thirty minutes before the rehearsal began. And so it went . . . the first trumpet player had a case of kidney stones and wasn't coming to rehearsals. There was great resistance to Ives's *Fourth of July* because it was felt that marching band sounds were beneath the dignity of the Orquesta Nacional Espagnol. Slashing, stamping, beating, sweating, the general rehearsal ended in a flare of tempers, and the orchestra's refusing to give me seven more minutes necessary to finish the third movement of the Berg. Apologies came later from the O.N.E., and we entered the Teatro Real that night (May 9) with a feeling of good will. The hall was filled, the audience enthusiastic, the orchestra played above their usual standard, and an ovation was received at the end. This was only the second all-contemporary concert given by the O.N.E.

My next destination was Basel, Switzerland, where I was to work with Pierre Boulez in his Course in Interpretation and Conducting of Contemporary Music, June 16–July 5, 1969. In the meantime, I learned of Boulez's appointment to become Director of the New York Philharmonic; and I received a telegram from Carlos Moseley inviting me to assist the Philharmonic on its next transcontinental tour.

The Boulez course was smashing, and I learned a great deal from him about analysis, rehearsal procedure, and certain conducting techniques. He impressed me very much, and I found out how fabulous his ears were and how well loved and respected he was by the musicians. At this stage in my career I thought I knew how to rehearse. Watching Boulez was a revelation. *Nothing* was left to chance; and he maintained that if you do not come to rehearsal with a specific plan, you are like "a dog looking for a bone."

In a preface to the course, Boulez said, "It is high time to demystify the term 'specialist,' much too convenient to rid oneself without qualms of the musical present; remove the myth, too, from the personage which all too often the conductor plays, to the detriment of historical progress, denying (renouncing, rather) his principal *raison d'être.*

"To be avoided thereby: the astutely camouflaged amateur as well as the rigidly narrow-minded professional. Both of these evils, equally formidable, open out onto parallel disillusions, identical setbacks, similar catastrophes. They divert the senses, they reject participation; they engender confusion, provoke misunderstanding; they slow down agreement, warp the perspective, dry up the vital flow of communication ( . . . neither dictator nor artisan! neither oracle nor valet! neither Messiah nor sexton! neither angel nor beast! . . .)."

The three-week course was intense, with orchestral rehearsals every morning and evening except Sunday; and theory classes every afternoon including Sunday. There were three public concerts. The repertoire: Schoenberg, *Erwartung,* Op. 17; Berg, *Drei Stücke,* Op, 6; Bartók, *The Miraculous Mandarin,* Op. 19 (complete); Stravinsky, *Chant du rossignol* and *Symphonies of Wind Instruments;* Boulez, *Éclat;* Berio, *Tempi concertati;* Stockhausen, *Punkte;* Webern, *Sinfonie,* Op. 21; Varèse, *Arcana.*

Boulez's professional demands were severe; from the knowledge of the scores, to the tuning of the orchestra, to balancing the general sound, to specific percussion timbres, string harmonics, etc. His ears were incredible, and he could shock you into a near state of paralysis. For example, I remember while I was conducting Berg, he suddenly stopped me and said that there was a mistake in the viola section. I acknowledged this in a general way, but he went further and asked, "Which viola made the mistake?" This was but one of many examples of his standard of perfection.

I got along well with Boulez, but didn't always agree with him. For instance, I challenged his decision about who should conduct *Erwartung* and asked to be given the opportunity to conduct it in rehearsal. My complaint was that the piece has an emotional content that must be considered as carefully as the purely mechanical problems, which are formidable and which he understood better than anyone. I conducted the piece, and the orchestra seemed to appreciate my efforts. The emotional content of the work was never discussed.

My Stockhausen experience in Madrid had prepared me for what was about to happen in Basel. The orchestral material for *Punkte* was just as bad as before. I told Boulez about the problems I had had, and he said, "We will see." At the first rehearsal of *Punkte* I was asked to conduct, and it was Madrid all over again. Great commotion—What to do? Original version? Second version? Boulez speaks: "Bring the music to the afternoon class." So all of the parts were brought to the afternoon class and stacked on Maître's desk. Whereupon, he anounced that we would all check the orchestral parts simultaneously. The questioning began: "Is there a fermata in the fourth measure?" Answer—no. "Is there a repeat in the tenth measure?" Answer—no. "Does the oboe have F# and G in the fourteenth measure?" Answer—no ... etc., until finally Boulez decreed that this was the wrong version. An amusing photo of Boulez, with his arms stretched out, was sent to Stockhausen's publisher UE, Wein. On the back was written "Warum?" and we all signed it.

Boulez and I got along best in the classes (without orchestra) when we would sing the orchestral music while conducting, cueing, etc. He had a large vocal repertoire of sound effects, but so do I. The result was that the entire class would break up. I remember an amusing moment regarding the *Höhepunkt* in the last movement of the Berg. We were singing along and at this moment his emphasis was a beat earlier than mine. And in my bullheaded way, I began emphasizing this beat over and over, until finally Boulez took a chair, held it in front of him, and said "OK, OK."

On another occasion he was grilling us on rhythmic curiosities in each of the three movements of the Berg. Two of my colleagues failed miserably. Fortunately, I had worked that noon on just those problems, and without score, did my "number." This secured an assignment to conduct the Berg, which concluded the course. I remember gesturing to Boulez to take a bow with us. He was warmly and enthusiastically greeted by the audience, the students, the orchestra, and it was a genuine ovation.

After the concert there was a huge reception for everyone. Boulez sat at the official table and was heralded for carrying on the tradition of European conductors, for becoming Director of the BBC and the New York Philharmonic, *and* naturally for his contribution to composition. Afterwards, Boulez thanked me for my performance.

Shortly after this, I made my conducting debut in Paris with the Orchestre Philharmonique de l'ORTF and the Orchestre National. I also conducted in Vienna and Rotterdam for the first time. In October 1971, I headed for The Hague and the Residentie Orkest for an unforgettable musical experience producing *Carré* by Stockhausen for four conductors, four orchestras and four choruses. Five performances were scheduled: one in the Kurhaus at Schevenigen and four in Paris at the Salle Wagram, a 19th-century boxing arena. We rehearsed morning, noon, and night, beginning at 9:00 a.m. and sometimes ending at midnight, using the rehearsal plan of the premiere in 1960. There were 18 separate simultaneous rehearsals for each of the 4 groups and 7 tutti rehearsals, making a total of 25 rehearsals for each conductor, but a total of 80 rehearsals (not counting chorus preparation) for the entire project of 77 instrumentalists, 48 choristers, and 4 conductors. Now, that's the way Europe did things in those days! It makes the meager American rehearsal scheme for large-scale contemporary projects pale.

My co-conductors were Gilbert Amy, Michel Tabachnik, and Lucas Vis. The four conductors, if not rehearsing musicians, rehearsed themselves. The mutual effort made in solving the many problems of *Carré* was rare as well as terribly enjoyable. Our first efforts were to sit at a crowded coffee table and discuss, argue, and try to maneuver our way through a work that presents so many new conducting problems. While each of us felt confident about conducting our own groups, it was no surprise to me that, with four conductors, reaching an agreement was sometimes as difficult as deciding foreign policy in a government.

What was difficult? To begin with, each conductor not only gave directions, but had to *receive* directions from his fellow conductors. This is against the basic nature of a conductor and provides immediate grounds for disagreement. Although we were very critical of each other, we overcame this hurdle because we were all friends and had been together in Basel at the Boulez course.

Then, the work *Carré* (square), as the title suggests, is to be performed in a prescribed space (about 25 m by 25 m). Each conductor, with his respective group, is placed on a platform at the points of the compass. The audience sits in the center area, enabling the sound to surround it. At these distances, the conductors have a considerable visual problem in order to coordinate their gestures.

Stockhausen's concept of a "music in space" considers "spatial dispositions by the sound, by direction of the sound (alternating, isolated, fusing, rotational movement, and so on)." This presents brand new challenges for the conductors. Besides the aesthetic qualities and innovative methods of vocal and instrumental execution, the conductors faced a complexity of performance that I had never seen before and found truly revolutionary. It was essential that each conductor practice his score with a stop watch and that all four conductors together utilize stop watches during their private rehearsals as well as during the normal rehearsals with the musicians.

The composer wished us to be able to cast the music around, back and forth, like tennis players. Stockhausen says about *Carré,* "This piece tells no story. You can confidently stop listening for a moment if you cannot or do not want to go on listening; for each moment can stand on its own and at the same time is related to all the other moments. . . . My heartfelt wish is that this music may afford a little inner calm, concentration and breadth; an awareness that we could have a lot of time, if we simply take it—and that it is better to come to one's self than to lose oneself."

The night before the dress rehearsal, we conductors had our own dress rehearsal in the Kurhaus at Schevenigen, that incredible "elephant" of a resort hotel on the North Sea. In the silent *Saal,* there we were, gesturing to each other, making vocal sounds in order to conjure some semblance of the combined orchestras and choruses, each on our own specially built platform.

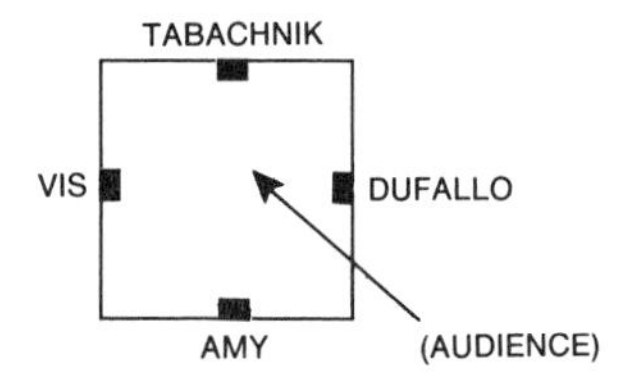


At these distances, we found it difficult but not impossible. We decided that there was only one more thing necessary to acquire, lest our final rehearsal become a total disaster, and that was an internal communication system. We had anticipated complete chaos without one, since musicians can be noisy under normal conditions, let alone with this complexity. The usually simple task of announcing where to begin or end became an enormous task. Our final rehearsal went smoothly and we used our microphones and earphones to great advantage. The most significant accomplishment was that disageements among the conductors were solved in a fairly calm and quiet manner.

We performed the next night to a packed, enthusiastic house which gave us high hopes for the subsequent four performances in Paris. Our only concern now was facing Stockhausen with our results. We flew to Paris leaving behind good times and hard work and took with us what seemed to me a new level of mutual musical understanding.

In Paris, the morning rehearsal began with the usual problems: microphones not ready, equipment late, etc. Stockhausen arrived, dressed in a costume that made him look even larger than he is, and his mood was one of profound seriousness. We met, and I was impressed by his determined face and his erect, lumberjack posture.

We ran through the work, Stockhausen in the center of the room taking notes. During the intermission we received our set of corrections, none of which were too severe or too lengthy, and were mainly concerned with percussion timbres. The composer was pleased, and so after another run-through we were ready for the evening.

We performed *Carré* twice each evening for two nights. The first night, Stockhausen spoke between performances and we demonstrated examples. He praised the performance and called it better than his own premiere. He sat with eyes closed through each performance as if in an trance. All performances were received well and the reviews were ravishing. The work is extraordinary, if not a masterpiece. These concerts with the Residentie Orkest were part of the Semaines Musicales Internationales de Paris (SMIP). With the Dutch and French premieres behind us, the four of us would eventually perform the English premiere of *Carré* in September 1972, in Royal Albert Hall, London with the BBC.

After an appearance with Domaine Musical in Paris in February 1972, my work with the Residentie Orkest continued in May 1972, when I was invited to conduct a marathon of three concerts. The first was part of an homage called "The Country of Cage": Brant, *Signs and Alarms;* Brown, *Event: Synergy II;* Ives, *Second Orchestral Set;* Varèse, *Amériques;* and the Cage *Concerto For Prepared Piano.* Then came two more programs with the music of Ligeti, Schat, Stravinsky, Halffter, Scriabin, and Davies. John Cage was present at these concerts, and David Tudor prepared the piano.

# *Pierre Boulez*

BORN MARCH 26, 1925, MONTBRISON, FRANCE

**April 10, 1987** **Paris**

I arrived at the Centre Georges Pompidou in Paris at 5:55 p.m. Five minutes early. I descended the underground staircase to the Institut de Recherche et Coordination Acoustique/Musique (IRCAM), which looked like an entrance to an elegant "subway." I asked for Pierre Boulez, and moments later he went whizzing by the reception desk. I called out "Pierre," and he stopped like some product of centrifugal force that had its power suddenly cut off.

He ushered me up a spiral staircase to his office, where I waited about fifteen minutes while he tended to other matters. While waiting, I reflected upon the fact that I had conducted Boulez's music in New York, Buffalo, Aspen, and Milan. And I also tried to review in my own mind the questions I was going to ask this formidable man!

RD Two nights ago I spoke with Karlheinz Stockhausen at his home in Kuerten; late last night I spoke with Mauricio Kagel at his home in Cologne; and this evening I am speaking to you here at IRCAM in Paris. In the many discussions that I have had, there's a myth . . . or a reality. . . . *But*, Darmstadt looms, in some minds, like a word out of the past that threatens some composers already in their maturity. In other minds, the Darmstadt initiatives have significantly contributed to the expansion of musical investigation; and there is also an expression of love-hate in regard to these initiatives. What about those initiatives which you have written about and helped to shape?

PB I never really felt concerned by Darmstadt. Darmstadt occurred quite late in my development . . . I mean, my development was already there! In France, I discovered the Viennese School in 1944, '45. The first time I went to Darmstadt was for one day in 1952. The real "first time" for me in Darmstadt was in 1956. After *Marteau*. At that time it was a very pleasant first meeting point; Stockhausen was there, Nono was there at the beginning . . . along with many others. It was interesting to meet them and exchange ideas; but it had absolutely *no* influence on my own development.

RD I don't think that is what I was getting at. What I am talking about is what you did there and what you produced there in terms of initiative.

So often when composers now speak of Darmstadt, they refer to "Stockhausen-Boulez" or "Boulez-Stockhausen"; and there is, I suppose, also this mixture of envy, approval, disgust, etc., in regard to what you and, let's say, Stockhausen achieved. In the recent collection of your writings,[1] your early Darmstadt lectures (1959–1961) on "Fundamentals," which are very succinct, present a very positive view of your aesthetic position.

PB When I was there, I gave classes in composition which took place for only ten days each year. I went there in 1956–1957, not in '58, then again in '59 and '60, and then quite regularly until 1965. In 1966 I began to conduct in Bayreuth, so I had no time anymore. During this rather short period of ten years, I was especially involved in giving these composition classes, rather like I am today at the Collège de France. They were more like courses in theory than practical composition classes.

Then I gave some lectures; and I think it was in '61 or '62 that I last really played a major role. Before Steineke died, I gave the Musikdenken Heute series, which was ultimately enlarged and published. The second important fact for me was that I shared the conducting of the Darmstadt Group with Bruno Maderna. The orchestral concerts were generally given by the Frankfurt Radio Orchestra, because it was the closest city. And other orchestras also came as guests. But Steineke's idea was to take some musicians from the Domaine Musical in Paris, some from Germany and others from Italy, so as to form a kind of international group. Maderna and I were alternately in charge of this international group during the same season. We shared concerts. He made a program . . . I made another. I don't remember how many programs there were during a session, but there were certainly no more than three or four.

RD That history I'm a bit familiar with. To put it a little more severely, taking one topic at a time, your dissertation on "Taste: The Spectacles Worn by Reason?,"[2] for example, leaves very little room for anybody else to have an opinion! And so it goes. Everything that you touch I think you explain quite thoroughly, in terms of what you believe. But it also produces, it seems, a notion that "this is the only way." And that is said by many of your colleagues.

PB Well, if they reacted in that way, they cannot be very secure about their own work. If one produces a reaction by taking a strong stand, I am not against doing so. If other people object to your opinions, that is not your fault. They are objecting to these opinions because they think *they* are right. But if they react to a situation twenty or thirty years later, it

1. *Orientations,* collected writings by Pierre Boulez. Ed. by Jean-Jacques Nattiez, trans. by Martin Cooper (Cambridge, Mass.: Harvard University Press, 1986).
2. Ibid., p. 44.

doesn't interest me very much, because I am now in another universe. The universe of Darmstadt lasted ten years for me. And now, who cares? I know that in Germany, for instance, some of the young people criticize the Darmstadt period. I think they have better things to do than just criticizing the Darmstadt period.

RD It does get down to specifics about that old world of serialization, doesn't it? You have been quizzed about this a lot.

PB I myself was also very critical of it. And I think my very first critical articles were published back in 1954 and 1955. So one cannot say that I was really uncritical. One of the lectures I gave in Darmstadt was against the teaching of composition, because I thought teaching was really not very necessary. Teaching is only necessary for a very brief period of time. A "school " of composition cannot really exist.

RD I've been asking a question to some of these composers, and I think it produces interesting results. What could you say was your musical genetic code? Stockhausen answered that he invented his!

PB That's very easy!

RD And others reflect on a situation, a teacher, this or that aspect of the past, etc.

PB Of course, you always invent your own genetic code yourself. You invent your personality. If you don't invent it, nobody will do it for you.

RD But you are finding things along the way.

PB But of course. Everybody is influenced by everything and everybody. Of that I am absolutely certain. But the personality that is able to transform these influences into something so personal that there is no longer any reference to them, then of course that is *personality.* You invent yourself; but at the same time, you "invent," because you take substance from other people and then you find *your* substance from that. And after that you are not influenced by other people directly, because you have found your own territory. What you take could be anything you hear, even in the most remote areas you don't want to explore very much. If you happen to hear something or look at something, some things can influence you in a very unexpected way. Therefore, I think you are like birds of prey . . .

RD Like vultures?

PB Eagles or something . . . Yes, perhaps vultures in this case!

RD When you were a young student, you worked with Messiaen. One sometimes hears that influence in your music. Is there anything else?

PB Yes, of course. There is the Viennese School and all the Stravinsky works, Bartók works, even Varèse. . . . There are all kinds of things you listen to in the beginning that are very important. But I don't think Messiaen could have written *Marteau sans maître* and Stravinsky could not have written my *Sonatine* for flute. Nor could Schoenberg have written my Second Piano Sonata. Sources can be obvious, but the way they are transformed is less obvious. And they are even less obvious to the composer himself, because he really does not know very much about this very personal process of transformation. It is part reflection, part instinct and even part irrationality, all of which play an important role.

**RD** . . . and intuition?

**PB** Yes, intuition. Something irrational.

**RD** . . . because that is something Stockhausen spoke about as needing to be methodically challenged. Beyond that, we also spoke about his concept of pieces and form in a bigger life-sense. You share that view to a certain degree, don't you? I mean, instead of movements . . .

**PB** Yes. Certainly yes. I am more and more for a piece—even if it is distributed in movements—where the movements are tied to each other so strongly that they form a continuity. But that is not new either. For instance, look at *Erwartung* by Schoenberg. Of course it is a theater piece; but you find even in Mahler that one movement is already an entity in itself . . . the first movement of the Sixth, for example, or the last movement of the Fifth. There are works that oppose the kind of accepted form of simply contrasting movements, and the definition of each movement according to separate prescriptions. I think that in music, the kind of narration, its continuity and development, forced the composer to think much more in terms of a work as an organic whole.

**RD** I don't think everybody shares that particular view. But I think you and Stockhausen are very verbal about it. This leads us toward *Répons,* which is your latest work that is now in a mobile state.

**PB** Yes, and it is becoming longer and longer because of very pragmatic consequences, and because of my reflections about its composition. Pragmatic because, if you have a special setting, a very complex building up of everything, you cannot really write a small piece, a short piece. It has to take up a whole evening. You cannot escape that. So, faced with this problem, you must conceive of it in terms of a long work.

**RD** But you designed that ambiance . . .

**PB** I wanted it that way. In order to have some musical facts given to the audience, I needed this kind of geography, or topography. And then, thinking of that, I was forced to consider the basic practicality of *writing* the score itself. Because you cannot write for musicians who are thirty meters away from you as if you were conducting them just under your nose. Under your nose, you can really conduct very quickly, and they will follow. Any kind of cue will be immediately reacted to. But when people are twenty or so meters away from you, they cannot play the score and look at you at the same time, even knowing the general feeling. First of all, they don't hear that well because they are far away. Then there are all the transformations within the work, and so on and so forth. So you must account for, in the writing itself, the distance and the fact that they cannot follow. Either you can give entrance and exit "cues" or you can give a beat. But it must be a beat which is very slow so they can follow it. Even if they miss seeing a beat, they won't lose your tempo. Especially through the environment in which they listen.

RD So speed is a big factor . . .

PB Yes, a big factor. But I didn't always want to write at slow speeds. So I had to find a conjunction between these two entities: the center and the periphery. I had to join these two entities by some very practical devices of writing which could give me a slow tempo, a quick tempo, whatever I wanted. In the beginning I had difficuly in considering and realizing it, because I could not write just *any* kind of music for both situations. And then there arose the problem of the length of the piece. If it is too short, you don't understand why you created all these earthquakes to end up with only a *mouse!* That's impossible. You have to have at least, maybe not an *elephant,* but at least a *camel* or a *horse* . . .

RD . . . a sizable something!

PB Yes, a sizable something, exactly! Since I conducted a lot of Wagner between 1976 and 1980, and Mahler before, I was much more aware of this kind of length of form. And if you use big forces, you cannot simply write a short piece. It is out of proportion. There are two exceptions in classical contemporary music: The Berg *Altenberg Lieder,* where you have big forces for very short pieces . . . they are marvelous pieces, but it is very difficult to grasp them . . . therefore they were not performed for a long time, and Stravinsky's *Le Roi des étoiles* (The Star-faced King). You have a male chorus, which has to be big, and then you have a big orchestra, practically like the orchestra for *Rite of Spring.* But the piece is only six minutes long! By the same token, you cannot make a very long piece without using large forces. There is maybe one exception in modern times, and that is Strauss's *Ariadne.* There you have a rather restricted orchestra, but you have real virtuosity. You must be very virtuosic to use small forces for a long period. I think there is a kind of relationship between the mass you use and the duration. In contrast, for me, when I was beginning *Répons,* I had just finished the first four parts of *Notations,* which uses very big forces for rather short pieces. Two minutes. Three minutes. But there is a series of them. That's different. I also wanted to think in terms of small form in this conception. On the contrary, with *Répons,* I had to think of form and forces in quite a different way.

RD How long do you imagine *Répons* will be eventually? It is now approximately 42 minutes.

PB Now it is 42 minutes. Sometimes 43 or 45. When I am very excited it is 42, and when I am less excited it is 44. It will be approximately 30 minutes longer.

RD Since *Répons* uses the 4X computer . . . this is just an aside . . . but in talking to Kagel about things of the Middle Ages and isorhythmic discoveries, he said something I thought was interesting. He said that he saw no difference between trying to use isorhythmics and trying to use the computer today. He felt that there is a constant in the human being which has to be discovered. The secret has to be revealed.

PB Well, of course, the esoteric element was always something which was ingrained in art generally: the *esoteric* of proportions in architecture and in painting. This occurs especially when science and art are going hand in hand. In the Renaissance period in Italy, for instance, when art and science were very close to each other, there always was a kind of speculative aspect involved in the study of perspectives and proportions. I think the *esoteric* is nice, but it's really more a matter of self-contentment than anything else. For example, the esoteric element in Berg's *Lyric Suite* does not really add anything to the music. Not at all. He put in all those secrets that nobody could decipher just for his own satisfaction. Finally, of course, it is deciphered, because he has given us the key. It is a piece which is very dramatic, romantic, etc., etc.; but I mean, with another love, with different initials, he would have found a way to introduce *those* initials!

RD And so what!

PB Yes, exactly! For him, the essential thing was to introduce initials; and after that it isn't very important. In other cases, you can have references which are more or less mystical in numbers. I once read a study on Obrecht, the Dutch composer; and everything is a question of one proportion related to another in the architecture of the church where he was performing the *Missa* or something . . . I believe it! But the kinds of relationships between spatial measurements and temporal measurements have very little to do with each other. You find that in all civilizations. It reassures one about order in the world! But if you observe scientific studies, the further you go, the more complicated the measurements become. And they don't fit at all into these kinds of very simple proportions which generally constitute the essence of the *esoteric*. Things are usually much more complex than the *esoteric*. Things are much more complex than the *esoteric* would like to have it. Therefore, I am amused most of the time by these kinds of considerations, because I always think of Hamlet . . . "There are more things in heaven and earth, Horatio, than are dreamt of in your philosophy." I think that's right. One can never really stop contemplating the complexity of the world. And one knows very well that one's own complexity will never match the complexity of nature!

RD Yet one goes on.

PB Yet one goes on. But I don't need any encouragement to go on! Or any encouragement of that nature!

RD Allow me to call your attention to how I met you. I heard the American premiere of *Marteau* in March of 1957 in Los Angeles. I recently called

Lawrence Morton[3] (who sends regards) in Los Angeles to ask him some questions, because I wanted to establish when we first talked. He is now eighty-two, and I haven't seen him in a long time. I understand you were just in Los Angeles for a big celebration called "From Pierrot to Marteau." Anyway, I told Lawrence that I was at your concert in 1957. I met you, but we didn't talk. He reminded me that you came back to Los Angeles in 1963, when you conducted and played on April 3 at the Monday Evening Concerts. We discussed that program briefly, and afterwards I recalled how impressed I was by it. I believe it began with the *Three Japanese Lyrics* . . .

PB Yes, Stravinksy.

RD . . . followed by *Trois Poèmes de Mallarmé* of Ravel. Both works were composed at about the same time, when Ravel was living with Stravinsky in Clarens (1913). Then came *Trois Poèmes de Mallarmé* of Debussy, followed by your own *Improvisations sur Mallarmé,* I and II. I was quite astonished by your conducting of the *Improvisations,* because of your new technique, which was extremely adroit. You then played your notorious *Structures,* Livre II for two pianos with . . .

PB Yes, with Karl Kohn.

RD . . . and finally there was Webern, Opus 24. It was a remarkable and well-built program. During that period you were also lecturing in the Los Angeles vicinity. At a party at Leonard Stein's house we had a rather lengthy discussion about improvisation and improvising. I had been working in this area with Lukas Foss and in that conversation, you dismissed our work as a "parlor game." In my youthful exuberance, I had to point out that it was terribly stimulating while we were doing it!

PB Oh! I am sure of that! I always say that that is a good game for those who play it. About this kind of improvisation craze . . . I looked at Stockhausen's group at this time also, and I listened to it. And I listened to Globokar's improvisations. What always disturbed me about the musical thinking in this period was its overly simplified approach. Let us call it the "sine-wave" approach, which is excitement-rest, excitement-rest, excitement-rest. That is neither terribly exciting nor satisfying. I suppose for the people who perform it, it is like a drug. They are so delighted with themselves that they don't see the real world.

RD What do you feel about jazz players? Because they have a good time.

PB But that's different, because in jazz they have the methods, the tunes, the harmony. Practically all the language is fixed. Fixed with a certain mobility. You know, jazz for me is closer to the Indian way of improvising, where you have a lot of infrastructures; and then you can move these infrastructures in a kind of lively way, and you are on a certain

3. Former director of Monday Evening Concerts, Morton died May 9, 1987.

*track.* Good. It does not mean that every improvisation will be "out of this world." Improvisation can also be low-key. And there are also improvisations which can be brilliant, like Liszt did on the piano. Or like the musicians in the 18th century in Baroque times did up until Mozart. Now, the rules of the game are really more complex. All of what I have heard, when it was interesting, was some kind of remembrance of works which already existed.

RD That of course could be the biggest discrepancy.

PB It was always based on memory, which I can understand perfectly well, because inventing something requires quite a lot of depth and concentration. But when improvising you have conditions which are exactly the opposite of depth and concentration: you have to do it immediately in front of an audience. That is perhaps interesting as a psychological experience: to sort out one's psychology, alone or with other people, within the reactions of a group. That could be interesting. But for the people who are confronted with that, especially under the guise of a concert, it is simply false. Some people might call it cheating!

RD There were other directions that were also going on. As I remember our early talk, you made it clear that your concept of improvising, if you want to call it that, was this maze idea, where you construct the maze and a performer may take a *left* or a *right* or a *straight* choice of direction.

PB It is almost like a map of the city.

RD Or rails, or whatever the analogy is . . .

PB That's right. Because everything is constructed. You can choose the direction, but it always has a meaning. And the meaning is always more complex than just the "sine-wave" meaning.

RD You never worried about somebody starting out toward Paris and ending up in Bologna?

PB That does not matter to me. The work should be composed like *that,* which is very difficult as a matter of fact, because you must see and then foresee more than one solution. It begins to be very difficult.

RD I guess it is a question of ultimate spontaneity. Is it your feeling that you cannot have ultimate spontaneity and then a very serious piece?

PB I don't think so. Spontaneity is really always acting on a lower level than non-spontaneity, I find. Because spontaneity lives on memory and/or on limited resources. Resources limited in different ways. Limited maybe in their invention, because one is tired and doesn't know after five minutes what to invent. Perhaps one has one minute which is absolutely ecstatic, but the next four will not be interesting. That's a kind of statistical approach. And I find the statistical approach not terribly exciting. Too many *downs,* not enough *ups.*

RD I remember our meeting in 1963. I saw you conduct several enormous programs in London in 1967. And then I had the *Great*—and I can't say it enough—that great experience of working with you at your Course in Interpretation and Conducting of Contemporary Music in Basel in 1969. You had a repertoire that we worked on. It was Schoenberg *Erwartung,* Berg *Three Pieces,* Bartók *Miraculous Mandarin,* Stravinksy *Symphonies of Wind Instruments* and *Rossignol,* your *Éclat,* Berio *Tempi concertati,* Stockhausen *Punkte,* Webern *Symphony* Op. 21, and Varèse *Arcana.* Would you change that repertoire if you had to do that course again?

**PB** No, I would not change it. I think it covers the whole territory of contemporary music. Maybe I would add a piece like Berio's *Sinfonia,* something like that. I could do my own *Notations* or I could do other pieces by younger composers; but the main pieces, in the historical sense, have not changed since then. I chose *Erwartung* because there is only one singer; but musically I could have chosen *Die glückliche Hand* just from the orchestral point of view. But the problem is with the chorus and soloist . . . I mean, that's impossible to do. I could have chosen the orchestral preparation of *Glückliche Hand* instead of *Erwartung.* And Stravinsky . . . I don't know . . . there was *Rossignol,* and I could have chosen *Noces.* Something like that. But I would not have changed the composers. Certainly not. Their place is fixed historically.

**RD** They are the pillars.

**PB** They are.

**RD** I think there is no living composer or musician who has had the privilege of accomplishing so many things and writing about them in terms of criticism at the same time. If one reads your essays, there is no doubt about what you believe; but there you were, first with Domaine Musical and then with the BBC and the New York Philharmonic, and your expansion into opera, etc. And now here you are at IRCAM. Do you feel your sense of personal accomplishment? Because you have put a lot of criticism in your essays in terms of the music world . . . did you get something done?

**PB** Yes, I suppose so. I got something done. I always have had two sides to myself: the reflective or theoretical side and the other side that wants to *realize* those reflections. I do not want to just think and not make things. For me it has always been very important to materialize things. Therefore, throughout my life . . . when I began with Domaine Musical it was quite early, and I had no experience whatsoever either in conducting or in organizing . . . but I began with that! And I forced myself to do it! I did not force myself, because I had the impulse; but sometimes it was difficult to find money, etc. It was a big burden for me. But I thought, "Yes, I must do it, because otherwise there will be nothing concrete left with which to be confronted."

**RD** That was "Domaine"; but then you were given the BBC. And you were given the New York Philharmonic.

PB Well, everything is not really given. They *hope* that you will do something with these organisms. That is the opposite of *given. You* are given to these organisms! Quite the opposite! Maybe, egotistically, I would find Stockhausen's point of view exactly the opposite of mine.

RD Yes, it is.

PB Everything is concentrated on himself.

RD Yes, his world is there in Kuerten.

PB Yes, totally. Now it is even visible. He always has these small groups around him, particularly his family. I am an "invader" in another way. I want to invade with a lot of people around me! I cannot say that one is better and the other is worse. Everyone has his own methods of working. If you compare Stockhausen and me, we are both very realistic, because we *propose,* but we *make.*

RD That is true. The results are there. Can you speak just a little about New York? First of all, the succession of music directors of the Philharmonic forms a peculiar trio—Bernstein, Boulez, Mehta. You came to New York and you did all these things. . . . I mean, it was the time for it . . . the "Rug Concerts," the "Perspective Encounters," and the "ghetto concept" and all that . . . how do you look back on it now?

PB It was an interesting experience for me. I was just in what the Germans call a "Zwischenstufe," which is "between two steps." A kind of step between two parts of my life. Certainly, when I began to conduct, I was interested first in new music—to promote contemporary music. And then there was the Darmstadt period. But composition was at the center of that. Then I was asked to go to Holland, Germany, and to England especially, and I discovered my abilities in conducting, which I had never before explored because there had been no opportunity. This led me to develop some ideas; and I thought I could convince *more* people by branching out than I could if I just stayed in a small area. At the same time, I had come to know what an orchestra with a capital "O" is, as a musician, not as an administrator. I came to know the orchestral literature much better than I ever did before; because when you rehearse things, when you look at things so very closely, then you learn quite a lot and you understand the mechanism of the orchestra as such. I say, *musically.* I am not speaking about the mechanism of the orchestra as an institution. However, if I had allowed myself to be tempted too far, I would have ended up conducting only, and dropping composition. The New York Philharmonic asked me to begin in 1970–1971; but because Szell died, I was obliged to start sooner, in 1969. What is not really known is that I saw President Georges Pompidou in the fall of 1969; and in 1970–1971 I met the people here in Paris about creating IRCAM. So when I began in New York, I already knew that my time there would be limited.

RD That's interesting! Because it was very surprising when you were with the Philharmonic . . . I was assisting you for a while. . . . All of a sudden, IRCAM was announced like an explosion. And it was rather sudden. But now you say it happened because of earlier planning.

**PB** You can imagine that such an institution is not built up just like that! As a matter of fact, the first idea for IRCAM came up in 1966. The Max Planck Institute, a scientific institution in Munich, wanted to create an institute for music. The initiative was not taken by musicians! It was taken by scientists. And they came to me, because the Institute of Music was to be for pedagogy, musicology, performance, and contemporary music. They asked me to prepare the organization for contemporary music. I was of course very interested. I was already giving concerts in London, but not living there permanently, so I was free. So I began to develop that. Finally, in 1967, it did not materialize, for various reasons. But the plan for an Institute for Contemporary Music was there. When Pompidou asked me in 1969 to do something, the plan had already been made. I had only to develop it in a different way. For instance, I did not want to develop instrumental music very much. That was not what I foresaw for IRCAM. And so in 1970—I remember it was on the day de Gaulle died, therefore I remember it—it was in November 1970 when we first met in Paris to seriously discuss the plans for the Institute. Remember, I had already seen Pompidou in 1969. So therefore, when I started in New York, I already knew that IRCAM would open in 1976. In the end, it was delayed by Pompidou's death and it opened only in 1977. But in 1970 I already knew that I would be leaving New York in 1976.

**RD** So when you accepted the New York post, you already knew you would be leaving it at a specific time.

**PB** Yes.

**RD** Do you have any bad memories of New York?

**PB** No . . . no bad memories.

**RD** It's a funny city.

**PB** It's a funny city, certainly, but all capitals are funny cities. They have their funny sides. I would have liked, retrospectively, for the press to have helped more generally in making the institution of the New York Philharmonic a kind of cultural institution.

**RD** I think you are absolutely right.

**PB** And not only, let us say, an entertainment institution. But they did not help very much. Some of them did, but certainly not the main ones. Therefore, it was very difficult to fight against the mainstream. After the 1960s, when there was some shaking up of society, there was a return to the right in the '70s. And it is still very difficult to make an institution an integral part of cultural life and not merely an element of entertainment.

**RD** How to solve that one is a big problem. You speak about it in one sense, but that means changing the sophistication and development of an entire society! How can you do that?

PB I cannot myself begin the education. We do the best we can to be attractive and to make people more conscious of contemporary music; but I cannot do more than I can! And I must say that we have achieved something here. Young composers come here to work. It's a very solid institution. We have an audience which is very loyal. Of course, not 10,000 people. I don't think it will ever be 10,000 people. But we have a very strong following, and that's the best we can do. The essential thing is to create works, after all. If the works are good they will survive. And then they will find a larger audience. You cannot trust the judgment of only people of our time. You must also think that if there is no connection right now, there might be one later.

RD There was just one statement in the IRCAM brochure . . . that you had lots of projects, some of them still continuing. But you are zeroing in on the electro-acoustic, live situation.

PB We are trying to develop that more, because it was . . . not neglected . . . but it was simply impossible to do it earlier, because the machines were not powerful enough. And now that the machines are powerful enough, we are developing that, because it helps the musicians quite a lot. The real time calculation. And also during the concert . . . I mean, *live-electronics* does not mean only the transformation of instrumental sound; it also means having music programmed. It could be synthesized at the time of the concert. So you don't have to worry about synchronization if you are playing an instrument. The machine can really follow what you are doing. That was the problem before, when you were playing with tapes—it was absolutely inflexible. So it killed the gesture of the musician. Now, due to all the progress that has been made in the field of technology, I hope that we will again be in a situation where we will have great flexibility in a very rich world of sounds generated in real time.

RD And this of course is something that is one of the last frontiers for you at the moment.

PB Yes, for the time being.

# *Karlheinz Stockhausen* *

BORN AUGUST 22, 1928, MOEDRATH (NEAR COLOGNE)

**April 8, 1987** **Kuerten, West Germany**

I was on my way to Europe to conduct some concerts, including a studio recording in Holland of Debussy's *Martyrdom of St. Sebastian* and Per Nørgård's cello concerto *In Between.* I flew from New York to Frankfurt on April 7, 1987. Stockhausen and I had corresponded in the last months, to no avail; he was always too busy to see me. I sent him my schedule and arrival time by train in Cologne, since I was to see Mauricio Kagel in any case. When I arrived at the Hotel am Dom Europa in Cologne there was a message: "Please come to dinner this evening (April 8th) at 7:00 p.m. at my home in Kuerten. To get there call taxi 'Thorn,' 02202/55555, he knows where I live. I have no private telephone—looking forward to seeing you—Karlheinz Stockhausen."

I took the 18:18 train to Berg Gladbach and was met by taxi driver "Thorn." After a twenty-minute drive into the country we arrive at Stockhausen's gate, which "Thorn" opens. Stockhausen is there to greet me in a bright yellow sweater. The house is a modern structure—very attractive. We proceed to Karlheinz's study where everything is in immaculate order—no clutter anywhere. We begin to talk for more than one hour. Then comes dinner with Suzanne Stephens, which begins with thirty or forty seconds of silence and bowed heads—a period of grace. During dinner Stockhausen continues to speak about his music. There is no doubt that every moment counts. He seems to live his life like sitting on a "timer."

His personality is at once formidable, exact and amenable, genuine and deeply concerned. We talk about music the whole time. About 11:00 p.m., the taxi driver "Thorn" arrives. Karlheinz gives me four autographed volumes of his *Dumont Documente.*

RD It's an enormous pleasure to see you and be here at your place in Kuerten. I want to begin by saying that, in this project of talking to, by now, twenty-four composers, Americans and Europeans—you seem to be somebody that is referential. The Darmstadt initiatives, the Darmstadt accomplishments are usually referred to as those of Boulez-Stockhausen.

* With the permission of Karlheinz Stockhausen.

*Universal Edition*

Or Stockhausen-Boulez. And it's produced a shock wave, in terms of composers of your generation and slightly younger. I think there's a myth, and I think we should talk about that myth; and there's a reality. The myth first: your idea of total organization and one way to do things seems to be the stop-sign that people mostly react against. Is that true?

KS No. In 1959, for example, I gave one of my yearly composition courses at Darmstadt. And in this course, there were so many different composers who composed in their own style, and I supported their individuality. Like Sylvano Bussotti, La Monte Young, Aldo Clementi, Friedrich Cerha, Gilbert Amy, etc.

RD That's quite a lot . . .

KS There were about twenty composers in my course. They're all written up in the archives of the International Music Institute in Darmstadt. Lots of Americans, by the way. I have always said to my students that every composition should be original, and composers should try not to repeat themselves. And the meaning of the multiplicity of composers is just that everyone is a different source. For twenty-one years I have taught in Darmstadt; and I have really fought for the invitation of John Cage, Earle Brown, Christian Wolff, Morton Feldman, to Darmstadt. The former directors Steinecke and Thomas didn't want at all to accept them. I said, "If you do not invite (for example) Cage, then I will not come again." And as a matter of fact, it was the only year that I didn't go, because of this problem. After that John Cage and David Tudor were invited. I had worked since 1955 regularly with David Tudor. He lived in my house. Also Cage lived in Cologne. I found him a place to live, and we were good friends. In 1958 he translated into English my first lectures which I gave in America. And I arranged the first performances of Cage's work in Cologne at the radio. His Concerto for Piano (1957–58) was first performed there with David Tudor as soloist, together with several other of his works. For the Donaueschingen Festival I tried the same. The only time when Heinrich Strobel, the former director of the Donaueschingen Festival, performed Cage's music was after I convinced him that he should—against the advice of Stravinsky and Boulez, who were very influential at that time in Baden-Baden and Donaueschingen. I have always tried to support the variety of approaches because I think that we live in such an incredible historical moment of opening into all directions, that certainly one man cannot do it all.

RD I guess it was that way when you faced the post–World War II period. There was the *big bang*. Is there another *big bang* now? Is that what you're saying?

KS I'm saying: every composer now has so little time, because there are so many new problems since 1953, that the approach from many different sides to all these open problems is necessary, like in science. That is why I have always supported the multiplicity.

RD Well, let me not lose sight of this. But it is strange that this reaction I spoke about ranges from affection to hostility. I think any figure of your stature would inspire such feelings. I think the musical affection stems from a great respect for the accomplishment of your thought. However, you seem to give the impression that your way, while allowing others, is the only way.

KS No, that's a misunderstanding. Here in Germany I have supported most of the composers who were not performed. I was the one who fought for Boulez and his performances at the West German Radio; I also found him a place in Cologne to make the first version of *Marteau sans maître*. Then he found out during the rehearsals in Cologne that it didn't work, and he had to rewrite it. He started working on it again and wrote the new version in Cologne. The same is true for Nono. We were friends for years, and I arranged the performances of his music at the WDR Cologne. I also helped Kagel to work in the radio in Cologne and to get performances. With Ligeti, it was the same. And I always supported him. He got his first performance through me. I still think the same way. That just this multiplicity, which in Germany has been the product of my influence for years and years, is so fruitful. Russian composers who came for the first time to the West were here in this house visiting me. I gave courses in Darmstadt for twenty-one years, from 1952 until 1974, analyzing also the compositions of other composers.

RD I mean only to say that there is the myth that your impact, by virtue of your composition and your teaching, was overly dominant. And along with this, the publication of *Die Reihe* seemed to underscore this dominance. That is what was coming across.

KS But that's so long ago! The anecdotal aspects will diminish in importance, because all that counts is the works. I mean, who thinks today about the competitive feelings during the time of Schoenberg? There are a few documents, and many people resented the existence of Schoenberg's works and his influence. But nevertheless the gossip has become unimportant. What remains are the works, and very soon one will see that probably there is something deeper in the feeling that is expressed through hostility. Perhaps several colleagues do not want to accept my *works* as they are. I think the personal things are not so important after a while.

RD No, the personal things are not important. But there might be other things, and they have to do with origins. For example, you let people know that the texturally composed cluster, moving in glissando, and the phenomena in general was of your origin and not Xenakis, not Ligeti, or Penderecki. Was that important to you? These discoveries of sound?

KS Yes. And they still are.

RD Do you claim them as yours? Or do you need to?

KS Sometimes I write texts like last year for the program of the world premiere of my composition *Evas Zauber* in Metz, describing what I have discovered. Then such a text becomes an objective document. For example, I described the research which I have done in 1985 and in particular throughout the year 1986, for three new compositions: the programming for new synthesizers, that I have learned a new method of programming for different types of synthesizers (though I had worked for over thirty years in the field of electronic music, this was new to me). I also described in the program book how challenging it is to feel like a student again, and that there were five comparable revolutions in my life. Each time I discovered a whole new world of sound and of new methods to compose the sound.

RD You mean live and electronic, or just electronic?

KS Both, because I have described how the influence can be shown between the discoveries in the field of electronically produced sound and traditional instrumental playing. For example, in 1985 and in 1986, I have written several works for basset-horn and for alto flute, and the musicians who performed these works had to relearn their fingerings. Suzanne Stephens is working since about two months on a new piece, called *Xi*, which I wrote in December 1986. It is the Greek letter "xi," which means undefined value (*Unbestimmte Grösse* is the German meaning). Suzanne is practicing every day for several hours the fingerings of these microtones. And for the flutist Kathinka Pasveer it is also a new technique which she has studied last year for two months.

RD Well, I am a former clarinetist too, and I know about that instrument. What is your interest in the idea of breaking the scale into quarter-tones or smaller intervals . . . is it philosophical, is it technical . . . What?

KS It sounds marvellous.

RD It's just lovely to your ear.

KS More than that, because the microtones allow a transposition of chromatic figures (I call them *formulas*) into larger or smaller frames. And this is an extraordinary experience, to hear a melody which you have heard several times in a composition, suddenly compressed or expanded. I have done this with an electronic synthesizer for years, but I never knew how to achieve it with traditional instruments.

Sometimes now, in *Xi,* a minor second gives me the impression of a major sixth, because there are at moments up to nine steps in a minor second. And it is fascinating to hear intervallic movements within a small interval. Our whole perspective of listening changes through this experience. And then, every fingering produces a different timbre. It sounds very clear if you place a microphone above the keys of such an instrument, and then play these compressed melodies and hear all of a sudden the most beautiful changes of timbres, which are the result of these "wrong" fingerings.

RD I know those. Those were wonderful to discover!

KS Especially if they are amplified. So, I have written many different combinations for alto flute and basset-horn in a composition called *Ave.* When the pitches of the two instruments come close to each other, they produce all the interferences—the combinative vibrations—not only different tones, but also "beats" which you can count in ritardandos and accelerandos.

RD . . . if done right . . .

KS . . . just by the fingerings.

RD Really! That's sensational.

KS The rhythm slows down when two pitches come close to each other and speeds up when they separate. And then, when they play together a very small interval, it's such a delicate sensation! So the music becomes extremely refined.

RD Xenakis also has examples of two instruments "beating" when playing close intervals. There's only one other composer who has a similar, it's not the same, but similar idea today, and that's Ligeti with his illusory rhythm. In the sense that rhythms can pile up . . .

KS I do the same.

RD You do the same?

KS I started composing what you call "illusory rhythms" in *Carré,* which is a key work for that.

RD *Carré?* Which I did.

KS I know.

RD How is that?

KS There are a lot of string sections where every string player has a different kind of ritornello. And I have carefully measured the length of these ritornellos which are regularly repeated. When they are superimposed, they produce a combinative rhythm with its undefinable micro-rhythm which gives this beautiful feeling of polymetric superimpositions. But because of the fact that the players play the same pitches, you do not hear this as usual like different *voices.* You hear it as the inner life of a *time spectrum.* And when you calculate very carefully the different lengths of these ritornellos, they produce accelerandi and ritardandi during longer time spans. These phenomena are new, you see. This is all the result of *The Song of the Youths,* composed from 1954 to 1956, where I started with such ideas . . .

RD Ah, yes, the *Gesang der Jünglinge,* of course . . .

KS . . . where I started all these micro-compositional processes by using many, many tiny elements, superimposed, in order to create just one *timbre* with an inner life. So, this was a new concept of the *composition* of sound. The "inside" of a sound. It's like an organism.

RD Something beyond the original material itself.

KS It's not the timbre that you *find* or that you *make,* but you *compose* it really like a living being. And this has steadily increased, and now I'm doing it every day.

RD We had a talk in Paris in 1971 at Gilbert Amy's home after our performances of *Carré* at the Salle Wagram. I remember several things about that talk. You were waxing philosophical about intuition . . . and I've seen it in your articles still, . . . but you were telling me that you wanted intuition in your hands twenty-four hours a day, if possible.

KS Yes. More methodically . . . yes! A challenge!

RD That's one, and we'll talk about that. Two: you said at that time that evening, that you wanted to travel to space with your mind alone. And three: you said that you had been through a suicidal tendency . . . I have it in my diary . . . for six or seven years.

KS Yes, but it happened in 1968.

RD Those were the three things I remember in my diary. What about this idea of intuition?

KS Well, at that time it became very urgent for me to find a method as reliable as a mental method to compose *intuitive music.* And what I mean is the following: usually one collects ideas by scribbling into a sketch book—describing the inner meaning with words, with little diagrams—and then brings all these different intuitive bits together at the moment of composing. But in 1968, I found a way to change my mind before and sometimes while I compose. By this I mean, taking very few elements of the skeleton of a new composition which I have worked out mentally, and then sit down and clean the mind completely, make the mind void and literally wait until I innerly hear longer stretches of music. Until I was about forty years old, these intutitive moments were rather short.

Traditionally, a composer has worked at the piano, like I composed my very first works; and then he collected small bits of intuitive music and brought them all together in the process of composing. But if one finds a way to open oneself for the ever-present music which is suprapersonal, then it becomes possible to imagine . . . not imagine . . . to *hear* longer stretches of music and develop completely new mental processes which are the result of these intuitive sound visions or music visions. And it needs a special way of living, which I didn't have before. One has to learn special methods of meditation, special ways of emptying the mind. One has also to learn to perform together with a few others in such a way that one does not read the music, but can close the eyes in order to concentrate fully on the nature of sound, which is developing in the air while one is playing with other musicians. Then, one can develop an instant reaction to the sound which is in the air, and shape this sound like a sculpture of vibrations, together with several players. Therefore it is necessary to find friends who are on the same wavelength and who have not only the gift of improvisation, but also the gift of intuiting music by listening to a musical organism which is shaped at the moment of the performance.

RD That's a very big demand, though, Karlheinz, having about three people doing that at once. Because I am a former improviser myself, and I know. And at best one reflects sort of the background of what you knew about Berg and that sort of thing. But to intuit something on the spot is a big demand.

KS Yes. But that is what has come.

RD Why?

KS Because the result is a kind of music which cannot be written. It's different.

RD Has it anything to do with the Eastern thought at all?

KS No. I have been in India last year, together with my son Markus, the trumpeter, my daughter Majella, the pianist, with Suzanne Stephens, the clarinettist, and Kathinka Pasveer, the flutist. Unfortunately the art of improvisation of the Indians—they are the only musicians who really improvise—is overestimated, because it's very academic. It's very fixed.

RD You were very fascinated by that whole Eastern . . .

KS Not by the Indian music, this is a misunderstanding. I was fascinated—and I'm still fascinated—by the timing of Japanese music, in particular of Gagaku music and Nōh music. My fascination with Eastern music has never had anything to do with improvisation, but with timing, which you can also find in Sumo wrestling or in the tea ceremony. It is the wide scale between the extremely fast and the extremely slow, and in addition there is the concept of instant change without transition. This is so important in the Japanese way of living and particularly in their best music. I mean the extremes of almost no change and then instant change; and then again no change for a while. So there is the sudden changing of direction, which is different from the speeding up and slowing down like in all European languages. European rhythm is based on our way of speaking. But the Japanese have, in certain disciplines, a way of timing which is extra-lingual. It is something else. It has nothing to do with the way of speaking. I have described this in a text, *Memories of Japan.* These experiences remain important forever as an enlargement of the musical time scale and of the concept of form.

RD Yes, the concept of form. Let's see . . . I don't want to just jump . . . we were talking about our conversation in Paris, and you did explain intuition. The second point I made was that you had told me that you wanted to travel in space with your brain alone. Is that still something . . . ? You said that.

KS Can you remember if I gave an example?

RD Well, to a star, to a place, with your brain only. You said . . .

KS It's not my brain.

RD Maybe it's frivolous . . .

KS No. I think the word is wrong. But the rest is right. It's not my brain which can travel.

RD Or thought . . .

KS It's my spirit!

RD Yes, spirit. All right, that's it.

KS That's something different. I think we all will do it in deep sleep.

RD Well, have you . . . how far along are you in that department? Your spirit . . . I mean, obviously you've thought a lot about it. I don't know anybody that . . .

KS I move a lot in space.

RD Example . . . ?

KS In deep sleep. And then I experience places which are extra-terrestrial. They have a totally different environment and allow a different way of moving. I can remember them. And whenever I close my eyes, I can move with any speed. I can be in New York within . . . no time.

RD You mean right now you're in New York?

KS Well, if I close my eyes . . . yes . . . I am standing at the corner of 42nd Street . . . right. There is the gallery Bonito at the left side.

RD And you feel you were there. And now you're back.

KS Yes.

RD Mm. Not just a blink of remembrance.

KS I see it.

RD You were there.

KS I see it, I feel it, and I am aware of it.

RD I can do that, too. That's not very extraordinary.

KS That's what I mean.

RD I mean, I can say, "Well, now I'm in Schiphol Airport. I am in Ramp 32."

KS Very good.

RD But what's significant about that?

KS Once you know that, then you know that you are not fixed to your body.

RD But don't you think most people do things like this?

KS Clear! But they don't take it seriously. They think it is only an illusion. They don't take it as a reality.

RD What application do you want to make of this for your own personal work? What do you want to do with it?

KS Well, it simply shows that all my aims, my goals, are not related to my body . . . to the time of my body, to the length it lasts. Everything I decide, I do in a way that I can gradually become independent of my body.

RD So, this opens up this whole spiritual aspect of your work, your life. Was it a reaction to over-organization, or was it organization that produced this?

KS This is also a misunderstanding. Already the very first works which I composed, like *Kreuzspiel* or *Spiel for Orchestra* or *Punkte for Orchestra,* were at their origin purely spiritual concepts. I wanted to make "star" music; I wanted to make an outer space music. And the organization was simply a process to realize this. So, the mental process came always second. There was, at the beginning of every new composition, an inner vision to discover a world which I had never experienced before. And it needed the kind of supra-personal state in which I was for a moment . . .

RD . . . not technical . . .

KS . . . before I dived back into this planetarian situation, where I thought, "How can I ever, with the means and with the notation and with the technique of this planet realize this?" And then it became a translation. So I think, all my works are more or less translations into the possibilities of what I have learned here and what is available. Most of the time I have invented new notations, and I have always gathered new instruments or I built new instruments, to approach at least to some extent what I had innerly experienced.

RD Well there's no doubt that the stunning quality of how your music goes from moment to moment—*Momente* if you wish . . . has been a revelation. There's no doubt about it. How it just simply progresses. And to hear you speak about it in terms of that word "spiritual" and having conducted your works, I find some of that same quality even in *Carré.* I find that it is not just a matter of conductorial administration. There are moments . . . I have a very favorite moment, but there are moments in there that transcend technique. And they take you to an area that is not in the obvious—sentimentality or emotions that are 19th-century—but they are certainly transporting. That's what they are.

KS Mmm. *(affirmative)*

RD They are very transporting . . . and yet you have an enormous resistance in some of the world about your sincerity in these areas. What do you have to say to that?

KS You have said "in some of the world." On this planet, a religious approach to art is now considered by most of the artists to be naïve. And I'm reproached by most of my colleagues and by those who write about music, that I am naïve. I have, since I was born, the same feeling as is described for composers like Schütz or Bach, naturally also for Mozart, who was a very religious person, and—in a free spiritual way—for Beethoven, also—to some extent—for Brahms, Schoenberg, Stravinsky, and for many other composers.

So, there is this tradition that one is—as a composer—rather content with oneself if one has added to the multiplicity of compositions a few small models of the universe; and to feel that one is a collaborator of the greatest of all composers, who has composed the universe of universes. This is perhaps what today most of my colleagues don't want to accept. I remember from discussions dating back to the early '50s (and they continue until today with some of my colleagues whom I like, and who seem to like me, too) that they always say to me, "Well ... let's not talk about this." With interpreters, who are atheists and nihilists, it is the same. The atheist and nihilist spirit is predominating. It was very fashionable after the war when I studied for a year in Paris, but every morning from six to seven o'clock I attended the early morning Mass. I lived there in a student house.

RD Are you Catholic?

KS No longer.

RD Were you then?

KS Yes, I was a Catholic. I left the Catholic Church in the early '60s. This has a particular reason, not because I'm opposed to Catholicism, but because I was not able to follow the rules. That is another chapter. What I am trying to say: already during the discussions with my colleagues then, it became very clear that the general fashion was a kind of Sartrean nihilism. Also in our country, the general philosophy was what they call neo-existentialism and nihilism, atheism.

RD And you oppose that.

KS I have told you what I experience, and how I am most happy when I compose. When I don't know where to go next—this has happened even today several times, sitting at this desk before you arrived—when I don't know what to do, I close my eyes and pray for a moment. Or I think nothing, and I say silently, "Help me." And then I wait until I have something that I can agree with, which makes me think, "This is new and this is good." So I have this attitude, and my whole life has been guided like that.

RD Do you think this is unique, or do you think others really have experienced that? I mean, can you imagine a Mozart, a Brahms ...

KS I have mentioned them, yes. That's the same. That's a long tradition, I said. But in our time this has faded away.

RD I think you're right.

KS Most of my colleagues don't accept this spiritual attitude. They don't accept that a composer can function like a *receiver,* like a *radio receiver,* and transmit music; because most of the composers think that the music is theirs ... their own music. And most of the interpreters even think nowadays that the written music is only something they can use to express themselves. Because they think they are most important. But I think the highest ideal for an artist is to be a vessel.

RD Well, those are Stravinsky's words. He was, he said, the vessel for ...

KS And to transmit, to translate. But this then, is an entirely different approach, because you know that the *mental* is a machine for translation. It should be very precise; it should be trained every day, hour after hour. And one should study all the signs of sound and of sound organisms. But, nevertheless, one should consider only as being valid what has come from the spiritual world into oneself. And this can very well be noticed if one sharpens this conscience.

RD It takes enormous discipline, doesn't it, to realize that?

KS It takes a different approach, not "discipline"—it's a different approach. I have practiced this my whole life.

I was a child who lost his parents very early and had nobody. But I had something else: whenever I didn't know where to go, I closed my eyes and stood somewhere in the road or on the street—or, during the war, in a field where bombs were falling—and I wouldn't move until I heard a message. And then I did what I heard, and I did not doubt it. I have taught my six children the same. Some can do it, some others more or less, because they are constantly under the influence of people who laugh about these things. That's the essence of my composition. So, there's nothing mystic about this. It is a technique, to wait until one hears; and if one hears, then translate it, find a way to notate it and then try to be moved oneself. And if one is not moved, one should wait.

RD Is that in the tradition of Western thought, though? Or is it more Eastern thought in your mind?

KS I have no idea. I think in the East there are no composers.

RD Well, then how about now the last big work—*Donnerstag*. Is this a prevailing attitude toward that big piece?

KS You mean *Licht*—yes, since I began composing *Licht*, it has increased because it needs a lot of courage, to trust. Most of the time, for a composer the mental fabric is a kind of safety belt. He wants to have a system; and the system should guarantee that one doesn't make too many faults and that one is original. So, most of the composers define themselves by negating something else. They say, "I do not want this, I do not want that." And then what remains is what they are. Finally you end up with a composer who is only blue—like Yves Klein—and who composes only in one style. Most of my colleagues do this. They are stylists. Or they are mannerists, because they are basically afraid of destroying a style which they have established by their previous composition, and to start anew, start fresh and not be afraid of dealing with entirely different processes, different materials.

RD Well, that's always been your premise. . . .

KS I think the essential is the spiritual message. The spiritual materialization of what is the essence of a work. And it's not the facet, it's not the style. But many composers, like painters, are stylists. They keep to their style, so that they can be identified. You hear three bars of Stravinsky on the radio and you say, "Aha! That's Stravinsky." I think this is a traditional concept of the artist and of the art altogether.

RD Well, if you hear three bars of Mozart, you know it is Mozart . . .

KS No. That's not true.

RD Really? Three bars of Mozart?

KS No. It could be Dittersdorf, it could be Salieri as well. I have conducted in the last two years several Mozart works, and recorded them. Most of what is called "Mozart" are clichés. There are very few characteristic details which are typical Mozart, which make it become Mozart.

RD That's interesting. But you do maintain that Stravinsky in three bars would be noticed?

KS Yes. He changed the style towards the end, and again it became his style. He was a stylist. I don't think that he was a really original composer in the sense that he would consider every work as an extraordinary challenge to create a new world.

RD So your interpretation of originality is that you start afresh each time.

KS Yes, you should be a source. Because in a lifetime of a composer, there is very little time. And one should not get caught by this feeling, "I want to be recognizable. I want to express myself." Because if one expresses oneself, then this is a closed system. "Oneself" is so very little.

RD But you have this idea that your life is to be a part of your composition?

KS Yes. I have said this several times.

RD How is this manifested in your mind; I mean, why do you need that concept?

KS I was reproached in the '50s that my works lasted three times longer than anybody else's work. Then, when I composed *Kontakte* lasting 34½ minutes, they said, "This is ridiculous. It is much too long." This was printed in articles and said by colleagues who came to me. . . . *Gruppen for Three Orchestras* lasts about 25 minutes in one movement. Boulez had composed nine movements, to make his *Marteau,* and so I said, "Why do we need these movements, sections within a section of life, within a composition?" I was already very early opposed to the concept of "movements" and of "pieces." The word "piece" shows that it is a *bit* of something else, of an entity.

RD Except *Momente* has a kind of shuffling that . . .

KS No, *Momente* simply says that musical events can exist on their own and don't always need to be developed in order to make the listener understand from where they come and where they go. *Moment-forming* is a lyrical concept which conceives individual events as being self-sufficient. This is a concept different from the Beethovenian development concept.

RD Is it discontinuous-continuity?

KS What? *Momente?*—No.

RD It is not. Well, that's another misconception.

KS Yes. In *Momente* it simply says that only certain *moments* influence the present *moment.* These may be one or several from the past or from the future, or both. There exist certain moments which are so much filled with memory and expectation that one can hardly identify them as individual *moments.* Only a few *moments* have nothing in common with other *moments.* This is all happening in *Momente.*

RD Yes, but looking, really, above it all, can't you allow the idea that discontinuous-continuity is really what it is?

KS Yes, but here is a lot of developing continuity within certain long *moments.*

RD I mean continuity but the discontinuous quality . . .

KS It's both together.

RD It's something that came out of Darmstadt. I remember the phrase . . .

KS It is discontinuity and continuity, depending on where you are.

RD Yes . . . exactly.

KS This is no longer a one-dimensional concept of composition.

RD Well, that's what was fascinating to me about it . . .

KS It happened already early that my works became increasingly longer and unbroken processes, long spans of bridges in time. *Momente* lasts 113 minutes; *Hymnen,* 2 hours and 10 minutes as an unbroken process. Then *Kurzwellen* with a duration of about 55 minutes, *Stimmung* 72 minutes, *Mantra* about 65 minutes, *Sternklang* about 2½ hours, *Fresco* 5 hours, *Inori* 70 minutes, and *Sirius* with the duration of 96 minutes (it took me about three years to write and realize it).

In *Sternklang* I translated star constellations into pitches, rhythms, and dynamics. In *Sirius* I composed *"The Year"* with the 12 months and the 12 human types of the 12 signs of the *Zodiac.* Then I came to the concept to compose *"The Week."* So, it does not seem strange to me that it will take about twenty-five years to compose *"The Week."* The individual compositions have simply taken on larger and larger dimensions through the years; and I can now say that maybe in the next generation of composers, one can consider this as a model: *Licht, Light.* After *Light,* I would like to compose *"The Day"*—the twenty-four hours of the day, giving musically a new meaning to each hour of the day. And after that I want to compose *"The Hour";* and then I want to compose *"The Minute,"* then, as the last work, *"The Second."*

RD Really!

KS Yes.

RD Down to the second!

KS Yes. I want to superimpose time. The tendency in my work is, on one side, to stretch the time beyond the European concept of the duration of "a *movement*" (which was, as an average, about a quarter of an hour), of "a *piece,*" "a *work,*" "a *composition.*" (Traditional compositions of longer durations—like the B Minor Mass of Bach or the *Passionen*—

are basically *suites* of more or less short pieces, based on the concept of the *suite,* change of mood, change of character, of orchestration.) On the other side, I want to compress time, suspend it.

RD But what could be interesting about a *second* of time?

KS The verticalization of time. To compose 600 layers recorded with a 600-channel tape machine, and then let every listener choose a combination of layers within one second. One would understand that everything that seems to be meaningless in a too dense superimposition within a second of time, can be studied carefully by listening to innumerable combinations of layers. So, one can listen again and again to the inner life of the second . . . which will be an extraordinarily complex organism, like a human being, like a cosmos.

RD But that's really getting to a very responsible area . . . you are going to decide a second? You know, you decide a week, you decide a day . . . there's enough seeming space in between. But to decide a second . . . !

KS I will just treat the time as being vertical instead of horizontal. Because time as a horizontal concept is an illusion. Therefore I want to end up with the smallest referential time unit, which is like one heartbeat. Within this I want to have a whole universe of music which is composed more vertically than horizontally.

RD I see. It takes very quick listening!

KS No. You can listen for hours, choosing always different combinations of layers within this second. The second may have 600 layers, and you combine layers as many and as often as you like.

RD And when it moves horizontally, what do you have?

KS Each layer has a different *Gestalt.* You can listen over and over again to these different *Gestalten.* It's like analyzing, with a microscope, cells or molecules within an organism.

Do you know my composition for choir and orchestra "*Breathing gives life . . .*"? In the fifth solo I have written the following text for the bass singer:

> When the mesons,
> the putty which holds the atoms together, decompose,
> are born the muons
> with a life span of two millionths of a second,
> before they can conceive an electron.
> And during this life
> can a muon make a distant journey.

RD That brings us to one more question, and it may be the last: if you were to define your musical genetic code, what would it be?

KS *Formula.* I have found, as the next step after many years of working with *series* (sets of proportions which determine the organism of a musical composition), the principle of composing with a *formula.*

The *formula* is something very complex. Since 1970, I have used *formulas* for different compositions. But now I work already since ten years with one *triple formula* which I call a *super-formula.* It has three *formulas* superimposed as time layers, and each *formula* has different characteristics. The *triple formula* is like the genetic code of the music for *Licht.* It has helped me to compose already twelve hours of music. And I want to compose with it approximately twenty-one hours of music.

RD So you refer to it, and it is always there.

KS Always.

RD So that your genetic code has nothing to do with past individuals or past times.

KS No. It's the *super-formula,* which I conceived in 1977 in Japan, in a temple. And since then I use it for *Licht.*

RD Is it called *Formel?*

KS Yes.

RD It's referred to, and I read it.

KS My composition with the title *Formel* from 1951 has already a kind of a nuclear *formula.*

RD That's right.

KS In 1970 I have picked up this term again; it has a significance like in mathematics. We use the word *Formel* for a mathematical *formula.* But we also use in fairy tales the word *formula* as a *magic formula;* when you say, *Zauberformel,* it's the same word "formula." For example, in some fairy tales a person might speak or sing a magic formula and as a result, a mountain opens up, or a frog transforms into a human being, a human being into a bird. We call such a *magic formula* a *Zauberformel.* I use the word *Zauberformel* also in my text in *Licht. Formula* therefore has this double meaning: to evoke magic, and at the same time to function like a mathematical formula which can engender a whole world of figures, *Gestalten,* forms.

RD So your DNA is sitting there for you . . .

KS . . . yes . . .

RD . . . and you have created it . . .

KS . . . yes. And I interpret it every day, every minute. I have used it all day today!

RD Well, is it endless in possibility?

KS Up to now I didn't find it to be too limiting. I don't need anything else, though sometimes I need to become a magician myself, to discover a secret application mode for my *formula.*

RD I see.

KS It is quite compact, I must say.

RD Well, is this a step beyond Schoenberg?

KS Oh yes. A *Reihe* (series) of Schoenberg has only pitches, intervals. A *for-*

*mula* instead has everything: lights and shadows, sounds and noises, speeds, tempi, improvisations, scales, echoes, ante-echoes, modulations, degrees of variation. It has all these different characteristics; and it contains pure silences, coloured silences, a typical character for every note, and many more qualities.

RD Well, this brings us really to a recapitulation. In other words, these are your mediating factors that you use. And going back to the original, this—I guess—what you've just said, this total compactness, total organization on your part, I think, either represents an envy on the part of other composers or it represents a kind of world of lack of feeling.

KS Never mind. Such psychological interpretations will disappear. What will remain are the works. I have now written over 180 individually performable works.

The *work list* begins with no. −11 (minus eleven) because in the beginning I did not number the first eleven works, and I have arrived now (1987) at work no. 60. Many of these contain sub-works, which I have published separately and which can be performed individually as self-sufficient units. These *works* will remain, and what we have said will be less important in the future.

I didn't finish my development of the concept of unity. The next generation of composers *might* envisage that a composer creates from the first day on, because of this spiritual concept of unit, *one work* during his whole life. And everything he composes is a limb of this spiritual body.

RD It's an extraordinary idea.

KS One might even try to live like that, and never to excuse oneself for what one has done before; never try to eliminate something one has done earlier; relate everything one is doing now to what one has done before and to what one might do next. It is a spiritual concept . . . of one life.

RD You think it would allow much *other* life itself?

KS Oh, my God, yes! We are so limited by our body. And by our mind. And by our time. And by the means of our time. We can only do what we can imagine now. Even if I sometimes dream about the future, I feel extremely limited. I have written a lot of texts, last year and also this year, about what kinds of auditoriums I would like to have built, what kinds of instruments, compared to the ones which are available now, what kinds of performance practices I would like, which performance techniques I would like to develop, etc. But all this might happen within 100 years, or later. It's impossible to imagine now what I would like to do if I would come back to this planet again. Or if I arrive at another planet which is more developed. Then the conditions will be entirely different.

RD Do you have notions of reincarnation?

KS I have lived several times on this planet before.

RD Oh, you have?

KS Yes. I have recognized places.

RD Really?

KS I walked there feeling completely familiar with everything without ever having been there in this life. I could go to certain corners and places knowing where I was.

RD And you recognized them?

KS I went there . . .

RD . . . and nobody told you . . .

KS No, I was alone. Or I sometimes wanted to go to places and I didn't know why. And when I came there I said, "Ah! I see! How interesting—my country!"

—DINNER—

RD [Before I left, I had one final question.] I didn't ask you how you felt about Boulez or Berio particularly . . . Boulez/Berio/Stockhausen . . . do you acknowledge that trio?

KS I've said in a book which has appeared last year in Italian (Stockhausen, *Intervista sul genio musicale,* with Mia Tannenbaum), having been asked about the friends of my generation, "Certainly Boulez and Berio are the ones who are real musicians and whom I estimate highly because of their accomplishments and their musicianship."

# *John Cage*

BORN SEPTEMBER 5, 1912, LOS ANGELES

**April 8, 1986** **New York**

RD I'm sitting with John Cage in his New York City apartment. It's Tuesday, in the afternoon. John, I thought I would like to speak in three general ribbons of thought. One has to do with a little bit of history; the other might have to do with performers or with conductors who have conducted your music; and then composition, or about how you compose, and other composers. I think I met you face to face, for the first time, in The Hague, in 1972, when I conducted your *Concerto for Prepared Piano and Chamber Orchestra* with the Residentie Orkest. It was part of a project called "The Land of Cage."

JC I must have been sixty.

RD My program also included the music of Brant, Brown, Ives, and *Amériques* of Varèse, and you were there. Much later, I remember doing the world premiere of *Atlas Eclipticalis,* simultaneously combined with *Winter Music and Solo 45,* with Joan La Barbara and the Residentie Orkest of The Hague. This was at the 4$^{emes}$ Recontres Internationales d'Art Contemporain, in 1976, in La Rochelle, France, and you were also present. Some of the other pieces that we did at the Aspen Festival without your presence included: *In a Landscape* for harp and dance in 1972; *Score (40 Drawings by Thoreau)* and *23 Parts* (for any instruments and/or voices); *Twelve Haiku Followed by a Recording of the Dawn at Stony Point, New York, August 6, 1974* and *Bird Cage,* both in 1982; and *Three Dances for Two Prepared Pianos,* in 1983. I also did *Score* at the New York Philharmonic–Juilliard School, Celebration of Contemporary Music, in 1976. But first ... John, can you speak a little about the early '40s and the musical atmosphere you were confronted with?

JC I had worked in two directions before I came to New York. One was with percussion music, to which I turned after my studies with Schoenberg, because it was clear to him, and to me, that I had no feeling for harmony. He had said that I would never be able to compose, because I didn't have such a feeling. I would come to a wall, he said, through which I could not go ... And so, since I had already promised him to devote my life to music, I said, "Well, I'll beat my head against that wall." He said I would never be able to get through, and then, I began,

*Photo Copyright Roger Gordy*

literally, beating on things, not because I was angry; but, because I'd been told by Oscar Fischinger, a film maker, that everything in the world had a spirit . . . every object in the world had a spirit, and that spirit could be released by setting it into vibration. So I spent a great deal of time simply going around touching things and hitting them and so forth, whether they were musical instruments or not. I knew that if I worked in the world of noise, rather than musical tones, that I was free of any problem of not having a feeling for harmony. So I worked first with percussion. Then, when I came to New York in the early '40s, it was difficult to organize a group of musicians, because everyone was so busy with his own schedule. So that, I turned to the prepared piano, which I had discovered in 1939 or 1940, in the West. In New York City, it was easier to work alone at the prepared piano than it was to organize a group of people to play percussion instruments.

RD You may have been asked this question before, but do you remember the impulse that led you to prepare a piano?

JC Yes, it was to make an accompaniment for a dance by Syvilla Fort, who was a black dancer, very beautiful. She later directed the school associated with Katherine Dunham. On a Tuesday, she asked me to make the music for a dance that she had made, that she was going to perform on Friday. There was no room in the theater for anything but a piano . . . there were no wings . . . percussion instruments would have suited her dance, but there was no room for them in the theater. I tried to find a twelve-tone row that was African and I couldn't find one; I decided that what was wrong was the piano, not my efforts, because I was conscientious. So I began changing the sound of the piano. I already knew of Henry Cowell's work and often held the pedal down while he played on the strings at the back, and I had seen him put his hands on the strings . . . both muting and plucking them and stroking the strings in *The Banshee.* So I put a pie plate on the strings and it bounced; then I put a nail between the strings and it slipped. Then, I found a wood screw and it stayed in position and I was able to repeat a sound. Then I knew I was going in the right direction and very quickly I was able to write *Bacchanale,* which was my first piece for prepared piano.

RD That early piece, *In a Landscape* . . . I remember somebody's asking, "Is that really John Cage? It sounds so tonal."

JC Actually, it's not really tonal. It's closer, don't you think, to modal . . . because it establishes a gamut, as does the prepared piano, and then it stays with it rather than modulating to another key.

RD Do you remember what you got from Schoenberg besides the statement that you were not a composer of harmony?

JC Oh, I learned very many things from him. He had a great musical mind and he placed importance on harmony, not only for the combinations of sounds; but he said the importance of harmony was its structural use, so that you could distinguish one part of the composition from the other by means of its tonality.

RD Was he referring to a system at that time . . . the twelve-tone system?

JC No, he was referring to structure . . . musical structure. And since I was working with noises and there was no harmony, the first thing I did was to establish a rhythmic structure. I called it micro-macrocosmic, because the small parts had the same relation to each unit that the units had to the whole. That's characteristic not only of my work for percussion; my concern with dance also gave rise, perhaps, to the notion of rhythmic structure, because that was the common denominator between sound and movement. After the twelve-tone, early chromatic pieces, all of my music had a rhythmic structure until the '50s.

RD You certainly hear it in that two-piano work, *Three Dances*. It's full of rhythm . . . you want to jump out of your skin!

JC It's not the rhythmic patterns that I was concerned with, it was the phraseology and the relationship of the parts . . . well, like I say, that's one thing I learned from Schoenberg, the importance of structure. I think you'll be surprised that I feel he would have agreed with my use of chance operations.

RD Why do you say that?

JC While I was studying with him, he sent us to the blackboard to solve a problem in counterpoint, even though it was a class in harmony. And he said, "When you have a solution, turn around and let me see it." And I did that, and he said, "Now give me another solution to the same problem." And I did that. And finally, after about nine solutions, when he asked for another one, I said, "There aren't any more." And he said, "That's also correct. Now, what is the principle underlying all of the solutions?" I had always admired him, but I admired him then more than ever—he simply ascended—because I was unable, at that time, to answer him. After some years of using chance operations, it occurred to me that the principle underlying all of the solutions is *the question we ask*. It was in fact his question, you see, that produced all of the solutions. And he would have accepted that. And he would have liked the idea that in order to use chance operations you have to know all of the things that might happen when you ask a question that is answered by them.

RD You were known to say, I think about that time: "We can compose and perform a quartet for explosive motor, wind, heartbeat, and landslide."

JC That was close to my work with percussion instruments, but it was looking forward to a technology that was then being talked about. One didn't know whether it was going to go in the direction of tape or film, or what. But, we knew that there would be machines that would make it possible to use sounds of any kind in succession; and Debussy had already said that any sound can be followed by any other.

RD And how about Varèse? Were you talking to him?

JC I was very impressed by the work of Varèse.

RD He was talking about a machine at the same time.

JC He was looking forward to the same kind of thing that I was looking forward to.

RD Were you friendly?

JC I admired him very, very much and I loved his work. I worked, for several years, with Lou Harrison . . . we gave concerts together in San Francisco. Finally, we were so pleased with our work and with the musicians who worked with us that we decided to issue a record. The record was a recording of a piece of Lou's, and the reason we chose that rather than a piece of mine was we let the audience fill out blanks, and the most popular piece was the piece that was recorded. The piece was Lou's Third Symphony. We called it the first recording of music of organized sound. I had defined music in order to include noises in it. I had defined it as the "Organization of Sound." Then I read an article by Varèse that was published in *Commonweal,* and he called it "Organized Sound"; and I thought it was the same idea, but two words were better than three. So we sent the first copy of this record to Varèse; and we got back a telegram saying, "Please desist from using my term 'Organized Sound.'" We were very sad about that and apologized; however; there was nothing to be done because we had already released it. However, a few years later when I came to New York to live, the first duty I felt was that of going to visit Varèse, because it was his work that had preceded mine. I was married then, and Xenia and I went to see Louise and Edgard Varèse. He and I sat at one end of the room and Xenia and Mrs. Varèse at the other end of the room, and we talked separately. Shortly, Xenia came over and said, "We must leave." I knew that if she said that, we should. My father told me, "Your mother is always right, even when she's wrong!" So anyway, we left, and out on the street Xenia told me what had happened. Mrs. Varèse had said, "The reason we sent that telegram (and she must have been the one who decided to send it) was because we didn't want your husband's work confused with my husband's work, any more than you would want some . . . any artists's work confused with that of a cartoonist."

RD Oh! How did you take such a remark?

JC Well, Xenia took it as an insult and said that we should leave, and we did. However, later, as the years passed and as my work with prepared piano continued, the two composers who were always interested in what I was doing before it was performed, even at a concert, and who would come to my house to hear anything, if there was something to hear, were Henry Cowell and Edgard Varèse. And Varèse would walk all the way from the Village over to Grant Street and Monroe and climb the six flights to where I was working.

RD That's remarkable!

JC I should tell you one thing. When I began to work with magnetic tape, Varèse came to the studio on Eighth Street where I worked and we had a conversation about harmony, which is relevant to what we said about Schoenberg. And we both, in a spirit of pleasure, mirth, and so forth, agreed that anything that was worth knowing about harmony could be taught in half an hour . . .

RD Could we touch upon Cowell as a teacher?

JC Cowell was a wonderful teacher, so open-minded . . . and so really interested in elementary, basic, but original ideas, such as in his classes on harmony. Instead of the exercises being just for thirds or sixths, they were sometimes for other intervals . . . fourths, fifths, and seconds, and sevenths . . . I mean to say, as harmony. It was from such thinking that his clusters emerged. And then he showed me the manuscript of his book on rhythm. And since it wasn't available, I copied the whole thing out by hand.

RD My God! That must have taken time. Now, from a sonic point of view, or a composition point of view . . . he piqued your mind?

JC You know his book *New Musical Resources?*

RD Yes . . .

JC It was in that sense that he was so exciting. He opened doors, and it was his mind that was so open. I asked him how it was that he did so much teaching, how he had time to prepare for all the classes that he taught. He said he never prepared anything. He said he simply went to the place where he was supposed to teach and taught what there was to teach.

RD The painters, John, in that particular period after the war, or maybe a little before, seem to figure in your work, or in your thinking. . . .

JC I think that was because, as a result of Hitler, so many artists in Europe came to New York to live, to escape from Europe. That was, of course, the case with Schoenberg, too. So that at one time New York was as brilliant as Paris had been. And somehow that idea of being brilliant stuck, or caught; and out of that, having European artists here in New York, came Abstract Expressionism.

RD And, of those painters, there was Rauschenberg . . .

JC I met him at Black Mountain.

RD He said his work was in the gap between art and life . . .

JC I would say that my hope is that art would serve to introduce us to life; and more specifically, that music would teach us to listen and to enjoy the absence of music. Or, as I've said, silence. I've pointed out that silence is not the absence of sound, but is, rather, the presence of . . . the absence of intention, ambient sound . . . *(traffic noises)* . . . I don't intend those sounds that we're now hearing.

RD I remember telling you . . . it was a coincidence . . . that when I did *Score* in the Aspen outdoor tent, it was to be followed immediately, as per your instructions, by a tape of *Dawn at Stony Point.* I pointed to the recording operator to start the tape and unbeknownst to me, he wasn't where he was supposed to be backstage. He had forgotten to be there. So I pointed, very carefully, five times, hoping . . . nothing happened; but during this time . . . from the distance, the sound of a dog barking was heard . . . a motor started up . . . a jet went by . . . etc., and the people, listening intently, thought that was it. And in fact, it wasn't so dissimilar, in sound, to the tape. Unintended . . . as you say.

JC Right.

RD And other painters . . . you were close to Rothko and Pollock . . .

JC And Motherwell . . . I also knew Gorky and Matta . . . Matta was a brilliant figure. I met Mondrian and Duchamp. . . .

RD Of the painters we mentioned, would you say that Duchamp made an impression on you and your work?

JC I admired Duchamp so much that I didn't impose myself on him. And then when I wrote the music for his part in the Richter film, *Dreams That Money Can Buy,* I still didn't do that. It wasn't until . . . I think the early '60s, I noticed at a New Year's Eve party that his complexion looked more like that of a painted portrait than that of a live person; I deduced that he wasn't going to live long. So I went up to Teeny and I said, "Do you think that Marcel would teach me chess?" And she said, "Why don't you ask him?" And he said, "Do you know the moves?" and I said, "Yes," and he agreed. Then we were often together. I would go to Cadaquès in the summer and stay for two weeks to be with him every day. I didn't play chess well, because I was primarily concerned with just being near him, but I didn't ask him questions.

RD He once said: "The observer should be challenged to take in several meanings at once, and that these meanings can operate on quite different levels of understanding." That sounds like a piece of yours, *HPSCHD,* Harpsichord. I saw it at the Brooklyn Academy some time ago . . . it was quite an extravaganza. Does that kind of simultaneous, multi-level concept still persist as an idea? Since you're working on an opera, is that part of the thinking?

JC Oh, yes . . .

RD Your new opera . . . *Europera,* I guess you call it. . . . Can you say a word about it?

JC It's not only new, it's the only one I've ever made. I'd rather speak about it in two years when it's finished. It's a collage of European opera, but it's a collage in a detail sense and in terms of every aspect of it: the costumes, the sets, the instruments, the voices, and so forth. It will be done in Frankfurt.

RD In the pieces that I have done, for instance, in *Atlas Eclipticalis,* the method of getting notes, or "raw material," I think you told me, came from a constellation observation, is that right?

JC I have maps of the stars, and I'm able to put transparencies over them and to define the pitches by means of them, and also durations and spaces . . . I mean where they are. You see, in "Atlas," the notation is in space equal to time. And the time can be any time, but once the time is established, the conductor acts, as you recall, like a clock. And the position of the tones in space lets the players know when to play their instruments.

RD I remember that process . . . that's one piece I want to speak about in terms of performance. In rehearsals, I remember, we started with small groups of musicians. I think the total, eventually, was eighty-eight, is that right? That's the full complement. And in the small groups, I rehearsed placing the pitch in relation to a staff, which was not the usual staff, because, to the eye, it reflected both major and minor thirds.

JC Right.

RD And in those rehearsals of small groups, we seemed to have a great deal of attention. You were there, you explained what you wanted and what this was about. And if you remember, when we got to . . . this was very curious . . . when we got to the larger group, there was a certain trouble with the musicians accepting the particular tasks that they had to do. What do you have to say about that, John?

JC Well, one of those Dutch musicians, not on that occasion, but in The Hague or Amsterdam, told me, "To play your music, one has to change his mind. You can't expect 100 people to easily change their minds." I think there, in La Rochelle, the kind of mutiny that resulted on the part of some of the musicians was the result of the heat, because you remember it was terribly hot. And realizing that it was so hot, I had suggested that when they had a rest, they could leave and get something refreshing to drink. Some of them took that liberty and turned it into license; and not only drank something refreshing, but something inebriating.

RD I remember the second oboe player very well! Along into my second minute of being a "human clock," I saw him pull a bottle of wine out of a wicker basket and take a long "pull" on his bottle. He continued, alternately, to play and drink. Eventually, he poured a glass and put it in front of me on my desk. . . . There I was . . . slowly imitating the revolutions of the minute hand of the clock . . . one revolution every eight minutes . . . and glaring at him and staring at the wine glass with my arms suspended in "time." About twenty minutes into my "human clock," a stage man discovered the glass and crept to the podium with a glass of Coca Cola, which he exchanged for the glass of wine. I was bewildered, but I persevered . . . and the oboist went on drinking and I went on "clocking," realizing I had absolutely no control of the situation! Talk about something unintended! . . . I think I lasted . . . was it two hours and forty minutes?

JC Right.

RD I never thought that my back would recover. The disruption of the performance led you to say that they had turned it into a "social event."

JC Well, turned it into a "theatrical occasion," I think is what I said; because when such things happen, the attention turns away from the sounds to the interactions of the people, and that introduces "theater."

RD We had encouraged them to do just what they *had* to do; and you said, "like ripples, it will happen, it will happen" . . . meaning, I guess, the effect would happen, like you touching something. . . .

JC No, what I asked them to do was to make sounds independent of their feelings, which was how I had composed it, and to try to let a sound move out from its own center, rather than from the musicians as instigators.

RD My recollection was that some were leaving and not coming back . . . and we ended with a greatly reduced orchestra.

JC You mean they actually left? Well, I think that's very unprofessional. I think the Dutch musicians, in general, in their orchestral behavior, behave badly. I have never seen them behave well. I can say that too, of the New York Philharmonic and also the Los Angeles Philharmonic and the Chicago Symphony Orchestra. The two orchestras, in my experience, that are well behaved are the Boston Symphony and the Cleveland Orchestra. Others might be well behaved, but I haven't had the occasion to notice it.

RD How did you feel after that evening? We went out together, but I didn't ask you that question.

JC I thought that you had done very well and that many members of the orchestra had done very well; but some of them, so to speak, vandalized the situation . . . turned liberty into license. I've continued, though . . . I've written now quite a lot of works for orchestra. I've just recently finished another work which will be done in Tokyo, in December. I've continued making demands that are unfamiliar to the musician . . . even more so than in *Atlas.*

RD I remember *Atlas* as being something that I had not much control over, other than the time.

JC I didn't want you to control it any more than I wanted to control it myself. But I did want people to behave in the same way that we were behaving, that is to say, selflessly. Instead, they took the occasion to become selfish.

RD The other piece, *Score* . . . now there's another method for getting "raw material" . . . we just talked about your map of the stars . . . here were these drawings which you used for orchestration also. . . .

JC They were drawings of Thoreau and I had taken them apart so that the marks, the notations would read from left to right as musical notations do go—from left to right.

RD Are these things from . . . sort of outside the composer, like the twelve-tone system is from outside the composer?

JC No, it's again a part of my intention to work unintentionally.

RD In other words, give chance a "chance"?

JC No . . . give the sounds an opportunity to be themselves rather than just expressive of my feelings or my notions of order.

RD Yes, but is there an implicit order in the "drawing" or that "star constellation" that you really don't worry yourself about?

JC No, there's no implicit order, because the part will go to one instrument, rather than another, because I don't specify the instrument . . . so I have no way of knowing what's going to happen: and furthermore, I permit the player to see the limits of low to high as any that he chooses, not just those of his instrument.

RD I was a little perplexed as to how to conduct it, in the beginning.

JC All of the conducting of *Score,* of *Atlas,* and so forth . . . has to do with time, rather than any control. It's quite different from ordinary conducting. In a way, it's simpler.

RD Than beating patterns, you mean?

JC No, I mean to say that very often a conductor does what he does in order to express something, or in order to make some relationship that is in the music clear. Boulez is particularly good, I think, about making things clear in the musical sense . . . I mean to say, in the sense of relationships.

RD And in no case did you want those objects called "drawings" to influence the conductor? Because, what you give a conductor to look at, in *Score,* are those drawings of Thoreau . . . ordered like a Haiku principle on the page . . . but the conductor sees them as a whole . . . an animal, or a straw, or the birds, or whatever. . . .

JC They could influence you to speed up or to slow down, which you are free to do, but even if they influenced you, you wouldn't be able to control them . . .

RD I found that out . . .

JC So that the changes you would make while you were being influenced would not occur . . . they were in your mind!

RD Very true. It was always surprising . . .

JC Of course. That's what it's supposed to be!

RD I did enjoy it very much, I must say. Are there other methods that you use to gain this "raw material"?

JC I object slightly to the term "raw material," because it's not going to be cooked. They are simply sounds, and I'm trying to keep them sounds. So rather than "raw material," I would say, "What do you do in order to let sounds be sounds?"

RD And if you have to use a map of the constellations, if you have to use drawings . . .

JC I'll go to any lengths.

RD Do you have a new one, a recent one . . . ?

JC You mean other than those?

RD Yes, that allows you to draw upon of form of non-order?

JC Well, I often write in such a way that would be appropriate to percussion instruments. What is characteristic of percussion, as you know, is that you can't have the same drum in one place that you have another, unless it's timpani. But if it's an unpitched percussion instrument, it would be quite different in each place, even a woodblock. So that, following that indeterminancy of percussion, I sometimes write for the insruments simply giving voice leading.

RD In terms of your relationship with those other two people that are associated with you, or you with them . . . Earle Brown and Morton Feldman . . .

JC The third one is Christian Wolff . . . And then the most important one was David Tudor.

RD In talking with Earle, he makes it clear that he wants to be known as someone *not* dealing with chance, and he will say, "John is dealing with *pure* chance."

JC Earle is still involved in controlling a situation, in bringing about a state of affairs that is in his feeling or even in his mind. We're very, very different in this respect. He feels that I have given up the most important thing that an artist could give up . . .

RD Which is?

JC Which is that "control." So that he stands, as far as I'm concerned, in a European position, rather than in this other position that I would connect more with our country.

RD Was there something that brought you together or kept you together?

JC Oh, yes . . . I was first with David Tudor and Morton Feldman and Christian Wolff. Earle was the last person to come into the group. And he was refused admittance by Morton Feldman.

RD On what grounds?

JC Morty didn't want anybody else other than me and Christian Wolff and David Tudor. And so, even my friendship with Morty was broken for quite a period. He became very, very angry. We solved the whole problem by raising money through the Foundation for Contemporary Performance Arts, and we gave a concert at Town Hall, which was partly Earle Brown's music and partly Morton Feldman's music. Then they became friends, and Earle became a member of the group. Once Earle became a member of the group, I regained my friendship with Morty. The whole thing was solved. The unfortunate thing about that whole series of events was that no concert was every given of the work of

Christian Wolff. And his work, as a result, is not noticed as much as Earle's and Morty's. I think Christian Wolff is the most important composer of his generation.

RD I recently had a conversation with Philip Glass in terms of his music and the music of Reich and Riley. He was very respectful of the work of Cage, Feldman, and Brown. He didn't mention Wolff, but he felt that he wouldn't be in the position he is right now without all of you. What do you feel about their work?

JC I've heard, naturally, pieces of all three of them and my first impression was that I preferred the music of Terry Riley; and the one that I heard was the piece *In C,* and I heard it in the late '60s, when I was at the University of Illinois.

RD I did a performance of that and the piano player walked out, speaking of instances. . . . I had to play the Cs.

JC Then, I've had various occasions to hear Philip Glass's music, and I think the one that I've enjoyed the most was the *Satyagraha* and particularly the last act, with that ascending scale . . . and it reminded me of the end of *Socrate* of Satie, with the repeated note. And then, I've heard pieces of Reich that I've enjoyed. They bring about, it strikes me, a very convivial feeling in the audience. People who enjoy the music are all turned, so to speak, into a group. They're under the influence of the music. Wouldn't you agree?

RD I think, group therapy of a kind . . .

JC I myself, more and more, try to bring about a situation that is so mysterious or not so easily solved, that the audience is more apt to become a group of individuals, each having his own relationship to what is happening.

RD Rather than the "herd-instinct," yes, I understand. And what about the word "original" in terms of this discussion?

JC I'm very interested in originality in the sense of invention. My father was an inventor. I think of it as being my responsibility.

RD You think of yourself as being original?

JC Oh, yes, because I try to make a discovery and sometimes I succeed.

RD Are there any other people in the music world today who do that? Can you name a few?

JC I think Nancarrow is one; Christian is one. David Tudor certainly is one. Morty, and in a sense many people are . . . [*telephone rings*] What was the last question?

RD We were talking about originality and who was original.

JC Oh, and I finally decided that everyone was original! That's of course neither true or false. Many people are not concerned with originality, even though they may be original. They are often concerned with expressivity, or as Earle is, with control and expressivity. Christian is more apt to be concerned with something about society and with using music for political reasons. Morty is concerned with expressivity and with beauty and with a rather personal notion of what beauty is. I'm not concerned with any of those things.

RD Do the two words "ambiguity" and "relativity" mean anything in terms of your composition?

JC Very little. . . . You see, I want each thing to be at its own center, and if you speak of relativity, you're speaking of two things . . . and ambiguity is close to that. I like the thing that Joyce did, of making things happen . . . multiple significances. But I think that's almost the *condition* of sound, that it has multiple references. It's very hard to give a single sound *any* meaning, mm? It's free of that and that's what is so admirable about music, that it is free of that constraint. . . . What I like is not meaning; but you see, any meaning can arise that arises in someone's head. I like it better if different meanings arise in different heads, rather than the same meaning arising in everyone's head. I don't like to push people around by means of music.

# Narrative IX

In the early fall of 1971, Goeran Gentile phoned me. We had met a month or so previously at a party given by Philip Johnson in Connecticut. Gentile had recently been appointed General Manager of the Metropolitan Opera, succeeding Rudolph Bing. At that party, we had a marvelous time discussing Beckett, Ionesco, Pinter, the theater in general; plus Stravinsky, Schoenberg, and composers who were involved in writing music-theater works and chamber operas. He also told me of his dreams to have (as he put it) a Piccolo Met.

The phone call was an invitation to come to the Met and see him. This was the first of many meetings throughout the winter and then the spring of 1972. In our initial discussion, Gentile asked me to begin a research of the chamber operas of the past, regarding libretti, cast requirements, instrumentation, length, publisher, etc.; and to compile a list of 20th-century works that might be suitable for the Piccolo Met project. He revealed his plan to work out a merger of some sort with the Juilliard School and thus have access to the 1000-seat Juilliard Theater. This ultimately did not materialize, ostensibly because of union problems that would escalate stage costs.

More important, Gentile disclosed the fact that he had secured a pledge of $400,000 from an anonymous donor, which was to be used over a period of three or four years in order to help support the project financially.

As months progressed, Gentile and a group of us, including Schuyler Chapin, Charles Riecker, Michael Bronson of the Met staff, and William Nix, who was in charge of the Opera Studio, would travel about New York City looking at theater spaces.

I completed my research on repertoire and sent the results to Gentile on January 29, 1972. During that same month, I met with stage director Tom O'Horgan, discussing our project and throwing out ideas about repertoire, casting, designers, etc. Gentile seemed keen about wanting O'Horgan involved. Soon after this, he announced that arrangements had been made with Jules Irving, director of the Repertory Theater of Lincoln Center, to use the tiny 299-seat Forum in the Vivian Beaumont Theater, Lincoln Center.

In my numerous late afternoon meetings with Gentile at the Met, we continued to discuss repertoire and began to arrive at a short list of composers who, for one reason or another, suited the first season of the Piccolo Met. Among those considered were Argento, Bernstein, Davies, Ligeti, Ohana, Purcell, Stravinsky, and Thomson. Gentile began to lean toward Virgil Thomson's *Four Saints in Three Acts* and Henry Purcell's *Dido and Aeneas* as two possibilities, and I supported both ideas. *Postcard from Morocco* by Dominick Argento had recently been premiered in America, and it was given great consideration; however, it was a full evening by itself and Gentile didn't appear to be totally convinced.

We went back and forth over the works of all of these composers, but never came to a complete decision before the tragic summer of 1972, when Goeran Gentile was killed in a car crash on the island of Sardinia.

I received word of his death while in Aspen, Colorado, where I was engaged in my third season as Artistic Director of the Conference on Contemporary Music, a position I inherited from Darius Milhaud. I was shocked, saddened, and full of sorrow. I also felt cut off, not knowing what was going to happen at the Metropolitan Opera as a result of this horrendous tragedy. Having no other choice in the matter, I simply concentrated on the moment at hand.

After the summer, upon returning to New York, I received a call from Schuyler Chapin, who had just been named Acting General Manager of the Metropolitan Opera, to discuss plans for going on with the "Mini-Met" project. It was being called the Mini-Met despite the rather pretentious "Metropolitan Opera at the Forum" label it acquired in our first and only season.

I was appointed Principal Conductor and Artistic Adviser. Chapin and I came to very quick conclusions. *Four Saints in Three Acts* was inevitable; *Dido and Aeneas* a must in terms of traditional fare. What to put with *Dido?* I suggested Maurice Ohana's *Syllabaire Pour Phèdre,* knowing that it wasn't the most original score; but the musical language, though eclectic, was contemporary, and the combining of the Greek stories of Phèdre, Dido, and Aeneas made a certain sense. Another thing in its favor was the contrast between something very old and something that could be called an American premiere. The consideration of using all of those wonderful young professional singers of the Opera Studio as a chorus in both works also seemed to make sense.

At any rate, the double bill of *Syllabaire pour Phèdre/Dido and Aeneas* plus a second production of *Four Saints in Three Acts* was decided upon, and we went ahead full steam. I conducted the double bill with Paul-Emile Deiber directing, Evelyn Lear and Thomas Stewart singing the opening two performances; and Roland Gagnon conducted the Thomson, with Alvin Ailey directing and choreographing. Ming Cho Lee designed a multi-purpose set, cleverly placing the orchestra above the thrust stage, since there was no pit, making it necessary to communicate via closed-circuit television (not the best of circumstances). Jane Greenwood designed the costumes for both productions. Marit Gentile was Program Adviser and William Nix was Special Administrative Coordinator.

The Mini-Met was launched February 19, 1973, in the Forum of the Vivian Beaumont Theater, Lincoln Center, and played to twenty-five completely sold-out houses through March 10. We didn't lose a penny. The budget was about $225,000—$100,000 from the anonymous donor (A.D.), $100,000 from the National Endowment for the Arts, and an additional $25,000 from the New York State Council on the Arts. The story goes that Rudolph Bing had expressed interest in the desirability of the Metropolitan Opera's operating a smaller house. The story goes that James Levine is presently interested in the possibility of a small house. The fact is, Goeran Gentile's dream, although only for one season, was realized.

But plans for the Mini-Met extended beyond one season. After her recovery from the accident in Sardinia, and in her new role as Executive Producer, Marit Gentile and I began to work very closely together and planned a 1973–1974 season, answering to Schuyler Chapin, who became the Met's General Manager. We proposed a triple bill consisting of Luciano Berio's *Recital* (for Cathy), with Cathy Berberian; *Ancient Voices of Children* by George Crumb; and *Eight Songs for a Mad King* by Peter Maxwell Davies. For this production, we wanted Robert Wilson to direct, and I brought Wilson and Berio together. It was like water and oil. They didn't get along at all. As for the second production, I investigated three very early operas of Mozart: *Il Re Pastore, La Finto Giardiniera,* and *Lucio Silla.* I also looked at Haydn's *L'Infedelta Delusa.*

Our projected budget was about $350,000, still knowing that our "A.D." fund contained $100,000 for the season. However, our parent, the Met, was in deep financial trouble, and on January 27, 1974, the New York Times described it as "one of the worst financial crises in its history." According to the Met's finance director, William H. Hadley, "You can't spend more money than you've got and survive." Brilliant! But what about the Mini-Met that had an "A.D." and wasn't asking for parental financial assistance? Hadley went on to say, "The Federal Government is absolutely committed to giving money only for new things, like the June festival or the Mini-Met." The *Times* went on, "This year, it looked as though the National Endowment for the Arts would hand over $400,000 to the Met. But $100,000 of that was earmarked for the Mini-Met, which has been canceled. Whether the Endowment will switch the funds over to the main operation remains to be seen."

Well, the Mini-Met was not exactly canceled, but rather "postponed," in the words of Schuyler Chapin. Undaunted, Marit and I began to plan yet another season for 1974–1975. In the meantime, the musical world wondered why the Mini-Met, with its successful beginning, was being cast aside, particularly in light of the fact that the Mini-Met had an "A.D." We all knew that the "A.D." was A.T., Miss Alice Tully.

At this point, Marit disclosed another idea of Goeran's, and that was a belief that while Charles Ives never wrote a stage work, his music contained theatricality. I took to the concept with great enthusiasm. Eventually, Marit and I sat down with Brendan Gill and discussed the possibility of a documentary-cabaret, based on Ives's life, his songs, and chamber music. Gill quickly came up with a libretto and a title, *Meeting Mr. Ives.* I gathered the musical material, with which I was very familiar. *Meeting Mr. Ives* was eventually produced at the San Francisco Spring Opera and the Holland Festival in Amsterdam.

We still needed a home for the Mini-Met and looked at many theaters, including the ANTA Theater, the Billy Rose Theater, the Little Theater, and Alice Tully Hall. We finally began to focus on the Harkness Theater on Broadway, between 62nd and 63rd Streets, which was being renovated. On April 8, 1974, I received a memorandum from Schuyler Chapin saying, "This is official notification that the Mini-Met will be on for the 1974–75 season. At the moment we have $100,000 from our anonymous donor and we should discuss final repertoire and budgeting within that figure before any attempts at expansion."

As far as repertoire was concerned, many different ideas came forth, including Kurt Weill's *Lindberg's Flight, Der Jasager, Happy Ending,* and Leonard Bernstein's *Trouble in Tahiti.* The outcome of it all was the projection of two double bills: *Meeting Mr. Ives* coupled with *Miss Donnethorne's Maggot* by Peter Maxwell Davies, and Massenet's *Le Portrait de Manon* coupled with Chabrier's *Une Education manquée.*

On December 19, 1974, I received the last letter regarding the Mini-Met from Schuyler Chapin. In it he said, "With all of the problems that we face at the moment, it seems wise at this time to postpone the Mini-Met from this spring into next season and to concentrate on a Bicentennial presentation of the Ives material." Chapin left the Met in July 1975.

During those last years, neither George S. Moore, president of the Metropolitan Opera Association, nor Anthony A. Bliss, who was re-emerging as a force in the Met's administrative structure, seemed to care about the Mini-Met . . . and so it died.

# *Jacob Druckman*

BORN JUNE 26, 1928, PHILADELPHIA

**June 22, 1986** **Milford, Connecticut**

It was Sunday afternoon. I drove to the Druckman home in the country in Milford, Connecticut.

RD Jacob, I have a vivid recollection of our first meeting. I was rehearsing your *Dark upon the Harp* in the Aspen Tent, and after stopping, I turned around and there you were.

JD That must have been the summer of 1972, when I came to Aspen at your invitation.

RD Let's go back to your early training; would you say you had a conventional training as a composer?

JD Oh, goodness, yes! Probably the most thoroughly conventional training! I began studying piano when I was three years old and switched to the violin at the age of six. My piano playing was arrested at that point and remains at about that same level! But at age six, I began studying violin, wasted a lot of time with not very good teachers. In my teen years my parents went looking for a better teacher. Luckily, we fell upon a wonderful man, Louis Gessensway. He was a good violinist and a good violin teacher; most importantly, though, he was a composer and the most wonderful teacher of traditional techniques. When I told him I wanted to study composition, he at first refused, saying I was too young. I guess I was fifteen or sixteen. I went out and found a student of Hindemith's, whose name I can't quite remember, who was teaching at the Settlement Music School in Philadelphia and was teaching from the Hindemith *Craft of Musical Composition.* So I studied that for a while and wrote a woodwind quintet, which was a direct steal from the Debussy *String Quartet.* At that time I was in love with Debussy and Stravinsky. The quintet worked pretty well—I had it performed by colleagues who were good, and took the recording (mind you, a wire recording) to my teacher. Gessensway heard the quintet, was impressed, and said, "Okay, let's study." So we began working on 16th-century counterpoint. My "graduation" from 16th-century counterpoint was the writing of an entire Mass in five voices. When I got to that point he said, "OK, now we can talk about harmony. This is a major triad!" And we went on from there, with Bach chorales. I can't imagine how many fugues I wrote in the style of Bach!

RD Then you moved on, studying with Copland and Wagenaar and Persichetti and Mennin. What did you gather from these people?

JD Copland was fascinating . . . He had a way of looking at all of music, finding everything interesting and putting himself in another person's shoes; very rarely did Copland say anything that seemed to be prejudiced by his own particular point of view. He was really a very extraordinary teacher. I started when I was eighteen or nineteen and began writing very vigorously. Some of those pieces I've allowed to be published. A Duo for Violin and Piano—I was nineteen or twenty when I did that. My First String Quartet was before that, but that was a little too juvenile.

RD What was that piece that I recently helped to record? The *Divertimento*?

JD Oh, yes. That was written during my first year at Juilliard. I guess I was twenty-one . . . I'd spent a few years out of school. I quit high school, much to the dismay of my parents. I was playing jazz and making what at that point was a lot of money playing the trumpet. To the dismay of my violin teacher! Dance bands and nightclubs and playing behind strippers!

RD I had a similar experience myself! I was a clarinetist; but by the age of twelve, I was playing the baritone saxophone . . . my first baritone sax was made in Czechoslovakia; and since I'm a sub-Carpathian Ruthenian, there is some destiny there!

JD There was a lot of energy and a sort of clarity in those early pieces, including the *Divertimento* and the Duo for Violin and Piano and a set of songs called *Laude* which I wrote at the beginning of my stay at Juilliard. But somehow in that time I began to get swallowed up by a neo-classic Stravinsky influence that was so overwhelming, I didn't seem to be able to pull myself out of it. I don't know how much of that was due to my training. I think certainly the kind of training I was getting at Juilliard was conducive to a neo-classic, very conservative attitude. I really went into a kind of decline for several years.

There is another part of my conservative background—I was a student of Renée Longy. With her I studied the old French method of score-reading. I still do it. That's why I can't write scores in C; I think in clefs. When I write for instruments, the horns are in mezzo-soprano clef, etc.

RD Along the way did you have to confront that opposite pole called serialism?

JD I was totally disinterested in it.

RD Were you aware of that music?

JD Oh, sure. We did a certain amount of serial analysis . . . one was not unaware of it in those days.

RD And the question of chance procedures, did that cross your mind?

JD Not really. I certainly knew about them. I never really took it seriously. It seemed anti-musical. I remember being absolutely infuriated at the idea of Cage and Feldman. This is funny, because Morton and I are now very good friends.[1]

RD But you did go to Paris along the way, and you at least tried to study with Honegger.

JD I was one of five Fulbright Scholar composers in France whose fates were really settled by a committee of French composers that sent us to this or that teacher. Two of us were offered our choice . . . Easley Blackwood and I were considered mature enough that we could make our own decisions. Easley went immediately to Nadia Boulanger, but I refused to go to her because I was so worried about the Stravinsky influence. I thought, all I need is Boulanger telling me that Stravinsky is God! I was already feeling that much too strongly. So I avoided her and went to the class of Honegger. As it turned out, I didn't get to study with him. It was the year before he died, and he was sick the entire year. He was replaced by Tony Aubin, who was a composer and conductor, more active as a conductor.

RD Is that when you began to be curious about and to deepen your interest in orchestration?

JD The whole notion of orchestral style came up through Aubin. He made me curious about it, and I am endlessly fascinated how the orchestra sounds different in different people's hands, what it is that makes for those differences.

RD How long did you stay in France?

JD A year. 1954–55. I was aware of Boulez, but I heard very few of his concerts. I thought Boulez was kind of nutty and doing strange things that didn't interest me at all at that point. It wasn't until later.

RD What about the Domaine Musical concerts?

JD They weren't called Domaine Musical then, they were called the Marigny Concerts. This was very much at the beginning, in 1954–55.

RD The whole Darmstadt experience that was going on. Was it attractive to you?

JD I didn't know about it.

RD Apropos Boulez and Stockhausen, how did you become aware of them?

JD After I returned, through recordings.

RD So, the choice you faced then was not the Darmstadt experience, not chance, not serial. The problem you faced was not to be too immersed in neo-classicism. What led you out of that?

JD The European avant-garde, first Stockhausen and Boulez. There was a

1. Morton Feldman died September 3, 1987.

recording of *Marteau sans maître,* with *Zeitmasze* on the other side. Shortly after that was *Circles* of Berio. Those were the pieces that really attracted me, particularly the Berio, because with all its avant-gardism there was a sensual side. I felt the notes counted and there was a genuine musicality there.

RD Was Berio an influence?

JD Sure he was an influence. There is, I think, an equal influence along with Berio, and that was Cathy Berberian. Cathy's input was enormous. I think pieces like *Visages* are so much Cathy. I often had the feeling that she was present, to an extent, as co-composer. I was thinking of her when I wrote *Animus II.* Her voice was very much in my ear. Later on, I started working with Jan De Gaetani again. I had written *Dark upon the Harp* for Jan, in 1960–61. Since then *Lamia* was written for her, in 1975–76.

RD Theatricality . . . you mentioned Cathy Berberian[2] and Jan De Gaetani, both of whom are marvelous performers who have a sort of built-in theatricality, no matter what they perform. I suppose the manner in which they were able to introduce innovation in vocal technique is also something you capitalized on.

JD About this question you posed about theatricality, I don't think it's something that came from the outside. It is very much a part of my natural inclinations. When I was in high school nobody paid any attention to the fact that I was playing the fiddle and was in the orchestra; I kept getting pushed as a talent in the graphic arts. I got a scholarship to the Philadelphia Art Museum School. It was at the very end of Roosevelt's WPA Federal Theater Project, which sponsored a professional puppet theater company at the museum. There were two or three people who built the puppets, a director who was working with the group, and a seamstress. To legitimize their presence there, they took on two or three apprentices. I must have been fourteen or fifteen at that point, and it was fascinating, just putting on professional shows. I worked with the puppets, acted, helped build them. I was thoroughly involved with it, all the backstage stuff: when to pull the curtain; how long are the bows; all the live aspects of show business. The director was a role model for me, the cat's pajamas!

The very first sexual stimulation that I can remember, at probably age six or seven, was at a performance of *Carmen* that took place at the Robin Hood Dell in Philadelphia. I remember a lady in a black dress twirling around, and her underskirts were RED! That had a profound impact. It hit a nerve that I didn't understand at all! That again—theater! I was fascinated with dance (to the point of marrying a dancer),

2. Cathy Berberian died March 6, 1983.

and Martha Graham was a huge influence on me. I guess that was the avant-garde I was interested in—not the John Cage avant-garde, until later.

RD You carry it out in the *Animus* pieces, kind of inventing a performer's theatricality. I think of the *Valentine* piece where the bass player actually has a comic play with his instrument. But, when a singer who is not a percussionist is striking a bell or mucking about with the percussion instruments or standing behind a gong while the gong is being rustled about, to what extent do you see drama in that?

JD All the *Animus* pieces work best in a situation where one is expecting a concert performance and where the theatricality just peeks out a little beyond those expectancies. On the other hand, if one sets up a true theatrical framework, it is not enough to discover a subtle theatricality in the normal playing of the instruments. In that case the parts should be memorized and the *Animus* series should be directed, just like an opera.

RD How would it change? I mean, you have the singer coming down the aisle . . .

JD I think I would engage a good director to go beyond my imagination. I think of what Ian Strasfogel does with Ligeti's *Aventures,* which is a piece that has no specific theatricality in mind and yet which is terribly theatrical. Ian invents situations and stories and relationships. That may very well be necessary if a piece such as *Animus II* is to be done in a truly theatrical circumstance with a proscenium.

RD The alternative would be to do it just as a concert piece, with the surprises being "dramatic."

JD It really works, Richard, when it's done as a concert piece. You saw the performance that Jan (De Gaetani) did years ago in Aspen. It was really magical. What happens in that instance is a subliminal theatricality. You're not sure what you're watching. You see a concert performance of something, but there seems to be a complex relationship going on between the performers. There is a secret not really being revealed to you—you're intuiting. And on that level, it can be fascinating. But as soon as you put colored lights on it, it becomes insufficient.

RD Did Mauricio Kagel figure in your discoveries at the time you were working on the *Animus* series?

JD I didn't know Kagel until later.

RD I'm thinking of some of his theatrical works. In his case, there's a lot of absurdity . . . tying your foot to the trombone slide! And walking around. That's almost Ionesco-like.

JD I admire that kind of Dadaism. I also find myself very curious and very interested in the original Dadaists of the World War I period. And yet my work doesn't touch it at all. I don't know that there is anything Dadaist about my music, not even *Valentine,* which is a funny piece. *Valentine* is a very carefully constructed piece; and if you remove the theatricality, I think it's one of my strongest pieces of music, in terms of the notes. Also, it demands a terribly high degree of virtuosity just to be able to play the notes. And I think that in itself removes it from Dadaism.

RD Well, now, the way the notes are put down ... you said you weren't influenced by the serial stream of thought. What do you say about your organizational methods? How did you get to that?

JD Let me say, I was not interested in serialism in my student days. Later on I became interested in it. In fact, I explored it very seriously in my Second String Quartet, which is serial ... a Webernian structure, where the grand scheme, the macrocosmic scheme of the piece is a statement of the row, which is its own retrograde inversion, etc.... very tightly organized. In writing that piece, I was really tackling the question of serialism. I discovered in the composing of it that I was not in agreement with that aspect of serialism which presumes the equal importance of all the notes, and in effect is a codification of a way to avoid sounding tonal. Also, I found that just about every *other* aspect of the String Quartet became more important to me and it was a kind of ... perhaps it's theatrical. The Second String Quartet has twelve moments of unison, and those twelve moments of unison are statements of the row, across the twenty minutes of the piece. What became important to me was not the fact that pitch happened to be reflected in the microcosmic aspects of the piece, but rather the very event of unison became the excitement, like the racapitulation in sonata allegro form, or that moment of cadenza in a concert, that delicious moment of coming home. In a similar way, it was the unisons that became fascinating to me. The piece is all about that ... it's about the excitement of expectancy ... the anticipation of those moments of unison, I suppose that is theatrical in a way, but whatever it is, it has absolutely zero to do with serialism.

RD But along the way, the multitude of notes comes from serialism?

JD In the Second String Quartet they do, yes.

RD Let's move on to the orchestral music which I know quite well, having done a number of pieces. Are there any reflections on the red bloomers carrying over ...

RD Not bloomers! Petticoats! Bloomers are *your* imagery!

RD Sorry ... any carryover of that sensual imagery to your orchestral music?

JD I suppose if one wants to equate a kind of sonic sensuality with corporal sensuality, there is a connection. I love the orchestra, and I was a fiddle player. Having spent a lot of time in an orchestra, I have the familiarity that you get only as an orchestral musician.

RD Let me mention just a few examples. For example, I remember doing *Lamia* in 1976 with the Berlin Philharmonic, with Jan De Gaetani.

There you have Jan singing in a number of languages. These languages and your desire to use them ... what was behind putting that piece together?

JD I was already beginning to think of an opera. It was right after *Windows,* and I was fascinated with the notion of layers of memory: the fact that when we speak to each other there's a forefront of our attention, and also behind it so many other layers of thoughts, other memories, other emotions. *Windows* was about this: foreground/background. In the case of *Windows* it was nostalgia.

RD But I do remember the "window" in an actual realistic sense being that little thing in the orchestra where you heard something and said, "Now wait a minute ... that sounds like ..." But it really wasn't a quote, it was simply like a quote. So this was a nostalgia?

JD In this case it was a nostalgia, but it was the idea of the different levels of awareness that I was so fascinated with. I began to think at that point of an opera which would be a theatrical projection of those concerns. One of my early thoughts was that the opera should somehow treat the earlier operas based on the same story. Also I wanted a story that had such a force driving toward the end that it could stand a fragmented treatment. Also it would be good, I thought, if it could have something to do with magic, because that would give a raison d'être for the kind of fragmented treatment that I was thinking about. I talked to Ian Strasfogel about that very early on; and it was Ian who suggested the story of Medea, having been done so many times, and also about a sorceress ...

RD Being the force and the magic ...

JD And also having the terrifying ending that everybody knew! No matter what you did, it was like a maelstrom going toward it ... inevitable, that whirlpool leading to the nothingness at the end.

I was commissioned to write a work for the Albany Symphony Orchestra, and I decided to write a piece that would be a trial balloon for the opera. *Lamia* deals with magic ... all the texts in the different languages are about feminine sorcery. The singer has to change roles constantly. It also calls for two orchestras, as you know, so that the soprano sets things up with one orchestra and turns to the other orchestra and gets it contradicted. There is also a quotation in *Lamia* from one of the earlier operas on Medea; *Il Giasone* of Cavalli, Giasone being Italian for Jason.

RD Then, other works that I can remember doing are *Mirage* which I did at the Zagreb Biennale and also at Carnegie Hall. . . .

JD Yes, with the National Orchestral Association! That was a wonderful performance—1977, wasn't it?

RD Then you were there when I did *Aureole* with the Concertgebouw Orchestra. Do you remember anything about it?

JD I certainly do. I have a tape of it.

RD How did you find the Concertgebouw doing your music?

JD They were generally good. A funny thing . . . I hate the acoustics of that hall for my music. I know it's supposed to be marvelous and wonderful; but it is too reverberant and too velvety. I think my music sounds better in Avery Fisher Hall than it does in the Concertgebouw. And also better than it does at Carnegie. Carnegie Hall has a sound similar to that of the Concertgebouw. It is the sharp, incisive presence, the hard edge of sound that I like, not the 19th-century notion of seamless music where all those edges are blurred and rounded out.

The Concertgebouw strings were beautiful, but the percussion players were not as good as in a fine American orchestra. They don't yet have the tradition that we have here among the younger players.

RD In *Dark upon the Harp,* psalms are involved and I seem to remember jazz.

JD That's right. One of the movements is big-band jazzy and is, by the way, serial. So are several of the movements in that piece.

RD How do jazz and psalms figure together?

JD That was *the* piece I think of as my re-emerging from hibernation. Before that piece there is almost nothing for five or six years. I keep thinking of Schoenberg, who withdrew himself from composing for seven years! It's biblical! In fact, it was Jacob who had to wait seven years, was it not, for the lady he really wanted? I should know! But *Dark upon the Harp* was the first shaking-loose, the first time I really got rid of the Stravinsky neo-classic feeling. That piece is not at all Stravinsky.

RD If it wasn't Stravinsky, where were you going?

JD I think I was, for the first time, finding my own place. Things in that piece are not atypical of what I do even now.

RD The next time you came to Aspen was in 1976.

JD Of my pieces, I know we did *Lamia* together. And, *Delizie contente* was done that year. The piece for woodwind quintet and tape.

RD On your next visit, in 1983, you were there alone, as sort of *the* composer-in-residence. We did *Bo* and *Animus IV* and *Aureole. Bo* came from your interest in the "boat people."

JD I had seen a photograph in the newspaper of the so-called boat people, the Chinese who were being chased out of South Vietnam after the war. In the photograph, people were huddled in the bottom of a boat, looking wide-eyed and terrified at the camera. It triggered a memory of an experience I'd had in 1962. I used to do a lot of sailboat racing, and that year I was in the Newport, Rhode Island—Bermuda race. We had crossed the Gulf Stream, with its fascinating visual aspect. You're sailing along the cold green North Atlantic, and on the edge of the Gulf Stream, which is like a river snaking its way through the sea, is a sort of yellow-straw color, caused by the sargasso weed. On the other side of that line is a blue tropical sea. It's the strangest thing to see. We sailed across the line and into the Gulf Stream and as we did, we got into a doldrum that lasted an entire day. An entire day of sea so calm it was like a mirror, absolutely unruffled by waves of any sort. Sails just hanging down like wet laundry, the sky milky. We all began to speak in a half-whisper for no reason at all. We continued our banter, the seven of us on the ship, joking, but always in a very quiet voice. The memory of this silence kind of got mixed up with my strong reaction to the photo of the "boat people," and I decided to do a piece about the sea. I looked for a Chinese poem about the sea; with the help of a man who was the head of the Asian languages department at Yale, I found a perfect text from A.D. 300.

RD I understand that *bo* is the Chinese word for waves.

JD The structure of the piece is also very involved with waves. I was fascinated with the idea that you can throw a stone into water and create ripples, and while they are still going, throw a stone somewhere else. The two systems of ripples cross each other, the circles co-exist and do not disturb each other. I had the idea of a music in which there were such events.... So that we have these co-existent patterns. Because I only had three instruments and three voices to realize all of those materials meant that some of the instruments, particularly the marimba and harp, had to be playing in both events when the events happened simultaneously. It became an absolutely fascinating challenge, just as writing a tonal fugue is fascinating, because the horizontal aspects have to be fit into a harmonic, a vertical concept. It was a fascinating project because there was something to fight, something to struggle against; whereas an atonal fugue is meaningless, because it's like wrestling with smoke. There's no challenge to it. In the case of *Bo,* it was a different struggle—not having enough instrumental forces to play all of these complex events, yet managing to do it somehow.

RD And then *Aureole.* I remember doing it also with the Berlin Philharmonic in 1984.

JD *Aureole* is one of my favorite pieces. I feel a couple of pieces here and there are exactly right and I wouldn't change anything. *Aureole* is one of those; *Valentine* is another. I think maybe *Bo* also. Just about all the other pieces I would like to redo.

RD In terms of the nostalgia aspect ... *Windows* equals nostalgia, and *Mirage* equals nostalgia. How about *Aureole* as an orchestral piece?

JD *Aureole* has two levels of organization: one on the surface and one rather personal. I think the reception of the piece has suffered because I don't ever talk about the personal side of it in the program notes. I only talk about the surface characteristics. The surface notion is that I wanted to do a piece that was basically a single line. I had the image of those sparklers that we used to get on the Fourth of July. We'd light them at night and try to draw designs or write our names in the air as we waved them about. If we moved fast enough, you got the impression of a line, but a line shooting off sparks. I had the idea of a piece that would do just that. A fast-moving tune would shoot off sparks and collect a memory of echoes that either surrounded it as harmony or just took off and had a life of its own. The other part of it was something I never really got into in the program notes. Leonard Bernstein had commissioned the piece, and I wanted to pay deference to Lenny as a composer.

RD You mention the line at the center which is his Kaddish tune.

JD The Kaddish tune . . . why did I choose it? It was very much at the center of it, almost a subconscious choice on my part. I began writing *Aureole* in the summer of 1978, at Tanglewood where I was composer-in-residence. Lenny was also at Tanglewood and was in a terrible state because of the recent death of his wife. He just stopped all his activities and was very morose, and was using his beloved Tanglewood to try to pull himself out of it. Kaddish is the Hebrew prayer at the time of death. Lenny was very much around that summer. He used to stop in and have lunch with us at the composer's cottage or say hello occasionally, on his way up the hill to the main house of the Koussevitzky estate. And so the fact that the tune is the Kaddish tune is not a total coincidence. I also got involved with my own memories of the prayer. Every time there was a death in my family, I heard that prayer and it became a very terrifying memory. The personal connection I have with the Kaddish is to the rhythm of the words, which is a strong, repeated dotted eight and sixteenth figure. For some reason or other, out of protection for Lenny, I didn't want to say this in program notes. For me, the piece is a very serious one which involves not so much death, as the idea of survival. Often, because this is omitted from the program notes, *Aureole* is treated as a superficial, pretty-sounding piece.

RD We discussed the question of tempo, and that was at the very end. How, if you had the right ending, and used that tempo for the beginning, it would work. I had something in mind there that helped me to establish the tempo, and that was saying, "Son-of-a-bitch" (the rhythm of the ending), about as fast as I could. And it seemed to help! So before I started the piece, I would say, "*Son*-of-a-bitch, *son*-of-a-bitch, *son*-of-a-bitch."

JD I remember something funny about one performance. . . . Theodore Antoniou, who did it with his American Composers' Orchestra, somehow got it into his head that the rhythmic figure was a bolero. So he kept wanting to do it in that slower, bolero tempo! It was so slow, and so wrong; but we got it up to tempo for the performance.

RD Well, I must say that I have always enjoyed performing your music; and

I have done a great deal of it, both in this country and abroad. And I look forward to other performances that might come my way. Jacob, can we turn to conductors for a moment? In a *New York Times* article from 1977 . . . Raymond Ericson was talking about the fact that you had three or four performances in one week or something like that . . . but more specifically, you mentioned that the late composer-conductor Bruno Maderna had championed your music, which I guess was the *Windows* Pulitzer Prize performance with the Chicago Symphony. And you say here, "Later there was help from Pierre Boulez with his performances in this country," and myself, who had done your music here and abroad. Are there any other conductors who you feel have championed your music?

JD I don't think there's been anybody who has done as many pieces as you have done . . . that kind of championing has been very special on your part. Other conductors who have done pieces are . . . Slatkin and Previn and Bernstein and Mehta; I've had very good experiences with all of those people . . . Maazel, Commissiona . . .

RD And performers?

JD I am very interested in the performer. Pieces of mine will take on a particular character if I know for whom I'm writing. It will definitely affect the quality of the piece. I think many of the rhapsodical chance things that happen in *Windows* are there because I was so familiar with Bruno, and his love of "winging" it. That was the real excitement for Bruno—to leave things open.

RD You mention chance in the sense of the conductor actually controlling exits or entrances. That's about as far as you go in giving a liberty.

JD There are moments of cross-fading between the two sections of the orchestra that can happen as many times as the conductor wants to do it. That's about as far as I go!

RD So . . . here we have that particular element of chance procedure emerging finally.

JD Emerging and submerging! It hasn't been there in recent years. It's funny. Back in the '60s and early '70s, we thought, "Now we are really breaking the chains." We thought we were seeing new horizons and infinite possibilities . . . but that kind of event now, which is often based on repetitive ideas . . .

RD . . . or the famous "boxes" . . .

JD Or tape loops, as they call it. I find that what happens is that cognitively you don't listen to the detail, even though the detail can be fascinating. As soon as you discover the "boxes," they kind of go into the background. I often think of it as being like a gaggle of geese going down a road. Maybe for a while you can get fascinated with the frequency of honks or the rhythmic relationships of one honk to another. But after a while, you begin not to hear the detail, and your reaction is, "Ah, yes. That's a bunch of geese." It's the same way one hears an Alberti bass. You don't listen to it, you accept it as background.

RD And what about the procedure of the conductor establishing unmetered measures, like in *Lamia,* through the use of triangles that contain the number of beats, but with the duration of these beats indicated in approximate seconds?

JD All that amounts to is recitative, which has been with us for a long time. The notion of analogue (or "proportional" notation, as it is called), opening up new horizons is false. I don't think it does that at all. In fact, it's only with pulse and the expectancy of the next beat that you can do anything really subtle. I suppose there is the exception of a really great recitative. A solo singer or a solo performer can do fascinating things out of meter, out of tempo. But in the long run, if you are talking about subtlety, the difference between a samba and a bassa nova, or between a Viennese and a German waltz—the tiny differences of tempo and nuance that are so fascinating can only exist through pulse. In addition to pulse, one needs a kind of cultural expectancy and a common language. It can be wonderful when you break out of that common language; but the nuance is always in reference to a language.

RD So you are mostly preoccupied with that these days?

JD Yes.

RD You have mentinoned the word "horizons" just too many times for me not to pick up on that word and move on to your experience with the New York Philharmonic. During these wonderful four years you've been presenting the musical world with festivals. This is another aspect of your musical personality. Now that they are all over, how do you feel about their impact, their successes? They certainly raised a lot of questions.

JD Well, I think the impact has been strong and good. I feel very good about it. The strongest one was probably the first one, mentioning for the first time out loud the idea of the New Romanticism.

RD You were heavily criticized for it. Do you still hold the same opinion?

JD Oh, absolutely. I was heavily criticized from some quarters, but I think the general feeling is that it is true. Everyone is using the word. You can talk about Neo-Romanticism the same way you can speak about the Baroque Era. It has, whether criticized or not, become part of the language.

RD Has the definition of that term or phrase become more clarified?

JD In my mind, yes. In the general mind, no!

RD How would you phrase it again? When you first talked about it, it had a rather sprawling explanation.

JD It doesn't feel sprawling to me, but it is a broad definition. Basically, we are talking about the polarity between Classical and Romantic, or Apollonian and Dionysian. It's a matter of a delicate shift of balance between ideas that are always co-existent, the rational versus the intuitive. Obviously, both processes have to be there always. However, does one lean toward the intuitive, or does one lean toward the rational? For example, Boulez's *Structure* ... obviously he was trying to be as rational as he possibly could. He was trying to rule out the possibility of any choice based on intuitive notions. What I find fascinating here is that we used to think that the opposite of really structured serial thinking was chance music. And now my feeling is that they were two sides of the same coin. Both took the decisions out of the composers' hands and were, therefore, a kind of hyper-classicism.

RD It's interesting that you say that.... I've heard that from Lukas Foss, also. And I agree with both of you ... in sound, you mean.

JD Not only in sound but in intent. Because in both cases the composer is removing himself as arbiter, as intuitive personality ... and he's placing the responsibility somewhere else. What is that line from Lukas's *Paradigm?* ... one of the players says again and again, "Someone will be held responsible!" And his work came at exactly the moment I was talking about, as music was beginning a shift over to the new aesthetic. It happened in so many different ways ... George Rochberg suddenly quoting Bach and then becoming Mahler. Lukas's *Baroque Variations,* Luciano's *Sinfonia*—the nostalgia—the change between the Stockhausen of *Zeitmasze* and the later *Aus dem sieben Tagen,* where you play the vibrations of your neighbor ... talking about the Age of Aquarius. Everything that happened in the '60s, the revolutions, the student uprisings ...

RD These have connotations under the Romantic umbrella?

JD All of these things are very reminiscent of the early to mid–19-century Romanticism. I think these are all part of the same thing—Byron's "Western Wind," that spirit that got him killed, shot in some revolution ... what was it about? Nobody knows anymore. It doesn't make any difference. It was about revolution. And there were so many of those things happening in the 1960s. There are so many different aspects of it, where there was a shift over from rational, controlled distancing to something very personal—much in the way it was in the 19th-century.

RD Well, if you open any dictionary, "personal expression" certainly falls into the definition of Romanticism.

JD These are what I think of as the central criteria. And interestingly enough, just in the last year or so, I see young people, my students for instance, returning to a distancing and a neatening up of things, while the older generation is moving ever more toward the more romantic. Steve Reich, for instance, who began (after his serial writing days), with *It's Going to Rain* and the phase music, writing a music that was all very intellectual, structured, and not a pouring out of his guts and soul.

And now, beginning with *Tehilim* and *Desert Music,* he moves more and more in a kind of romantic way. Yet I see my students pulling in an opposite way.

RD Maybe that "accordion" is condensing itself. It's no longer 19th–20th-century; it's five years or less!

JD So we're seeing this fluctuation. At least, it is a countercurrent. I know so many people think, when I speak about the New Romanticism, that I'm talking only about those people who are writing music that sounds like 19th-century Romanticism, particularly the young Germans. I would certainly include that, but that's not the definition of it. For instance, I think Diamanda Galas is a total Romantic and she certainly sounds nothing like Tchaikovsky.

RD Well, that's behind you now. Your stay at the Philharmonic, in general . . . can you give us a "coda" about it all?

JD I loved the time. It was enormously consuming . . . I don't know how many pieces it cost me that I didn't get written!

RD At this point, I guess what occupies you the most is the completion of your opera. How is it going?

JD It's going well. I began very slowly, full of all sorts of trepidations, discovering what I didn't know about voices. Somehow, when you write a twenty-minute piece, the question of tessitura isn't as terrifying as it is when you think of someone singing over a period of two and a half hours. Also, just the idea of being able to predict the theatrical necessities: the way in which a piece plays, how many pages of libretto amount to how many pages of music, many things I didn't know but have now learned. I'm feeling much more secure about it, and it's going fine.

RD The last time we talked in any depth about the opera, you were at the formative stage of receiving a libretto and either adjusting that or working with it. I guess you're way beyond that stage now. The libretto is now rather fixed—or not?

JD No, we're not beyond that stage. What I decided to do, rather than ask the librettist, Tony Harrison, to make little repairs one at a time, was to save them until I had a clearer picture of my total needs. There are still revisions to go on the libretto. Originally, I gave very general directions to Tony. I did talk about the modus operandi of the drama and basically what kinds of things I would like to see and, in some instances, a detailed scenario, particularly for the prologue. But I didn't provide great enough detail; what I should have done was to have anticipated the musical needs and say to him, "I want Medea to do an aria all by herself when she's talking about this. And then Jason comes on stage and I need about twenty lines of duet." I did not do that. I just left him on his own. Harrison wrote some of the most beautiful poetry, much of it usable. But in some instances not usable, because of the musical necessities.

RD I wonder if any other opera composers run into that same problem?

JD It's a common trap but I didn't know about it.

RD Have there been any new conceptions about the role of men and women? You mentioned that femininity was a central issue at one point.

JD Rather the eternal war of the sexes, which became a more central part of the libretto itself. Tony Harrison is a very strong man, and he has his own thoughts about the piece, which come from a very wide knowledge of the Greek myths. We're going to be using more of the Medea legend than appears in the Euripides treatment. Most of the treatments of Medea are just the last couple of hours before she kills her kids. But Jason and Medea were married ten years. Ten years earlier he was a wandering sailor-adventurer on the *Argo,* and she was a princess who fell in love with this traveling salesman.

Tony made some wonderful contributions. The character of Hercules, for example, who was one of the Argonauts, and is often presented as the other side of Jason. Things keep happening similarly to the two of them, except that they are opposites in so many ways. Jason is a smooth talker, not a confronter but a mollifier, a seducer. Hercules is a confronter, a brawler, a fighter rather than a talker. He's also a drunk and he's gay. He is married, but in the legend he falls in love with a young cabin boy on the ship. While the *Argo* is on the search for the Golden Fleece, they stop at an island and the boy is seduced by the water nymphs. They call him, and he drowns in a pool. Hercules has gone after him, and Jason decides to haul anchor and leave Hercules on the island.

RD Now this is the invention of Harrison?

JD No, no. This is in the legend. Also, Hercules goes mad when he's older, and he kills his two sons, which is identical to the crime of Medea. The people of the town where he lost his mind dressed him as a woman. Each one of the Argonauts has a very strong and ironic fate. Hercules goes mad and kills his two kids. Jason wanders down to the rotting hulk of the *Argo,* and the rotting prow falls on him and crushes him. It's almost like an . . .

RD . . . erectile . . .

JD Exactly. He gets killed by a phallus! Butes, who is an older man on the *Argo,* who is the one so attached to his wife who keeps wishing he were home with her, is the only one who is seduced overboard by the sirens, and he drowns at sea. On and on go these strange connections and cross-currents.

RD How are they updated?

JD It's not updated but rather re-examined from the point of view of the 20th century. The character of Hercules is just wonderful. I'm writing it for heldentenor. And Jason is a baritone. There is one wonderful scene that Tony invented. In the Euripides play and in the legend, when

Medea is told she has to leave Corinth and leave her children behind, she kills her children out of spite and revenge. But before she does that, she sends the two children with a gift to Glauce, the young princess Jason is going to marry and for whom he is dumping Medea. What she sends as a gift is a poisoned robe and diadem. When Glauce puts them on, she is consumed by fire. Tony invented a scene in which Medea calls Jason in for a last time. She has the gown that she is going to send; it is the gown she wore when she married Jason. She summons him and opens the casket to show him the gown and says, "Do you remember?" And he says, "No, no, I don't want to remember." She keeps insisting, "Do you remember? Do you remember the golden light shining off my bosom, my golden breasts, golden thighs." And she attempts a seduction scene with the memory of their wedding night. But she fails. Jason says, "I don't want to remember." And she says, "You are going to remember!"

# *Gilbert Amy*

BORN AUGUST 29, 1936, PARIS

**November 29, 1986** **Paris**

Two days ago I spoke with Xenakis, and now I am on my way to Amy. Gilbert lives in the Courbevoie section of Paris; and he has invited me to lunch "en famille" at one o'clock. The day is clear and sunny.

RD Could you begin to speak about what interests you now in composition, what pieces you are writing, what developments you are very concerned with at the moment?

GA I had been involved until about three years ago with a big work for chorus, soloists, and orchestra, which is my largest work to date. It is a *Missa,* the Catholic Mass, on the Latin text. Part of this work was written during the time I was in the States in 1982; and it was finished a year later. Since this work has not yet been performed, I have not undertaken another new large-scale work like it. I guess I'm waiting, subconsciously, to hear a performance of it before starting another large work. Since completing the *Missa,* I have written an orchestral piece called *Praeludium,* which is meant to be the first movement or part of a large-scale symphonic work. Up to now, only the *Praeludium* has been completed. It was intended for the tenth anniversary of the NOP (French Radio Orchestra). Once I finished that movement, I stopped writing for the orchestra and became involved with smaller-scale works, like a trio for clarinet, violin, and piano, a piano piece, and other similar chamber works. I also experimented with a computer at IRCAM; the result was a piece called *La Variation ajoutée,* for seventeen instruments and computerized tape, made mostly out of synthesized sounds. That was not my first attempt with electronic equipment, but it was the first and only time with a computer. This year the work was partly rescored, and I made some improvements to the tape. So I am just "in between." It is not easy to define "in between." I think all composers of my generation, after having written a certain number of works, mostly instrumental, need to stop for a little while. I never really stopped; but I think right now I do not use the same process of composition as I did previously. In other words, I am trying to reconsider, with a critical eye, most of what I have written up to about 1983 or 1984; and I am attempting to

make a synthesis of what I find good, what I find less good, and what I can, frankly, reject. This might lead to another period in my life as a composer.

RD I think you aren't alone, as a composer, in undertaking that kind of reexamination; and we will get to that subject. But first, why did you decide to write a Mass?

GA There were two major ideas. First, a spiritual idea, and second, a more technical, musical approach. I have made several attempts to write large-scale works dealing with strong poetry or having spiritual focus. The largest one I wrote before the Mass was *Une Saison en enfer (A Season in Hell),* which was performed in New York last year at Cooper Union. It was written in 1980, for voices, piano, percussion, and electronics. It deals both with a very strong text on the one hand—Rimbaud's *Season in Hell*—and a large-scale musical period—forty-five minutes. The Mass was a next step, dealing with a text that belongs to my youth. I was brought up as a Catholic, so I went to Mass and heard the Latin. That was part of my culture. Many great composers, people as different as Bach, Beethoven, Verdi, Bruckner, Janáček, and Stravinsky, have written a Mass. I wanted to project my musicianship, working on a text that many musicians have used, and to challenge myself with that, too. In other words, I wanted a path, a road already designed—the text of the Mass. I could not choose the words, I could not choose to eliminate some and keep others. I had to retain the scheme of the Mass. It was an attempt, both spiritually and musically, to be in the hands of something which transcends me.

RD In one way it could be comforting to do such a thing, to know that one has parameters. Was it comforting?

GA In one way, yes. It was also a point in my life in which I wanted to know where I was in relation to those parameters. That's why I did it. And I did it in a very humble way, because I didn't eliminate anything from the text. I used the whole Common Latin Mass, with the Kyrie, the Gloria, the Credo, the Sanctus and the Agnus. Therefore I had a large but disproportionate shape: i.e., a very long Credo and a smaller Sanctus, and things like that which you have to accept. It is in the text! The other aspect was more musical. After a long period of not writing for the choir, I wanted to try to write appropriately for the choir, and to master the confrontation with the soloists and the orchestra. I used a very special orchestra, influenced by the *Symphony of Psalms,* using celli, basses, and a large wind section plus percussion, eliminating violins and violas. Stravinsky didn't use violins and violas because he felt they connoted too much of a Romantic expression.

Also, I have been dreaming, like many of my contemporaries, about opera. I frankly thought that the *Season in Hell* was a step toward writing an opera. I had made several unsuccessful attempts; and I thought that after writing the Mass I might have more mastery over vocal writing.

**RD** Do you have a subject area for the stage?

**GA** No, I have no precise subject. My position has always been to start from a musical compulsion. One *must* do it in that way, and then the subject will come later. When I started working on the Mass, it was unclear at the beginning whether it would be a "real" Mass. I realized only later that it *would* be a Mass. I think it will be the same with a subject for the opera. I will start with musical structure, with musical principles, and musical ideas, and then perhaps the subject will come.

**RD** You mentioned the dilemma that you might be in, and I said I thought you were not alone. In speaking to other composers, some of them also mention a kind of musical dilemma. In most cases it is thought of as a healthy dilemma, albeit worrisome. I find it seems to be more worrisome for those composers who, like yourself, were originally involved with Darmstadtian principles, with Stockhausen or Boulez, in terms of organizing musical material. Is part of your dilemma the questioning of that process?

**GA** Yes, I'm sure. To give you an example, I'm currently trying to write a piano piece. I haven't written for piano solo for almost twenty years. I have written many pieces involving piano, or even two pianos, prepared piano, etc., but not solo piano. I find myself troubled because, after twenty years, my way of writing music has developed in so many different ways, starting from the Darmstadt principles that you mention, post-serial technique, and then proceeding in a more specific, "individual" technique for each work. I don't know where I am within the piano writing. The question is, can I start a piece for the piano which would be related not only to piano technique and the genuine aspect of the piano, but also to the way I am writing music now? I find it very difficult indeed: and to tell you the truth, I'm not sure of any of the material I wrote last summer. In fact, if I use the critical eye I spoke of, I accept only a portion of all the works I wrote before the age of fifty, and I reject another part. There seems to be a crystallization, a condensation of all these problems in the piano piece. With no solution so far. A kind of "little crisis."

**RD** Maybe these are obvious questions, but does the crisis involve your doubt about carrying serialization further, or not at all? Does your crisis involve the formal aspects of your work, or the method of composition involving determinist versus non-determinist methods? Are you concerned with the phenomenon of complexity?

**GA** I think you put your finger on a few important aspects. They are part of my concern. Speaking about determinism, when I wrote pieces like *Cahiers d'epigrammes* (1964), I was very frank and definitive in the predetermined material. I accepted the solution which was "prepared" and would not criticize the complexity which was a result. Part of this complexity I now reject. I don't think it fits in with my personality. So now, speaking of that parameter—determinism—I accept one part of it and reject another. I try to shape my music in a much more "spontaneous" way. That is to say, I try to formulate in a more physical, more emotional way. Although part of the material is prepared in advance, it is not "frozen." I try to be much more selective in the definitive version of the different formulae I have created. If you want to prepare duration, dynamics, register, intervals, etc., everything is easy. But the control is utopic. If you want the composition to go in a more fluid way, in a more physical and emotional way, you have to be critical about that kind of determinism, without falling into vague improvisation.

RD But this is what you are interested in at the moment?

GA Yes. I find it difficult with the piano, partly because it is my instrument. For instance, unlike Xenakis, I am careful about what is "playable." I do not want a write a "utopia." Which is not to say that he writes only a "utopia"; but part of his writing belongs to utopia as far as instruments are concerned. I try to be humble on that side. The piano is particularly difficult, because you are confronted with polyphony for just one instrument. You are confronted with the harmonic block . . . and with the five fingers on each hand—all problems which don't occur in the orchestra or the chorus or other instruments. I feel very aware of it.

Let us look at another aspect. In 1985 I wrote a little trio. One movement was written for the sixtieth birthday of Pierre Boulez and based on material from different composers. Part of the material came from Boulez himself—his *Domaines, Le Marteau,* the Second Sonata. Part of the material came from Bartók's *Contrasts,* and part from Berg's pieces for clarinet and piano. And why? Because I was using these instruments—clarinet, violin, and piano. Using the same instruments, I was interested in trying to master musical material which came from other composers. Of course, in the case of Boulez, it was already dodecaphonic, so it was close to me. In the case of Bartók, it was difficult, because you have triads and major-minor. It is another world. So to bring it to my world, not just as a collage, but as a *function,* to make it work from inside, I tried to be aware of what might satisfy the ear. This brought me to a larger reflection and discussion about what the material itself was. What value does it have? Can you use someone else's material without changing "your" style? And maybe the "material" is not the important thing. I think perhaps it isn't. In fact, I think the important thing is the fluidity, the dynamic, the shape, the phrase, etc., and not the intervals. Because the interval belongs to everyone. Of course, the adjustment belongs to each composer; but there is another layer, other than simple manipulation of interval and manipulation of rhythm, which is the shape of the music. I think I was successful . . . at least, the feedback I got from that piece was strong. And it encouraged me to continue this personal discussion about the relationship of musical material and the music itself. It is difficult to put into words.

RD Did you feel at any time that you were stealing?

GA No. I can show anyone what I took, and possibly they will recognize it. But I think the conclusion of that musical confrontation is that the musical material is *abstract*. It becomes *your* material from the moment you reformulate it for your ear and do not simply copy it.

RD So you're not quoting.

GA No. This has nothing to do with quotations.

RD In other words, here are two examples of your being conscious of an historical past. We spoke about a Mass, and we spoke about using material that belongs to other composers. There is a slightly historical aspect—immediate or, in the case of the Mass, eternal.

GA Yes. I think it is obvious that we need to put our feet into something which has the power to remain, to stay. The whole landscape of art is so spread out today, with no line, no real line except for individual lines. This applies to painting or poetry or architecture as well as music. In music, for example, I note that there is no longer any school, nor any way that one can still speak of the "Darmstadt School" or the "École de Paris." You have individualities making their own road amongst many, many lines. But it is inevitable to rely on some historical foundation.

RD This is a very important issue and question. In your dilemma, are you working toward reunifying into one school, or are you simply observing the fractured quality of music? Are you interested in the strengthening of only your own avenue?

GA I think you made the point. I observe the fractions and certainly would not like to be at either the head or the bottom of one line, one school. I observe the fractions, observe the ambiguities which are sometimes dead lines, which do not belong to the art anymore. And I observe on the other hand, the issues which were not mine before, and which perhaps I could integrate into my musicianship. For example, three years ago I wrote a piece called *Écrits sur toiles,* which was based on letters by Rainer Maria Rilke. This text was not meant to be sung, but just "spoken" in the course of the work. It is a work using more or less the instrumentation of *Pierrot lunaire.* And I realized that I took material in a very obvious way from Stravinsky's *Symphony in Three Movements.* I wanted to cope with his harmony, his distribution of major triads and sixths, and bring it into my world. I thought this approach would produce an enrichment of my own harmonies. Why use only major sevenths and diminished fifths, as in the case of Webern's Variations for Piano, and not use triads, for instance? And to make it nonobvious, I reversed the harmony of Stravinsky. But it is there. And I was quite happy to be able to do that. I thought it was an interesting and a fruitful adventure as far as shape and structure are concerned. The triads have a significance in the ear, in the memory, that makes you remember quite well, because they have such an attractive image. I used this attractive image to build the piece, which is a kind of set of variations. I didn't use only triads—to the contrary—but that was the basis.

RD This thought is a reaction to dodecaphonic material, isn't it?

GA Yes, it is a reaction. It is also a way of incorporating what I call minimal, repetitive . . . that is, material which is due to repeat, so that the listener will follow a structure by some kind of repetition, not only in harmony but also in rhythm. It is somewhat like what Berg would call "Haupt Rhythmus," which also belongs to Stravinsky in many works.

RD But in no way connected to the so-called Minimal School?

GA No, in no way. But it was my personal way of trying to avoid the entropy of the dodecaphonic system, which makes every bar different until the end. Which I have experienced and which I think is over.

RD That's the point. There is an acknowledgment that the idea of serialism and how far it went was a major thought in today's music. But the interesting thing is how many of you composers are reacting to that thought and are now trying to solve that problem in so many different ways. It is fascinating to me how many composers are reacting to that one phenomenon of the mid-century, which is how music became so organized—or perhaps over-organized.

GA But don't forget, Richard, that some composers of Central Europe have been impressed by younger people who wanted to bring back Romanticism. That is not the case in France. It certainly is not my case. I think every feature which I mentioned, of using material that belongs to other composers as different as Berg, Stravinsky, Bartók, or others, has nothing to do with a "coming back to. . . ." It does not have the ambiance of Romanticism of any kind.

RD I know you are very positive about that.

GA Yes. This would not be the truth. There is another aspect. When I started in Darmstadt, the main focus was to have a nucleus and to make the nucleus control the whole work, okay? And now, almost twenty-five years later in my *Praeludium,* I made the simplest start that I think possible. I took an interval of a semitone and shaped it into a rhythm. And for this rhythm and this interval, I found a shape of instrumental colors (the Germans call it "Klangfarbenmelodie"). I took that nucleus from the *Five Pieces,* Opus 10, by Webern. Although it was taken from a dodecaphonic, if not serial work, I attempted to organize a non-serial work from that nucleus. Does that sound paradoxical?

RD Well, it does. You must explain how to organize a non-serial work with serial material.

GA I tried to start with that—I don't mean that one has to work with one single interval!—to utilize the shape of this nucleus, not merely as an interval, abstract from the rest, but as a potential for other developments: as a rhythm, as an instrumentation, and maybe as the beginning of a line. Like in Beethoven's Fifth Symphony, where you get a whole first movement from sol-sol-sol-mi, which leads to fa-fa-fa-re, and everything is deduced from this.

RD I would dare to say that your biggest crisis is harmony.

GA In one way, yes, because as we have no harmonic "function," our problem is to replace this absence of function with a "something."

RD Ah! But with what?

GA I think this "something" is more or less the use, the repetition and the modification of something which one can recognize—a recognizable object which is more or less underlying all one's composition. If you do not have at least that, then the composition is not perceivable.

RD Would this music be considered developmental or static?

GA Developmental.

RD If it is developmental, then I suppose this recognition is an important feature. If it were static, it wouldn't matter.

GA I reject the static in music. I think that every music, because it develops in time, should be developmental, even if it has a "static" aspect. I can imagine a piece that is apparently very static . . . but with peaks in either rhythm or pitch. I like music which I feel is developmental, where I can start from somewhere and go somewhere. I don't like and certainly don't compose what you would call contemplative music. I'm sure that repetition is part of developmental music, that pure repetition is non-developmental, and that non-repetition is also non-developmental. In other words, development starts where one can feel the dialectic between non-repetition and repetition. That is perhaps one of the main considerations.

RD That leads us to the next consideration, which is the formal concept. How to put development into a formal context. Now, you might use a Mass and its proportions as a rather comfortable way of approaching the problem. But if you are dealing more in the abstract, if that is the correct word, how do you personally approach the concept of formal organization on the largest scale?

GA This of course is the most important issue. You are right to say that in the Mass I had certain given proportions, although I could have decided to write a Mass of thirty minutes, or even a Mass of sixteen minutes like Stravinsky, or one on a larger scale. I decided that it would be on a larger scale—nearly an hour. But I accepted the division of the text as an idea of proportion. However, it was I who gave it its final shape when I decided, for example, to spend three minutes on one word instead of three seconds! I didn't take that division from the text itself.

On the other hand, I had to "accept" the division of five movements. Now, the big issue (which is also an issue for some composers of my generation) is whether we write precisely, in one simple big movement, or rather in several parts. Stockhausen, for instance, is the man of the one part, the one big thing without interruption, and the *Moment* form. On the other hand, Boulez writes in smaller shapes which are related to each other but separated. I have never solved that problem. In one of my largest works that you know, *D'un Espace déployé,* a work for two orchestras and a singer, I wrote it in three different parts: one medium-long, one short, and one longer. But at least there are three movements, even if they are to be played *attaca* or even overlapping.

I had a conversation with Xenakis after the first performance of *Praeludium* in May 1986. The piece lasts about thirteen minutes. He rang up and said, "I like the piece very much. I think this is very new. I like the shape. I like the energy. Don't touch it. If you make it longer, you will spoil it." He was warning me, because he read in the program that it was part of a larger concept! So I said, "but of course, Iannis, you tell me that because you are the man who cannot conceive several movements. For you, it's one landscape. It might be for different durations, but it's only one object." He agreed. And I said, "I'm not sure about that. I think I need some architectural balance. If something has too much energy, too much information within a short time, the piece doesn't satisfy me." If you play one movement of a classical symphony, you feel it is not finished. It's an interesting issue to consider. As a side issue, let's take the Schubert "Unfinished." I have always been attracted by that problem . . . "unfinished." We say it is an unfinished symphony because there are only two movements, and because he started other movements without finishing them. I have a facsimile of the manuscript, with the sketches for the other movements; and my opinion . . . this is absolutely personal . . . is that Schubert was perfectly happy with these two movements. And he was unhappy with what he had written later. I also feel that the proper balance of the work . . . which is why its shape seems so modern to me . . . is that it falls into its proper time. I don't know whether you have ever been struck by the two tonalities, B minor and E major, which are so far from one another. These two tonalities are not related. It is interesting to see in the manuscript that in the first sketches, with the particella, he finished the movement in B major, with a major triad. So he could go into E major in the next movement in a smoother way, rather than going back to B minor and having to have this sudden E major. He rejected that possibility as probably too easy and too formal. And because he rejected it, I think that he could have considered a rejection of the other movements, too. This is very personal. I'm not necessarily right. There are many unfinished works by

Schubert, not only because he was working quickly and had problems and did not live long enough. There is another way of seeing it, and that is that he was uncertain about the real shape.

RD You build a strong case, I must say.

GA This was not the case with Bruckner. Quite the contrary. He would start each symphony the same way and wouldn't dare think of taking out the Scherzo! I think we are far from this problem; but I feel very strongly about the difference. Why would I not?!

RD Let me ask you to relate this Schubert phenomenon with your thoughts of today, because I think it is a very interesting personal observation. You may be trying to apply this to yourself.

GA I spoke of Schubert because at the moment I am really concerned about the music which belongs to our world of Western culture, by virtue of the form. And I am concerned with why we call certain composers great. Perhaps it is because they destroyed the "given," old schemes by way of making them their own. Do you remember the issue that Charles Rosen brought up in his book about Classicism? He tried to explain that in Classicism, in the so-called Classical forms, the musicians who were using these forms did not think that they were using Classical form. They were simply using the most common form of the time. Rosen claimed he was able to date the moment when music split from Neo-Baroque or pre-Classical to Classical. But at the moment when their form seemed stable and became "Classical," this form was beginning to deteriorate under the hands of the greatest composers, like Beethoven, or even Mozart. This struck me. I am impressed by those composers who find a shape, find a form which is correct as far as a proportion is concerned. It cannot be followed by anything else. Take the *Prélude à l'après-midi d'un faune.* I often quote that piece as a striking example of a just balance. This is a nine- or ten-minute piece, and it is perfect. You cannot add anything. The circle is complete. You finish the circle, and you might play it another time, and that is all.

RD Could that be called organic?

GA Exactly. Organic. And I think my problem, and the problem of other composers, is to continue a work when you find that your ten or twelve or fifteen minutes are not organic. And to continue until you find that it is organic. In Impressionist painting we are familiar with a number of canvases. And sometimes we are told that a canvas of Monet or Cézanne is unfinished. We feel, however, that it is perfectly finished, even though technically it is called otherwise. I find it quite arrogant to second-guess the painter, who might have been satisfied with this "unfinished" painting. Since music is a more abstract world, we have no precise canvas. We can write ten or twelve or thirteen minutes, or forty; and we can feel free to do it. But one must find an organic solution related to one's own pulse. This was the contradiction in my conversation with Xenakis. I said, "My work is not finished. This is only one part. I must continue. It is like a house with only two rooms instead of four." And he said, "No, I perceive it as a whole." So you made the point. It has to be organic. And it could well be that in going too far, the organic aspect in one movement could be lost. That is a possibility. But we have no criterion.

RD Except a friend or an ear.

GA Exactly. I remember speaking with Celibidache many years ago. He listened to a performance of *D'un Espace déployé,* and I was of course anxious to know his opinion. And you know what he told me, Richard? He told me, "There are two bars too many in the second movement. You should cut out two bars in the second movement." This is the reflection of a man who listens to music as an architect, and everything is kind of computerized proportionately. It struck me very much. I think each ear has a different way of managing the organic . . . what is organic for one could be non-organic for another.

RD That is called cultivation! But now, about your musical genetic code, Gilbert . . . because you, for instance, are among very few composers who were considered protégés of a very influential man called Pierre Boulez. Could you talk a little about that influence, and the Messiaen influence, the French influence? At what point did you feel you were on your own?

GA This is exactly my genetic code, you are right. I was very struck by the Messiaen class, because suddenly I had the doors wide open on music. Except for my very early scores, I did not borrow too much from Messiaen's technique, which I found and I still find a technique of juxtaposition rather than of composition. I admire Messiaen very much, but my thinking is very different from his technique of composition, because of this repetitive and additive process. The Boulez influence was strong because it was, on the one hand, related to the school of Messiaen, let us say, for example, in the vertical and the sense of harmony. But on the other hand, it was turned toward complete opposition of this additive way of composing.

RD Toward what, though?

GA Toward much more contrapuntal and serial ways of composition, which have nothing to do with Messiaen. The air which was breathed at this time was really the only one I could breathe. I think when you speak about my genetic code, you are right. I still have in my music strong issues which come from that period, and which will probably be there all the time.

RD They are in the blood by now.

GA Yes, in the blood. And now, how did I get out of that? I think I started exploring new directions as early as 1965, with my *Trajectories.* I tried to be as organic, but to use more indeterminate parameters, to bring back a kind of "improvisation" and a type of musical energy which I had never used before, and which was not contained in the music of Boulez at that time to my feeling. From that period, I think, my music has reflected sometimes the music which Boulez was writing; but it also has gone in other directions. Maybe it was going back and forth—I don't know. Maybe it could be analyzed more precisely. I think I would never completely abandon certain colors which I have in the brain, and some ways of orchestrating and some way of shaping the intervals. It is difficult to say that, from my point of view. I think it should be the work of a critic.

RD But on the other hand, you do admit that you are facing a crisis. Could this crisis include a further break from this?

GA Sure. I think, for example, that to bring music material, sound material which you would not have used before, is already a step forward. Look, in the *Variation ajoutée* (1984, for computerized tape and seventeen instruments), I have tried to integrate some sounds from the bells. We call these sounds from the bells non-harmonic. They have a very complex spectrum; and by a certain process you can analyze this spectrum and it gives you the right pitches with envelopes and dynamics and everything, of one bell sound. This is like mountains, the superimposition of many pitches, very, very complex. I have used that kind of chart in some part of the work. I have tried to "apply" that to other instruments, to forget the bell and use only the chart which was on the paper. It was like taking a picture and then filtering it completely, and you have another result. This issue was completely against my principles. In my principles, I wouldn't use a single note which I have not "invented" myself. There, I had given notes. So this is another example like the one I gave you of the Stravinsky *Symphony in Three Movements,* or the Bartók *Contrasts* and so on . . . but in this case, I started with raw material, the bell. 1) You hit the bell and you have a complex sound. 2) You analyze this sound and it gives you certain results, a chart. 3) You use the chart as a given, a borrowed object. György Kurtag has a very nice word for this. He speaks of a "stolen object"—"objet volé".

RD Well, I asked you if you were stealing.

GA Yes. And I think the stealing process can be a vivacious process. Of course, only if you use it in a creative way! So, to go back to your point about the genetic code, this might be the most difficult thing to explore. The only way of exploring it is when you look back at old scores, and you say, "Did I write that? Yes, of course, this belonged to me, and this . . . well! I'm not quite sure!"

RD You say the critics should do that, but I think it's interesting to ask the question of the composer himself, because after all, you might have the best memory.

GA Yes, but it is sometimes funny when you have forgotten a work and you hear it again—and you hear it like a work of someone else. And all of a sudden there is a point that you say, "I am the only one who is able to write that!" I had a funny experience a few weeks ago. I was sent a tape of a broadcast from Vienna, a broadcast done on me, for my fiftieth birthday. There was some of my early music at the beginning, and I said, "What is that? I recognize my music, but it is transposed." And I made the confusion between two works—I thought it was one work when in fact it was another work. So at the same time, I recognized my music and I thought it was transposed, when in fact it was another work. After thirty seconds I recognized it. But for the first twenty-five seconds, I was completely puzzled. This is just to point out that there is a great deal of unconscious automatic rejection. I think the composer is not the best person to ascertain [his or her] authentic musical genetic code.

RD What about Stockhausen in terms of your early beginnings?

GA I was never really a pupil of Stockhausen. I was very impressed by his approach to form; but I was never really engaged in his way of composing. I admired him a lot and attended many of his courses, and even started to compose a piece at that time. But I think I was not absolutely convinced of his "*Ur*-Method," this way of having everything built on one single, abstract process.

RD We were together for the first time in Basel, Switzerland, in 1969 at Boulez's conducting course. There we began to have discussions about music.

GA Yes. I remember. I think that was a great period, because there were places like Basel where you could feel a little bit out of time. There was no other commitment except working on the scores and discussing and rehearsing the orchestra. It was fascinating for me.

RD I continue to carry with me a great admiration for Pierre's analytic approach to rehearsing, and his grasp of the material. I was terribly impressed by it all and remain so.

GA It was interesting to meet people, not only from different nations and cultures, but with different commitments: that is, those of us who were primarily composers rather than conductors, and those of you who were primarily conductors. It was so highly professional, and everyone was so confident about what Pierre was saying. I'm not sure that would happen again today the same way.

RD Following the Basel classes, we had a unique experience collaborating on the production of Stockhausen's *Carré.*

GA There again we were in a similar situation. Although the four of us were of different ages (Lucas Vis was younger than we were, and Michel Tabachnik was in the middle), Michel and I were also composers. You and Lucas were solely conductors. But it was total osmosis. We worked well together. I also think that the problems of *Carré* are unique, in the sense that it is not over-written. We had to shape it. Not like in *Grup-*

*pen,* where everything is more precisely written. We really had to shape within the bars of *Carré* and to make music out of it.

RD Yes, the freedom we had of time versus beat.

GA The musical landscape has changed a lot since then. I hardly meet a conductor now who is interested or involved in problematic discussions. It seems that individual ambitions have taken over the musical commitment that we used to have at that time.

RD And music has changed, too.

GA For instance, in that period a number of pieces were composed using two or three groups. Now that is finished. The conductor wants to be in charge of everything. And in a certain sense, it is also related to economics. There was a boom in the '60s, and everything seemed possible. Now we have to be economical. Pieces using extra players are rejected. If you are doing the Webern, Opus 6, you have to do the small version.

RD I also feel another thing about it. You point out the economy, and I think that is at the root of it. There are a number of composers in this "fracture" we spoke of who are writing toward "pulse" again. The temporary resurgence of Romanticism which you mentioned has allowed the average conductor to utilize his mentality again. The period we were talking about eliminated a lot of conductors, either because of the complexity of bar-line and meters, or the necessity to complete the musical gesture of a piece through indeterminacy. So that meant a very critical judgment had to be there. That eliminated a lot of conductors. I think the conductor today, if he can beat 3 and 4, or possibly 5, can get by in what is called modern music now.

GA Yes, you are right.

RD Is it the composer? Is it the time? Is it economy? Society? Maybe everything?

GA We certainly are not in the highest period of modernity!

RD I don't think so.

GA I'm sure that an economic analysis would be a good one to make, because modernity really belongs to the prosperous years. 1973 was the year of the first oil crisis and recession, and that started a big decline in the arts. Economy and the arts are closely related; but one must consider that the economic issues are in the first rank. All of music is economically based: the orchestra, the star system, the size of the orchestra, the number of rehearsals. Everything is against adventure!

# *Elliott Carter**

BORN DECEMBER 11, 1908, NEW YORK

**May 29, 1986** **Waccabuc, New York**

I am with Elliott Carter at his country home in Waccabuc, New York.

RD I'm intimidated in one sense, because you have written so much and so clearly about our music, our new music, for such a long time; and of course it's published, and you've talked about almost everything. It's hard to think of a question for you. *But* . . . I have some.

EC That's good, because I'm wondering myself what you could ask!

RD First question, and I know you have written a lot about Charles Ives, but do you think you would be a composer today if you had not met him in 1924–'25?

EC Oh, I think I certainly would have been a composer even if I hadn't met him. If you had asked whether I would have become the composer I now am if I had not known his music from that time on, I would have been more uncertain. Certainly works of Stravinsky, Bartók, Varèse, and Schoenberg were just as important to me then. My interest in polyrhythm dates from before that when I fell for the late works of Scriabin. Ives's *Concord Sonata* and many of the 114 songs (scores of which he gave me) were as exciting as the *Poem of Ecstasy, Prometheus, The Rite of Spring, Pierrot lunaire,* and *Intégrales.*

Of course, such modern scores seemed much more vivid, alive and imaginatively evocative than older classical scores or the heavily systematized routines of pop music. Because of this music's great imaginative power its future seemed assured, once the public caught up with it. I did not realize clearly, then, that these overwhelming works (including those of Ives, in a different way) had emerged from a background where many aesthetic standards were widely accepted—standards of craftsmanship, of unity and coherence, of musical thought, of style, and above all, of musical interest arising from an arresting vision on the part of the composer.

The apparent iconoclasm of the early modernists confined itself to

*With the permission of Elliott Carter.

*Charles Abbott*

extending the musical vocabulary in the interests of more immediate personal expression, while it continued the many older Romantic patterns, including that of first shocking listeners only to be later accepted by them. To me, today, this musical revolution gives the impression of having been a fierce attempt to make music vitally alive in the face of a rapidly changing cultural situation that was already beginning to fragment the professional musical world, causing it, among other reasons, to lose the authority it once had with the general public. At the same time there was a constantly increasing concentration on the performer as the center of musical life, and, hence, a growing cultural vacuum as far as creative work was concerned, expressed in Eliot's *Wasteland* or Schoenberg's *Erwartung*. The sense of alienation, so characteristic of that time, was not what struck me until much later—rather, it was the freshness and power of expression and the amazing flights of imagination of these early scores.

RD You were at Horace Mann High School at that time, and you were off to Harvard very shortly after that as an English major, is that right?

EC Yes. I went to Harvard, not so much because of Harvard, but because Dr. Koussevitzky at that time played a great deal of contemporary music with the Boston Symphony. They came and gave the regular subscription concerts in New York, and Dr. Koussevitzky conducted concerts of new music for the League of Composers there. I remember hearing Honegger's *King David* under those circumstances. Koussevitzky was really energetic in supporting contemporary music; and it was really for that reason that I went to Harvard, because I wanted to be where I could go to the concerts. I went every week during my undergraduate years; and as a graduate student I continued to attend whenever the Symphony played contemporary music. Koussevitzky, of course, had at that time a very remarkable man as his adviser . . . Nicolas Slonimsky, who used to give lectures at the Boston Public Library about the new pieces that Koussevitzky played. The conductor was a great friend and admirer of Nadia Boulanger, who also suggested new scores to him; so we got a large dose of Stravinsky. And as a matter of fact, Nadia persuaded Koussevitzky to play Roger Sessions's First Symphony, because she admired this work and many pieces by Copland and Harris. We heard a large amount of French music, of course, and very little Viennese music. While I went to Harvard mainly for the BSO, I was very disappointed in the music department, because there it was thought that what was going on in Boston was insane; and that Dr. Koussevitzky was crazy, playing all that modern music. And it wasn't long before I felt I was learning nothing, particularly in the field of music theory, harmony, and counterpoint.

RD But you had Walter Piston to go to.

EC Walter Piston didn't show up until the end of my time at Harvard, you see. Piston was studying with Nadia Boulanger while I was an undergraduate. And when I had to make up my mind what field to follow in Harvard, it was before Walter came back from Paris. When Walter returned, things changed a good deal for me, because I was very interested in his music, some of which I knew even before he came back. After I graduated as an English major, I joined the music department because of Walter Piston. Gustav Holst taught there the first or second year that I was a graduate student.

**RD** That was for your M.A. Do you recall what you gathered from those two men, as a composer? Does anything stick in your head?

**EC** Well, Walter Piston was very remarkable, very critical, and a very demanding teacher. He made us work at traditional orchestration and harmony and counterpoint which he criticized in detail. He was not as inspiring or as interesting a teacher as Nadia Boulanger, being rather cynical and difficult, but one that loved music and had an enormous interest in all aspects of music, especially in the instruments of the orchestra. When we had orchestration classes, he would invite players from the Boston Symphony to demonstrate their instruments; and he would discuss fingerings with them in enormous detail. That was very interesting to me, I must say.

This education assumed that a composer had to be well grounded in the "common practice" of harmony, counterpoint, form, and orchestration derived from composers of the 18th and 19th centuries. We wrote Mendelssohnian, Brahmsian choral works, Cherubini fugues, Clementi piano sonatas, at times boldly reaching out toward Chopin, Schumann, Wagner, Wolf, and Fauré. After this we were supposed to break loose and write our own kind of "modern music." These exercises left a permanent distate for what is now called Neo-Romanticism, as if this trend had not gone on among so many epigones throughout our century . . . even interesting for a time such avant-gardists as Ernst Krenek. It did make us very aware of the many wonderful things great composers were able to achieve in a style far more restricted than any system devised today.

**RD** And Holst?

**EC** Holst in those years had not only some kind of arthritis and couldn't play the piano, but he had very thick glasses and couldn't read scores, so that all the students had to play their pieces for him. I was never a very good pianist; but he would judge our music on the basis of what we played, and that made it rather difficult for me. I never played my pieces that well! It used to upset me deeply.

**RD** Shortly after that you went to Paris, like a number of artists.

**EC** As a result of Walter Piston! He said that I should study with Nadia Boulanger and sent her a letter of recommendation. The reason I went, of course, was partly because I could speak French, and I had traveled in France as a child after the First World War. I went to Paris in 1932 and began to study with Nadia Boulanger, and it was during that time that Schoenberg came to the United States. If I had not gone to Paris that year and had waited another year or two, I might have studied with him.

RD Well, Aaron Copland went to Paris; and I asked him why he went, and he said that was where the action was in terms of what he was pursuing.

EC What was important was that there seemed to be only these two teachers who were very authoritative about new music . . . Nadia Boulanger and Arnold Schoenberg. They were willing to consider contemporary music seriously and explain how it was done, why it was done, make something out of it; and I knew that Nadia would be helpful to me. What was disappointing about her during the years I was there, many years after Aaron, was that Nadia by that time had kept right up to date. Which meant that she was no longer interested in the "old" contemporary music, but in Poulenc, and those "neo-classic" composers who were then functioning, and in the Stravinsky of *Perséphone* and *Duo Concertant.* Indeed, by the time we got through with Nadia we all knew these and the *Symphony of Psalms* backwards, practically. She admired *The Rite of Spring,* but she was much more interested in the Stravinsky of the '30s while I was a student. Not only that, she was very attracted to the music that Stravinsky was interested in. We don't know who started all of this; but Stravinsky then was very concerned with Bach cantatas, and at the same time we studied, I think, almost every Bach cantata that exists. We sang them every Wednesday afternoon, two or three at a time, for three years. Nadia was also very interested in Renaissance and pre-Renaissance composers like Guillaume de Machaut. She used to point out interesting details for us and we used to sing them.

RD You mean you don't know who first started that interest in Bach?

EC Well, Stravinsky was interested in the very same things at the very same time, and one doesn't know who influenced whom.

RD As a composer, then, you came armed with something . . . was it mainly analysis and exposure to things or was it technique, or what?

EC Well, it was very interesting with Nadia, what she did with many students, for she made us reconsider under her very perceptive guidance the fundamentals of musical composition. Traditional harmony and counterpoint came alive through her remarks. Her vast knowledge of musical literature with which she illumninated the most basic matters gave them a significance few of us were aware of previously.

In addition to all of this, I was a member of a chorus conducted by Henri Expert, the leading Renaissance scholar of France. We sang large numbers of Renaissance, French, and Italian madrigals. In fact, I had done some of that at Harvard before I left. Even at Harvard in my last years we had sung Bach cantatas so that I was able to carry this further in Paris. In the last year of my stay, I ran a madrigal chorus with some of Nadia's students. And we sang French, English, and Italian madrigals which Nadia was, of course, very fond of. We went around to little churches in the suburbs of Paris and gave concerts. So I was very preoccupied with choral music in those years.

RD In 1936 you came back to the United States and I understand you were motivated by the demagoguery coming from Italy and Germany.

EC Yes, well, I was very much concerned. I had many friends in Paris who were German refugees. Paris, of course, during those years was overwhelmed with German refugees. Some of them I had known from before Hitler's time. I remember in the late '20s when I was one time in Baden Baden. I can't tell you why now, but I was with a group of fellow German students of my age and we all swore we would never fight in a war. Some of these people later turned up in Paris as refugees when I was living there as a student. Of course, many of us felt a great obligation to help these terribly disoriented people. This occupied a lot of my thought in those years. I was very concerned about their fate. The French, after all, were having a hard time themselves. There was the Depression, and what to do with refugees became a very serious political problem. Some were put in prisons or insane asylums, in all sorts of places, to get them off the streets. When I came back to the United States, I went to live in Cambridge for a while. It was nearly impossible to find a job of any kind during the Depression, and I thought that if I stayed close to the place where I had graduated it would be easier to find one than it would be if I went to New York. I was asked at that time to write some music for a Latin play given by the Harvard Classic Club. The score ended with a lively tarantella for men's chorus and small orchestra, which became, in a four-hand arrangement, a standard staple of the Harvard Glee Club. Finally, I orchestrated it for large orchestra and men's chorus, and it was given at the Boston Pops with great success.

RD Called *Tarantella?*

EC Yes. It was the success of this piece that got Lincoln Kirstein, who was a classmate of mine at Harvard, interested in commissioning me for a ballet. I think the Harvard Glee Club still occasionally sings the *Tarantella*.

It's somewhat entertaining. When I listen to it now I'm surprised it's as good as it is. The trouble with the big orchestration is that it's so brilliant at times you can't hear the men! I had the xylophone banging away, all sorts of things. Just like a young person would do . . . when you get a big orchestra, you're going to make it sound like a big orchestra! And I did it! It was rather hard on the singers sunk under the brilliant noise.

RD At that time and subsequently, you did a lot of writing as a reviewer and essayist. I'd like to call your attention to your article from 1939, called "Once Again, Swing." You speak about the Carnegie Hall concert in which there was an attempt to trace "swing" to African origins, and you talked about its being traced to the spiritual, in jazz, boogie-woogie, and Count Basie. I love some of the terms you've got there, like "gut-bucket licks" and "in-the-groove" and "jitterbug" and "a solid sender doin' some tall rug-cuttin' " and all that "jive talk."

EC During this whole period I was very interested in jazz and used to go to all those nightclubs on 52nd Street, with Fats Waller and other people playing. It was very exciting and has had a strong influence on my music.

RD You spoke of the black performers in terms of a comparison to gypsies in Europe.

EC Oh, yes. I was claiming that dance music in general tended in a special direction . . . this was something that even Nadia used to say . . . which persisted in many different kinds of popular music, that is the idea of establishing a dance rhythm that is rather strict and regular, while playing an improvisation against it. Apparently, even Viennese waltzes were performed that way during the high time. Certainly, gypsy music has that aspect, a kind of free rubato combined with a regular beat, which of course is the basis of my entire musical outlook as are many rhythmic devices found in African, Balinese, and Indian music. My recent *Penthode* is a development of the notion of one long melodic line, such as I heard in Dhrupad music as sung by the Dagar brothers of Northern India, with various regular rhythms in the background.

RD Could we speak just a little bit about Ives without going into all the detail which you've already written so much about . . . ?

EC Right in this house many years ago, both Dutch and German television came and interviewed me at great length about him. I took them over to General Putnam's monument in Redding. This was long before America was making anything out of Ives. Now, Americans are making something out of it and hardly anyone has come to interview me about him. Not that I want them to, as I have had my say in my *Essays.*

RD You went to concerts with him.

EC There are very strange things about this. What I have said in my writings doesn't seem to be corroborated by what John Kirkpatrick found in Ives's memos. For example, it seems to me I remember hearing *The Rite of Spring* with Charles Ives. On the other hand, John Kirkpatrick says Ives never heard anything but the *Firebird* of Stravinsky. Perhaps my memory is faulty, or perhaps the memos are not complete. I don't know that.

RD You spoke somewhat of his suffering under "undifferentiated confusion" in his composition; in that perhaps he wasn't able to digest his experience as an American and make it into a unified and meaningful musical expression.

EC Yes, I felt and I still feel (and this has, of course, become more and more a bone of contention) that a seriously considered work of art needs to have a fundamental unity, a fundamental musical vision behind it. To me, the element of quotation, at least certainly when I wrote about Ives anyway, destroyed the magic and the power of the music by suddenly referring to a literary meaning not relevant to the piece itself, but that came from another background. I'm not so sure. This is a much too complicated aspect to discuss . . .

RD The quotation aspect, you mean?

EC I don't know what I think of this . . . I don't like it as a rule. But certainly there are examples of music, especially old music, where quotations are used effectively, as in Bach or in all those variations before the time of Bach, quotations of plainsong, for instance, in Renaissance music.

RD It has a kind of photographic quality, doesn't it? A kind of snapshot of something, that quote?

EC Of course, the kind of justification to many, today, is that Picasso put pieces of newspaper in his pictures or quoted words sometimes. I'm not sure what my impression is about this. Stravinsky, of course, quoted a great deal. His later works do have a suggestion of Bach; they do have a suggestion of Guillame de Machaut; and also certainly of Tchaikovsky. And Stravinsky managed, in my opinion, at his best to take this material and make it sound like something of Stravinsky. People who quote literally suddenly give the impression that they are talking in another voice and in another person. It's like band music running through the hits of the day.

RD Well, certainly Ives had his own way. Those tunes, for example, played in a crazy kind of rhythm and polyphony! Nobody else can quote quite like that!

EC I used to feel that his method was superficial, that he didn't somehow transform his quotes into something his own, as Stravinsky did; because Stravinsky was, after all, the model I was thinking about at the time I wrote my articles. You must realize that although my admiration for Ives's vision and originality has never changed from the time I first came to know his music in the '20s, my opinions about his actual musical achievement have varied considerably as my own outlook changed. I no longer find my writings about him of the '30s and '40s a reflection of what I think now, yet those articles represent, I think, a real effort to think about his music at a time when few cared about it enough to formulate any considered opinion.

RD Well, I had the great pleasure of conducting the Fourth Symphony of Ives in the Concertgebouw with the Dutch Radio Philharmonic last season (1984–85); and I did it with two conductors. Let's consider the second and fourth movements which you referred to many times as perhaps the most interesting. The second movement in particular I found necessary to do with two conductors in order to accomplish that "splitting" or "tearing" away and collapse of one group from the other. While it can be notated (and people think that by doing it yourself you're making something out of it; but that's only notating something into a common rhythmic unit), in this case, I don't think that's any great accomplishment. But if you do it with two conductors, you underline the chaos of the two groups going their separate ways.

EC The second movement after all is the only orchestral work that we know of that Ives had a comparatively professional performance of. This was the one case in which he was actually confronted with actual sound; and as a result he worked very hard to make it come out as he wanted to hear it. I remember when I got involved in starting the first Ives Society, of trying to get that particular work into some kind of performable shape, the fourth movement was just a scribble. I mean, you literally couldn't decipher it. Suddenly, there would be six trombones playing one chord, and then nothing else like that again. I mean, it was all as if someone had jotted down little ideas that were not very coherent. And then later, somehow, some colleagues got busy and fixed it up!

RD The fundamental question is: how do you keep that fourth movement together while allowing the various layers of sound their rhythmic independence? How do you contain a chaotic mass of sound? Perhaps it is akin to Beckett's search for framing a form he called "the mess."

EC Well, I'll tell you a story about that Ives Fourth Symphony. In 1955, which was the year after Ives died, I was the vice-president of the ISCM; and at the festival in Baden Baden that the ISCM gave, where my Cello Sonata was played, Rosbaud conducted other works with the Sudwestfunk orchestra. I went to him and said, "You know, it would be a wonderful thing if you did the Fourth Symphony. It's a great, remarkable piece, and in America we never have enough rehearsal time for difficult music. It would be wonderful if you could do it, because you could really work out everything in the score." The orchestra thought this was a very good idea and corresponded with the people connected with the manuscript and tried to get it. It wasn't long before Oliver Daniel of BMI said it would be a scandal if the Germans did it before we did it in America. So at that point he raised money and convinced Stokowski to conduct it. But this happened mainly because the Sudwestfunk was so eager to do it first. The fourth movement was the stumbling block. Finally, Stokowski himself and, I think, Lou Harrison and Henry Cowell and others got together and worked out something. My own old editor, Kurt Stone, was the one who got the whole thing into a form for some kind of publication. But there was never any consistent overall view in all of this, as there was never enough money to do it. It was mainly just a rush to get it into some kind of performable shape before the Germans got there.

I find that the orchestral pieces, indeed a great deal of Ives's music that is published, contain lots of invention by other people. There's an awful lot that has been somehow doctored up and changed. Now Ives himself, of course, did change a great deal during his own lifetime, as you know very well.

I used to think that there were two sides to Ives. There was a transcendental side, and there was a side of hatred and anger at the musical profession. And a desire to make trouble in it, to make difficulties and be disagreeable about it, and to make a mess. Of course, some of his songs are about that, and he says it in his essays. There was a real fist-shaking at the musical profession which I thought was in the music. This explains a lot of the confusion in it.

RD And certainly against the politicians! You did, however, credit him with a streak of originality.

EC Oh yes, he was original in many ways, but also conservative in others. The whole question of primitivist composers is a very puzzling one. After all, Moussorgsky is a composer that Rimsky-Korsakov must have thought didn't know what he was doing. A lot of the time he thought he had to fix the scores up to present their ideas more effectively. Now we think Moussorgsky really did know what he was doing, and the same thing is true, in my opinion, of certain works of Janáček. Sometimes it's hard to know when there is real ineptitude or when there is the intentional desire to give the impression of not being "cultivated." There are an infinite number of ways . . . sometimes it can be very interesting; but sometimes it can be very disturbing when you get the impression the man didn't really know . . . Ives is a very original composer who sometimes seems to me to have not realized his intentions completely.

RD Well now, so much for Mr. Ives. Can we move now to your music?

EC I have, of course, been talking all along about my own music. Everything we have discussed is embodied in it in one way or another.

RD I know most of your early music, and I've conducted the *Pocohontas Suite* and the early songs . . . *Voyage* and *Warble for Lilac Time,* and of course, other works like *The Minotaur* and Symphony No. 1. What made for the compositional change after that, because that's certainly a different kind of music, isn't it?

EC I've been asked about this hundreds of times in my life.

RD I'm sure you have.

EC There's a long London Weekend TV about this made in London that was shown in March. A camera crew came here to this house and took pictures. And in the end, the producer, Alan Benson, said, "You made it perfectly clear that when the War was over, nothing really stopped and nothing had been changed; that a state of anxiety remained in the society. The end of the war was not the end of the situation." And he said that explained why the music changed so much. We had hoped for a new society and a new world emerging and suddenly realized that it wasn't going to happen.

RD Some composers of your generation speak of the war without saying that it influenced them; and here we have an example of you feeling that the war did change your thinking.

EC Not only the war, but the Depression before it caused a vast change in me and in the American musical scene. One of the results, as we said, was the ever-increasing erosion of commonly agreed on aesthetic standards—what used to be called artistic excellence, by musicians, and by critics who finally convinced the public of it. That is how all the works of our present repertory came to be included in it. This now has been taken over by 'public relations' with little regard for quality. Although sight and sound gags have always played a part in musical life, the authority and expert awareness of musicians has usually been able to convince the public of what came to be thought of as worthwhile. Now *PR* predominates and every effort is being made by it to turn new music into a fashion industry (if not so profitable, as that of painting or clothes.)—and as changeable, impermanent, and, ultimately, as unsatisfying.

As for me, I did after all start by being intensely interested in the advanced music of the '20s and that prior to the First World War. Then I came to another period that turned away from all that, and then I returned to my original interest. In the middle period I tried to be more communicative, I had a hope that the public would be different from the stuffy public that we're usually faced with in music. There was a whole series of philosophical ideas about music reaching a different kind of people and a different kind of audience and being more accessible. But suddenly there was the realization that things were not going to be like that; and it was important to go back and write music that was important to me.

RD Well, would you consider that the *romantic* gesture or the *heroic* gesture was "out"?

EC Romantic and heroic gestures continued to be indulged in by many throughout the 20th century, particularly in Nazi Germany, where Hitler set the style, and in the Soviet Union. Acts of terrible cruelty pricked the bubble of this nonsense as well as that of the insistent repetition of slogans reflected in the minimalism of Carl Orff and Boris Blacher. Yet all this seems to have a perennial appeal to the inexperienced public. Most of the rest of us felt, after all that trumped up melodrama, that we had enough.

RD You were in Aspen in 1973, 1974, and 1979. And 1973 was also the year that Nicky Nabokov was there and also Isang Yun, if you remember, and Gilbert Amy. And we did your Cello Sonata and *The Harmony of Morning,* the String Quartet No. 2—the Pulitzer Prize winner, and I remember conducting the Double Concerto.

EC That's right. That was a good performance.

RD In 1974 when you appeared again, the German composer, Aribert Reimann was there, and also George Crumb. Performances included your *Pastoral,* and again the two songs and the String Quartet No. 1; and the Fesitval Orchestra gave your Piano Concerto.

EC Gerry Schwartz conducted and Sam Lippman played the piano.

RD Then, in 1979, the composer who accompanied you there was Peter Schat from Holland. We did his opera, *Houdini;* and your music included the *Brass Quintet,* the *Recitative and Improvisation for Four Kettledrums,* the Eight Études and a Fantasy, and I remember conducting *Syringa.* Could you tell me a little bit about *Syringa* and how it came to be, and about how you treat words?

EC *Syringa* actually came about by a strange circumstance. John Ashbery, whom I didn't know personally, called me up and said he'd like to come over and talk to me. He said that a great deal of his poetry had been written while playing records of my music. He admired it, loved the pieces, they were a great inspiration to him. When someone says that, a composer is naturally very pleased. So he asked, "I'd like to collaborate with you on some piece. What do you think you'd like to write?" And I answered, "Send me some of your poetry and I'll think." So he gave me lots of poems and I finally chose *Syringa.* One of the problems for me with John's poetry is, that it is very understated, using rather simple, vernacular language, to suggest very complicated intense thought; yet its actual statement is simple and apparently direct, and rather cool. I began to try to find a way of dealing with the subdued poetic tone in my music, because I think of my work as being more extravagant and exciting; so I decided that I would make *Syringa* into a duet. Since the poem was about Orpheus, I would have the character of Orpheus be alive, so to speak, and singing passionately in the background during a cool presentation of John's poem. I thought of the Ashbery text as being sung very much the way the *Combatimento di Clorinda e Tancredi* of Monteverdi is. You have a singer constantly reciting and explaining what happens. Some very exciting things do happen, but the voice is somehow removed from all that, telling you how the people acted and how the two of them fought, and how they discovered one another as man and woman and so forth. I decided I would try to write something like that, that would have the story of Orpheus as told by Ashbery in the foreground as a kind of recitative; and then in the background a series of arias sung by Orpheus, a bass. Then in order to have Orpheus sing as in an opera, I decided to use the language he was familiar with,

Greek. In fact we have a couple of poems said to be sung by Orpheus that are in Plato's *Cratylus.* And that set me off to compiling a text made up of fragments of Greek poetry. That piece has its problems . . . it's difficult to balance, it's hard to find a mezzo-soprano who has a good middle register.

RD The last piece that I want to mention is *A Mirror on Which to Dwell* which I had the pleasure of conducting in its Rome premiere in 1984, with Dorothy Dorow singing.

EC That performance was excellent, and Dorothy marvelous. I don't know how you got those Italian musicians to play so well. They didn't used to do that.

RD That was a group called Gruppo Strumentale Musica d'Oggi. If you remember, it was a concert of the music of Petrassi and Carter that was co-sponsored by the American Academy and the Aspen Institute Italy, of which I was acting-director.

EC They played the Petrassi *Sestina* very beautifully.

RD The other piece of Petrassi was *Grand Septour.* There seemed to be an incompleted compositional outcome in my mind about it. He used sort of aleatoric fragments in interrupting his music . . . in both pieces. How do you feel about Petrassi's music?

EC I love his music.

RD And these two pieces?

EC Even these two. I like especially the *Sestina.* The other one, I've only heard the one time when you played it. One of the things about Petrassi's music in recent years is that he has tried to get a sense of an unpredictable spontaneity in his music; and sometimes it becomes very fragmentary, as though he wrote little bits of music which, when assembled contrast surprisingly with each other. I think *Sestina* is one of his best pieces, with his usual wonderful sense of sound.

RD And of course, also performed in Rome was your *Riconoscenza per Goffredo Petrassi. . . .*

EC Yes, with that wonderful violinist, Georg Mönch . . . an outstanding performer!

RD That little concert I thought was going to be so easy to put together, and it turned out to be an Italian "opera" in a way. First of all, it was scheduled right in the middle of Mardi Gras, so at the last minute we had to change that. Then we had some conflict about Mönch and a pianist that were going to play your Duo.

EC It's very funny in Italy. Things are always happening like that, and then in the end everything usually turns out very well.

RD Elliott, as a composer, you claim that you are not a serialist.

EC Oh, I claim that. There are some people who find serialism in my music, but I don't start out that way. Maybe I end up that way. . . .

RD And your manipulation of pitch . . . is this a private system?

EC I can't say that I use an organized system of composing, certainly not one as limiting and strict as the common practice of the past two centuries. I do focus on a particular area of musical vocabulary for each piece much as a painter restricts his choice of colors for a particular painting. It's not a system.

RD It's not a system . . . In fact, you are against system, aren't you?

EC As David Schiff's book and my *Essays* explain, I order pitches, rhythms, and tone-colors according to the needs of each piece to be written.

RD You spoke about the idea of freedom and control in your music; and you said that you achieve that freedom and control through polyrhythm quite often. Is there something else that you use to achieve that freedom; or, what do you mean by freedom in your music? Freedom for the performer?

EC For the performer, freedom in the presentation of a work of mine is like the freedom of an individual in a democratic society or a game. The player, if he chooses to do the work, has to enter into the responsibility of understanding what the work is all about. In order to do that he has to follow the notes quite accurately and at the same time interpret them, make them come alive for the audience. The score is not only a string of notes, of course, but a human experience expressed in musical sounds, groupings of notes, phrases, etc., which he must re-create like an actor playing a part in a play.

RD Would you just say a word or two about the relationship of a musical idea versus a compositional requirement . . . you speak of that.

EC Schoenberg said that musical material has its own requirements, meaning that once a muscial idea is stated it requires some kind of continuation or development or carrying on of one sort or another. He felt very strongly that the composer had to be aware of the responsiblity that he had in stating an idea as to how it should be carried out. In my opinion, any musical idea is partly an idea of how it will be carried out and partly how it will or will not be projected in time, let us say, depending on what the idea is. But certainly the musical idea itself, certainly in my case, is almost always a sequence of ideas and an idea of sequence, rather than one isolated event. The isolated moments come into focus while the piece is being written; but the original idea is how the music will flow. And then gradually, each little detail becomes more clarified as I write. But the compositional requirement, so to speak, is how this will all be combined together. And as I say, it's not necessarily through some sort of logical development or even carrying out motives; sometimes it's a clash between two or three or sometimes four opposing ideas which become a conception, which then requires of itself its own progression. It seems to me very important that music have a conception of progress or motion. I find a lot of music doesn't seem to be written that way anymore; and it bothers me. Older music had this concept pretty much built into it. One of the inventions of our time has been to find requirements that are not the standardized ones, and then to find ideas that need this non-standard progress.

RD You speak about it very succinctly when you, as you say, mustered your courage to ask Stravinsky how he composed. You related that wonderful tale of looking at his composition scrap-book. You saw in his book scraps of multi-colored paper which had ideas, motifs, and what-not on them, which got put into various arrangements. Then you mention a very interesting aspect of that process. You referred to it as cross-cutting. You then went on to trace this in Stravinsky's music. Would you speak about that a bit?

EC What was interesting about Stravinsky's music in particular is the fact that he really did cut ... he cut things short. Very often one would expect something to happen and it doesn't happen; or it's delayed until another place in the piece. And often there's a short stop that isn't a cadence or a dramatic pause. In the finale of his ballet *Orpheus,* as he said himself, there is a fugue cut in pieces. Orpheus' harp part comes in and interrupts the fugue, periodically. This is a very small example of something that happens in many, many Stravinsky works ... the cutting off. And in fact, it's even something that I remember in the arrangements Vittorio Rieti made of *La Sonnambula* for ballet. One of the things that he did was to cut off the last measure of an aria. Or the last two measures. As you hear it, you're left in the air, and the score goes on to something else. It became the standardized way of arranging old music during the Poulenc period, when they did funny things, left things hanging ... and it's quite pretty. I used to find it an interesting idea to do that.

RD Is it stretching the imagination to say that, while convention is in the air, along with the deisre to break the bar-line, this cross-cutting which you refer to in *L'Histoire du soldat* ... had some parallel in Cubism?

EC In movies!

RD How about Cubism?

EC Yes, in Cubism, too, of course. Cubism used a whole series of broken shapes derived from reality, as I see it (I'm not an authority). Its material often is like shadows of real objects under various sources of light, shadows of a guitar on a wall, for instance. These would be super-imposed, because two lights don't make a shadow in quite the same way. Cubists took fragments of this sort and put them together in a picture. The picture may have had a very realistic subject as its basis. But with shadows cut up in different ways, there is a whole new picture which is in a sense abstract; but it also isn't abstract, in that all the material is traceable to some realistic subject. Of course, Cubism is much more complicated than Stravinsky. Stravinsky, I think, really got this out of the movies—in old movies as I have pointed out, in Eisenstein's movies for instance, the cross-cutting of materials. Indeed, in order to show any scene in an ordinary movie, the maker has to cut a whole series of different shots from different angles so that you see something that gives the impression of a continuous action. Yet what you are seeing is a whole series of different shots from different angles. In the old days film makers were much more modernistic and cut the shots up in such a way, that there were very contrasting things happening very rapidly one after the other; and you were left to put these together in your mind.

RD Could we talk just a little bit about Varèse?

EC Oh yes! I knew Varèse for a very long time. I met him first when I was a high school student, in a speakeasy in Greenwich Village, where we students would go to get drunk! I followed his music from that time on. I heard the first performance of *Intégrales* and other early works. His music was always very compelling to me. I came to know him better, personally, during the time when his music was never played. To my utter surprise, he came to a reading of my old First Symphony by the New York Philharmonic, which did American composers the great favor to read over some of their music, although unwillingly. Varèse liked that score very much, which seemed to me unbelievable. In any case we became good friends after that, and we used to have Louise and Varèse to the house.

RD I met him once, at the New York Philharmonic French-American Festival. I did his *Hyperprism,* and he heard that.

EC Then, during the '40s, he was very disgusted because nobody played his music. He felt that the "Americanizing" composers were very against him and didn't want his music played, and he was discouraged. But he was continually writing a piece, *Espace;* and he was also conducting a chorus at the Greenwich Music House. Some of us got together at the New School and arranged a performance of *Étude pour espace.* I didn't think it came out very well. He had the chorus singing in different languages, which was a novel idea in its time.

RD Well. That was that dream about simultaneous radio broadcasts all over the world. I've had the privilege of doing all of his works that are accessible, including *Amériques* and *Arcana.* What about your relationship with Pierre Boulez, who seems to have been very outspoken about you as an American composer?

EC I met Pierre Boulez for the first time when I was in Baden Baden in 1955, the year when I was one of the vice-presidents of the ISCM. Pierre had his *Marteau sans maître* played, in an earlier version, and I had my Cello Sonata played, as I said; and we got to know each other. Then, when he came to New York with the Barrault company, we saw each other a couple of times. He has always been very friendly and interested in my music, and more so as time has gone on. While he was conductor of the New York Philharmonic, the Centre Pompidou was opened; and at the opening of that, IRCAM gave two concerts of my music. One was a small, chamber concert. On the other, they did *A Mirror on Which to Dwell,* the Double Concerto, and a whole series of things. Since that time, his ensemble has played every chamber work of mine, some of them a number of times. I've heard the Double Concerto four or five times over there. Then his ensemble commissioned my recent piece, *Penthode,* which it played on a tour of the United States.

RD What is your view of Boulez's music?

EC I must say that when in 1955 I heard the first performance of *Marteau sans maître,* I was very puzzled by it. I found it very difficult to understand and peculiar. I remember I was then with Petrassi, who at that time was the president of the ISCM. And he said this was something I would come to like later . . . and it's true. I now like especially the vocal settings. I'm also very fond of the big piece, *Pli selon pli.* I think that's wonderful. And also the new piece *Répons.*

RD How did you feel about the electronics in *Répons?*

EC Well, it's hard to say . . . we only heard one version of it. But the ending that we heard in New York with the electronic instruments playing, I thought very beautiful.

RD The concept of pluralism in today's music has been kicked around a lot. What do you feel about that?

EC If pluralism means that there are many different kinds of music being written today, it is hardly anything new. I think the composer can only do what he thinks is important for him to do. If he thinks it is important to make a career quickly and become famous, he should write junk and work at *PR!*

*Charles Abbott*

# *Aribert Reimann*

BORN MARCH 4, 1936, BERLIN

**December 5, 1986** **Berlin**

It's after four in the afternoon. I am greeted by Aribert telling me that a performance of mine will be on the radio tonight: his *Variationen* which I recorded in Berlin for the SFB in December 1976. This will be coupled with the excerpts from his opera *Lear,* sung by Fischer-Dieskau.

RD What is your state of "creative affairs"? Are you finishing a piece or are you in the middle of working on one?

AR I just finished a piece for mezzo-soprano and piano in October, after sonnets by Louïze Labé, from the 16th century, in old French. It's a long piece. I think it will take thirty-five minutes. I wrote it for a young mezzo-soprano and a young pianist.

RD Who are they?

AR Liat Himmelheber is the singer; and Axel Bauni, he's a young pianist from here in Berlin . . . I wrote this piece for him . . . and they will do the first performance next year in Hamburg and then in the music festival in Schleswig-Holstein. Now I'm writing a string trio commissioned by the Alban Berg Gesellschaft in Vienna. I should be finished with the piece by the end of this year, but in the last six weeks or so, I couldn't find the time. So I think it will be the middle of February when I finish.

RD The first piece is a song cycle?

AR: Yes, a song cycle. Louïze Labé was a famous poet in France in the 16th century, and the text is love sonnets—my first love songs!

RD That's a switch, because you have been attracted to the darker side of poetry.

AR I think now the lighter side will come into my life!

RD How lovely!

AR So the explanation will come after me! It was strange. I was finished with *Troades* last December, and then I was supposed to write a concerto for violin, cello, and orchestra. I started to write this piece and it was not possible. I was absolutely empty. I couldn't write another piece with orchestra. So I started to compose again in April, with this song cycle. Small cast. And then this string trio. I postponed the double concerto. When I finish the trio, I will write a piece for mezzo-soprano, piano, and orchestra. This is commissioned by the Festival in Berlin for Doris Soffel. She was the wonderful Cassandra in *Troades.* The piano part is for me, with the orchestra.

**RD** Voice, piano, and orchestra. That's unusual.

**AR** Yes, I love that! I think there is one piece by Mozart, the famous "Chio mi scordi di te." I played it at one time. . . . This comes through the connection with Doris Soffel. Sometimes we do recitals together, and I admire her very much. She is a wonderful mezzo-soprano. So from that came the thought to write a piece for her with orchestra.

**RD** What text will you use?

**AR** The text is by a German Romantic poet, Karoline von Günderrode. She was from the same time as Heinrich von Kleist. She became famous through a novel by Bettina von Arnim, called *Die Günderrode.* And in this book I found fifteen short poems, written not in rhyme but in prose. One called *Ein apocalyptisches Fragment* is a wonderful poem. Next year is the 750th anniversary of Berlin; and Bettina von Arnim lived in Berlin, and this *Günderrode* book was written in Berlin, so there's a kind of connection.

**RD** It's also Romantic, you say?

**AR** It's a Romantic poem, but the poem looks toward the future.

**RD** So you have two women poets. About ten years ago you wrote the Sylvia Plath songs. It's interesting to hear you be so confident about the pieces you are writing. I have been speaking to a number of composers; and I must say many of them are in a crisis. The rejection and the self-criticism that's going on in composers today is really quite severe. You are not in a crisis?

**AR** I was, in the first four months of this year. I thought I would never compose again. After *Troades* I knew it was not possible to write further in that way again. This was absolutely a final *Endpunkt* in my life. And then, something happened in my life and I started very timidly *(vorsichtig)* in April to compose again. I started three pieces, one a piano sonata, then the string trio, and then the song cycle. The cycle, the Labé cycle, took hold of me and so . . . *ja.*

**RD** What brought on your crisis? Was it your system of composing?

**AR** Yes. A little bit the system and also the kind of composing. . . . I think my pieces in the last years became darker and darker and darker. You know *Lear,* you know *Requiem.* Then after the *Requiem* it was the *Ghost Sonata,* and that's also very dark. *Troades* is absolutely the blackest piece I ever wrote in my life. I knew there was no further way for me. There were two possibilities. I stop. I make an end of my life . . . or, I try to . . . *ja,* I didn't know. Then life changed, and my music changed somewhat. And it was not a joke when I said I had the feeling that my music became a little bit lighter. There was a new beginning for me, an absolutely new beginning. I think this piece, I can show you later, this cycle may be a new start, a new beginning for my composing.

RD That's a wonderful thing to hear!

AR I am probing at the moment.... But I have now finished this cycle and I am very happy about it. It will be very important for me to make a new beginning with the string trio now, only the three instruments.

RD To reduce to that thought, yes.

AR I have the feeling that I have never composed before in my life.

RD That's very honest!

AR It's very strange, but that's the feeling I have! So next week I will start again with the string trio, and we'll see.

RD And the new way for you ... obviously it had to do with some psychological feelings. Has it to do with your manner of organizing? Is that new?

AR Organizing ... *ja,* a little bit, too.

RD Order?

AR Also. I think my language, or my style, if I have one, will be the same. I cannot now say I am another person, another composer. But I think the language will be more variable. I see new aspects and more life; more positive things, I hope. I think it's more psychological, more in my musical visions than in my organizing music or way of composing.

RD The influences of the past, just to retrace—you had admitted some sort of allegiance to the music of Webern in terms of its conciseness and maybe also to the expression of Berg ... did you acknowledge the twelve-tone or serial writing at all?

AR I have done it, yes. In my earlier years. My first opera, *Traumspiel,* has a kind of serialism or twelve-tone approach. But only part of it. I never could write very academically. It was against *me.* I tried it very often, but I stopped after *Traumspiel.* This will be performed now again after twenty-two years in Wiesbaden in February. So it will be very interesting for me to hear this piece again. In the *Ghost Sonata,* in the last picture (there are three pictures), I found a new kind of harmonic system.

If you divide up a fifth into two evenly symmetrical halves, you have—in the case of C-G—to flatten the E by a quarter of a tone. The resulting triad is between major and minor. I played around with this. I kept the sixth, the seventh and the ninth intact, and I reached new harmonic combinations. And I was able to *mirror* chords in a way not possible before: if you mirror-reflect a major chord, you get a minor chord. The mirror reflection of C-E-G is C-Ab-F. I constructed on this basis a complete harmonic system, and I could mirror-reflect the whole thing, and I turned the mirror around ("ich halte genau den Spiegel nach unten"), and there was no change. It always remained the same harmonic system. I discovered something very interesting. It's very difficult to perform, because the middle voice, the quarter tone voice, was written for the flageolet of the cello. And it was difficult for the cellist to play the quarter-tone flageolets sufficiently precise to leave the harmonic system intact.

But it's a very strange sound. Very cold. It's a sound like from another world. It was very connected to the texts of Strindberg. He speaks from another world, like the Jesuits' world. So I found this way to give expression to these ideas of Strindberg; I arrived at this harmonic idea.

RD Can you use this harmonic system in other works?

AR I can, yes. But I think it's not very valid to make a system on this, rather just use it to express something very much out of the ordinary. It was interesting for me, because I found a new way to a harmonic base, to a new kind of harmony.

RD That seems to be another theme in speaking to these composers. There is a search for their kind of *harmony,* and however they arrive at it, through whatever method, that word is being used a lot today. It is part of the crisis of composition.

AR This kind of harmonic system was very important for me in order to find a new way in the music—a new way *up* from serialism and a new way *around* the old harmonic system.

RD How do you mean—around?

AR A new harmonic system affects everything that lies outside of this system. You find a new harmony that has nothing to do with traditional harmony: it means that there is somewhere a central point, a harmonic central point, a harmonic focus for the music. This was very interesting for me.

RD Well, to make a discovery for yourself like that is very inspiring, I must say. You studied with Boris Blacher. Can you reflect upon this experience?

AR I began composing in my tenth year. I had no system, I must confess. Nor do I today. But I found a way to find myself. Blacher was the most important person for my development as a composer. When I heard his music for the first time I was fascinated by the mere fact that one can deal with tones; that one can say so much with very little, and that one can say everything with a few notes. And I wanted . . . I needed this . . . my music was very full, teeming with notes and unorganized . . . and so on.

RD As opposed to this economy.

AR Yes. It was important for me to reach a central point, a focus. And to understand the tension between two notes. That it is possible to construct and write a piece with very few notes. And to understand and lift into consciousness the element of tension between two notes was for me more important than anything else.

RD Can you speak of which pieces of his you heard?

AR Yes. One piece was called *Ornamente* for orchestra, and another *Orchesterfantasie,* and then his ballet *Otello.* I had the desire to study with him. And my studies began with exercises for one voice, for two voices, and then I had to write a piece for flute and violin. And for three months I wrote nothing but a first movement, a development section, and he always said: "That is horrible! That's absolutely music of the '20s, and nobody is interested in hearing that." Then I came to the point where it was actually impossible to write any further. I could compose no longer, and I decided to stop writing. It was absolutely impossible. So then I wrote poems.

RD Just words, you mean?

AR Just words. And then I went to Darmstadt, in that summer of 1956, for the first time; and I heard all this serial music. Some things were very important for me. I heard for the first time—it was twelve-tone but not serial—the *Lulu Suite* for orchestra. I heard it there for the first time in my life, and I was very impressed. So impressed that I could not get up. Other people were sitting with me in the box, and I could not get up. Other people went out, and I was still sitting. I think it was the most important impression I ever had in my life. And then I thought, I must compose again. But how? I saw I had no way. I came back to Berlin and I asked myself the question, do I continue to compose or not? I went to Blacher and brought him the second movement of this sonata I tried to write. He said, "Oh, this is much better." And then he said to me, "I had to bring you to a point of decision—a point in time at which you realize clearly—at which you know for sure—is it important or not for you to compose? Is it an inner necessity for you to write, or could you go on living without composing? And looking at this movement I know that you *have* to compose, and now we can start to work." It was a very hard time for me, because he taught me everything. He enabled me to find my own language.

I remember bringing him a piece for violin and piano. In the second theme of the first movement there were four bars. And I had the feeling when I wrote those bars, that's me! I don't know why, but I had the feeling—that's my music, and I cannot compose in any other way but this way. But I did not say this to him. I had the feeling that if I mentioned this, he would say, "That's kitsch!" or something like that. So, I showed him these four bars, and he looked at them while I said nothing and then he said, "These four bars will be your style." This was his genius in teaching. To be able to think like your students.

RD It takes a very strong personality to do that.

AR From that moment, I knew what I had to do.

RD And you tested it so honestly in not saying a word.

AR And if I wanted to go to him it was because I knew that only this man could put me before the decision "to compose or not to compose"—my way of composing is absolutely not his, but it was his way of thinking which led me to him . . . he not merely showed, he *gave* me my way.

**RD** Yes, I can understand that.

**AR** He brought me up.

**RD** That's a very clear explanation.

**AR** I never published this violin sonata because . . . this piece is still a mixture of many things—this experience with Blacher and my experience with Darmstadt. So I have this piece at home. But it was played . . . it was a piece to find myself, that's all. I think it was the most important year of my life, thirty years ago in 1956–1957.

**RD** That's a very touching and very real kind of story. In the course of composing from that time on, you have done so much for the voice. I think you said once that your mother had some kind of influence on you in writing for the voice. . . .

**AR** I think there were two things . . . of course my mother, and the students of my mother, because I started to accompany the students when I was eleven or twelve. And I sang too, as a child. When I was ten years old I sang in the *Jasager* of Kurt Weill, here in the Hebbel Theater, on stage. And at that time I started to compose lieder. Then I accompanied a lot. I need sometimes to compose for voice to find myself. Every voice has something special. It's very important for me when I write a piece for voice or an opera to know who is singing the first performance. So it was Fischer-Dieskau in *Lear,* it was Helga Dernesch in *Troades,* and it was Katherine Gayer in *Melusine* and the Plath songs. Each voice gives me a new color and inspiration to work with.

**RD** Don't forget Joan Carroll in *Inane* which I did . . .

**AR** Which you did in Amsterdam with the Concertgebouw Orchestra in 1975. Of course, she was something special.

**RD** Very, very unusual.

**AR** I would never have written *Inane* without Joan Carroll.

**RD** Speak a little about *Inane* because that was a unique piece.

**AR** *Inane* was, I think, the second piece I wrote for Joan. The first piece I wrote for her was *Verrà la morte,* and it was for soprano, tenor, baritone, chorus and orchestra. Joan was singing the soprano part; but the interesting thing in this part was it ends the piece with the "blues"; and I had never written any "blues." In my childhood, when I was fourteen or fifteen, I wrote some Schlager. But when I heard Joan, I thought, "Ah, that's a special kind of voice." And she gave me the "blues." She was an extraordinary person. She would sing the most complicated voice parts in contemporary pieces, and she loved to sing difficult pieces. She said, "It's boring for me to sing Mozart or Schubert. I must sing pieces that are very, very difficult and complicated." And on the other side, she could sing jazz, because she sang in New York nightclubs, like many American singers.

RD And she sang Lulu . . .

AR She was a famous Lulu! She sang Lulu in four languages. She was Lulu in Sante Fe, she was Lulu in Florence, in Munich, and I saw her in Düsseldorf. I met her in Stuttgart in the audience, when Anja Silja was doing Lulu. Suddenly a person appeared and she had on a little hat. And on top was hanging a feather. She came to me and I said, "I am Aribert Reimann." And she said, "Oh, you are Die Aribert!" She was wonderful! Then I wrote *Inane* . . .

RD Well, the importance of that piece lies also in the expression of a woman's problem, I think. There aren't so many pieces that concern themselves with abortion.

AR Yes, but in that piece it was not like it is today. A woman wants to have a child, and her lover forces her into an abortion, because he doesn't want to pay for the child. He does not want to recognize the child. And she is compelled to have the abortion and because of that she becomes insane. This situation is not the common one—usually it is the woman who wants to get rid of the child because her circumstances do not permit her to keep it—or else she gives birth to the child, while society applauds . . . In *Inane* we have this very special situation. She wanted the child and was not allowed to have it. The child was murdered. She was forced into this abortion . . . that is the story of this piece. It is a soliloquy of twenty-five minutes. You did this piece in Aspen in 1974 with Susan Davenny Wyner, and she was very good.

RD That performance was the American premiere. She doesn't sing anymore. She had an accident on a bicycle.

AR Oh my God. She was a very gifted singer, and it was a marvelous performance, I remember. I liked it very much. And it was interesting for me, because she was the first singer after Joan who had sung this piece. Who was it in New York?

RD That was Nadine Herman at the Lincoln Center Celebration of Contemporary Music in 1976, which was the New York premiere.

AR That was also interesting. Anyway, all these experiences with these different types of singers and voices are very important for composers when they are writing for the voice.

RD You had worked with Dieskau of course before *Lear*. . . .

AR Yes. I met him for the first time in 1958. He was looking for an accompanist to coach songs and opera and so on; and my mother knew his teacher in the Hochschule. Our connection started then, and now we are very good friends and we have done a lot of concerts and recordings, and I've written a lot of pieces for him.

RD Can you say how his voice impressed you?

**AR** It wasn't merely the beauty of his voice. It was the unbelievable spiritual aura of this voice. This voice is a singular phenomenon in our century. For me he is the greatest singer of our time.

**RD** I would agree with you.

**AR** And he can sing everything. He can do so many things with his voice, he can express and articulate whatever he wants. His voice can encompass a whole vision, a complete spiritual world. During the years his voice has changed a great deal. He always discovered new possibilities. The first piece I wrote for him was the *Five Songs* from Paul Celan, for baritone and piano. It was in 1959. He was a lyric baritone. Then I wrote the *Totentanz* for baritone and chamber orchestra, it was 1960. Then his voice changed. I never thought at that time to write *Lear* for him. In 1968 he asked me to write *Lear,* and it was absolutely impossible for me, because I admired this work but I thought, it's not to be set to music! We talked about *Lear* when you and I were together in Aspen in 1974. I told you about my plan for *Lear* and everybody said, "My God!" But then Fischer-Dieskau insisted and insisted and insisted. So in 1971 I composed a piece for him, *Zyklus,* also from Paul Celan, for baritone and orchestra. When I heard these songs with orchestra, I said to Dieskau, "I think I can do it. They are something like 'Lear' sounds." I had the feeling it had something to do with *Lear.* Then five years later I started. The *Variationen* you conducted here in Berlin are also on the way to *Lear.*

**RD** Why? How? How does a set of variations lead to a dramatic score?

**AR** I think these *Variationen* are dramatic too. The psychological situation of these variations, you know. They are not variations on a musical theme. They are variations on a circumlocutal *Zustand* (situation). And this situation has something to do with the person of Lear. It is the situation of an abandoned human being, of a man condemned to face total solitude, outdoors, he simply is somewhere out in the wilderness, and suddenly he experiences horrible things, and he is totally and helplessly exposed to everything that now rushes on towards him. This is the situation of Lear. Perhaps this thought came to me while I was writing these *Variations,* because I thought so much about Lear at this time. And all that entered my subconsciousness. Because of that these *Variations* anticipate the language of the orchestra, the harshness, the brutality. All that I wrote and produced in those days had already a great deal to do with Lear. With all the cruelties, all the brutalities of our era. While I was writing these *Variations* I suddenly had the feeling that I was in midst of these horrors which happen around us all the time.

**RD** So that is a kernel, a seed, of the concept of a *Lear* personality?

**AR** Yes. And I felt that I wrote *Lear* not because I wanted to compose Shakespeare, but because *Lear* is a piece for our present time. Many situations arising today are like in *Lear,* aren't they? I wrote this piece during the period of the terrorist crimes. Horrible things happen today, and we are all victims. . . . *Troades* goes beyond even that. This also is a piece for our time. Our time has a great deal to do with *Troades.* With the Trojan women.

RD How—in our time?

AR It's a piece against the war. Euripides wrote this piece against the war, as a warning. The women have lived through the war, they are brutally carried off as loot, and now they are waiting. They are to be distributed as slaves. This is the situation. And I also wrote this piece as a warning, in my own way—we cannot but act in the only ways given to us. Against that which goes on, ominously, in the background. And everybody knows; if there is to be another war, it will be the last; this war will mean the end of all of us. And that's why I wrote these pieces—as a warning. Everybody can speak only in his own way; I can speak only as a composer. Once in a lifetime one reaches the crisis point when one has to do it: therefore *Troades.* This war must not happen. And we all have to think positively that it must not happen. Everybody knows that it must not happen because it would mean war for the last time. And if we all think in this way it will not happen.

RD The theater is a very strong element in your need to compose.

AR Yes. I've done five operas and three ballets. My next opera will be in five years! I don't know what I'll do, but I will write an opera again. For me I can bring out me and my life the best way in opera.

Opera is necessary for me if I want to express myself to the utmost. And this I can do best if I have a human voice to deal with. The singing person on the stage speaks out for me as if I were to speak out for myself. Through the voice I speak out from the deep "within" in me—which doesn't mean that "in between" I don't write other pieces of music. . . . I like to write pieces without text and pieces for orchestra and instrumental pieces. But an essential part of my life, and what I have to say about it, goes into opera. What I think about our time I could say only through *Lear* or *Troades.* And perhaps my next opera will deal with things which concern myself personally—I have certain ideas—but I cannot talk about them at this point. I can do it on the stage. I believe that one can express oneself nowhere as completely as on the stage, through the voice and the orchestra. Wagner could do it only in using the stage, through *Tristan* and the *Ring.* This is something that goes beyond all limits, that bursts through fences and walls and transcends the merely musical element. There are areas which one cannot touch with other means; one has to resort to the stage. Every composer has his own feelings about the stage, his own kind of affinity. I personally feel the stage to be something very powerful. I hope, though, that I do not belong to those composers who write only operas. To write other pieces is very important for me too.

**RD** Do you have any last words?

**AR** I think to find new kinds of Melos, new variations of Melos is very important. Music without Melos is for me not music. I can express myself only in Melos that is, in melody, and then by means of the counterpoint belonging to this melody. Music always has to be shaped by emotion. And it always has to be expressive. Man is the expressive animal; as long as he lives he expresses himself, and this manifests itself in the tension between two notes—but this tension is not something accidental; it comes from the "within," from the soul. The reason for the tension between two notes is emotion, and emotion again is Melos. Without that no music has reality for me. Anything else is accidental tone-conglomeration, which has nothing to do with music. Music is a *situation (Zustand),* and also a situation of tension. And what drives me on to write music is the inner need always to recreate the situation of tension. And for this reason I am glad that I overcame my crisis and that I have now the feeling: I can compose again. . . . It is a kind of emotion which urges me on to write music. Music has to live to be able to express something.

# *Aaron Copland**

BORN NOVEMBER 14, 1900, BROOKLYN

**July 22, 1975** **Aspen, Colorado**

Excerpts from a panel discussion: The Arts in America, July 22, 1975, Aspen, Colorado
(Other panelists included: Earle Brown, David Del Tredici, and Kurt Oppens.)

Aaron Copland was Composer-in-Residence at the Conference on Contemporary Music, Aspen Music Festival 1975. We were celebrating his seventy-fifth birthday. This was in conjunction with the Aspen Institute for Humanistic Studies.

RD Having introduced everyone, I'd like to begin with Mr. Copland. Can you tell us what life was like in Brooklyn around the beginning of the 20th century?

AC Well, Richard, you're really beginning at the beginning!

RD Well, what better place?

AC It wasn't like Paris, I can tell you that! I don't think that if I had had a choice I would have picked Brooklyn as a place to be born, given the fact that I was going to become a composer. But I would say there must be worse places than Brooklyn. We had a museum and we had the Academy of the Arts where the Boston Symphony was playing. It wasn't an absolute desert, but I don't think anybody ever connected serious music with the street on which I was born. So I had the feeling, when I first became interested in Chopin Études and so on, that I was really discovering an unknown country. I couldn't talk to my pals about it, because they simply weren't interested.

RD Why did you go to Paris?

AC I'm not done with Brooklyn yet. You see, the generation I studied with in New York City belonged to a long tradition of German music. I studied with a very good musician, Mr. Reuben Goldmark, whose uncle was a famous composer of operas in Hungary. He was a man who really knew his stuff. But music to him was really German music . . . *serious*

*Material quoted from a panel discussion, "The Arts in America," July 22, 1975; Aspen, Colorado; with the permission of Aaron Copland.

*Charles Abbott*

music. And to the younger generation at the beginning of the twenties, the newer kind of music seemed to be coming from Paris. Germany had its great composers—Beethoven, Bach, Brahms, and so forth; but our older generation of American composers, MacDowell and such, had been influenced to such a degree by the German masters that we wanted to find a country where the very newest thing was happening. At the beginning of the twenties, that definitely seemed to be France. After all, Stravinsky was living there. We heard about Bartók, who was active in Hungary, and the Russian composers in Russia itself, Prokofiev and later Shostakovich. So, naturally, being a young man, and wanting to get close to the latest thing in music, I headed for Paris.

RD Can you tell us how you found Paris and what the atmosphere was like when you arrived?

AC That was Paris of the twenties, an historic moment in the history of Paris which has had many historic moments. But the twenties left a real impression, partly because of the fact that the First World War from 1914 to 1918 had stopped all artistic activities, I suppose. So when that was over, there was a certain re-burgeoning of all the arts, and that included music. The latest thing in music seemed to my generation to be the music of Debussy and Ravel. We knew about Schoenberg, who was working in Vienna at the time, as sort of a name, but not very much more than as a name. His music really wasn't played very much in New York, although some of his pieces had been played, especially *Verklaerte Nacht.* And so the fact that the latest things in the twenties seemed to be coming from Paris, in painting, literature, and music, naturally influenced me to want to go there.

RD Did Nadia Boulanger influence you or your music?

AC Nadia Boulanger was a very imposing personality. To be around her would normally influence you, yes. But I would like to correct the general impression that I've gathered from looking at notes on myself in various books. Most of them say that I went to Paris to study with Nadia Boulanger. That is incorrect. I had never heard of Nadia Boulanger when I went to Paris. I knew nothing about her. Fortunately, I learned about her through a student who had been taking harmony lessons with her, and persuaded me (with some difficulty because I had already studied harmony and wasn't interested in harmony lessons) to just go and see the way she did it. And I went and was very impressed. But I must add that I was somewhat wary of the idea of studying with a woman teacher. For some extraordinary reason, which no one seems able to fathom, the history of music to that point showed no women Beethovens and Bachs. That fact alone, you see, made me hesitate. It took some doing on my part to come around to the idea that I would write home to my parents: "I'm studying composition with a lady in Paris."

**RD** Mr. Copland, do you consider your music to be as American as most people say it is?

**AC** Well, I like to think they are right. I think one can exaggerate the importance of the "Americanness" of one's music. You see, I belong to a generation in which some of us were trying to establish the idea that serious music could be written in such a manner as to suggest the fact that it could only have been written in this country, in our century. That seemed an important thing to do, in my mind, especially when I was in my twenties and thirties. Once that had been established, it seemed less important. I know that it certainly doesn't seem at all important to younger composers of today. They aren't even thinking about being American. Possibly because we had established the fact that it could be done, it no longer seems interesting to do. And other things happened in music which also took away from the nationalistic tendencies of the composers of the '20s. A composer like Roger Sessions, for instance, who is my colleague and friend, had no interest at all in trying to write a music that was obviously American. But it happened to be a preoccupation of mine at the time.

**RD** Do you still see something of America around you?

**AC** What do you mean?

**RD** I mean, if you were looking as you looked at that time, in terms of a nationalism, if you were going to look again around you, do you find parallels?

**AC** No, I'm not that concerned. If I were asked to put some film with a typical American subject matter to music, naturally I'd try to write a music that suggested what you'd see on the screen; but if I were going to write a piano sonata, I suppose I would want a foreigner to know it was written by an American composer, but I wouldn't need to use American themes, jazz elements or whatever. By now I hope that I could write the piece and it would show the Americanness without my making any special effort about it.

**RD** I can remember the painter, the late Barnett Newman, talking about a question he was asked, "Mr. Newman, why do you paint?" His reply was, "Because I want to have something to look at." Mr. Copland, why do you compose?

**AC** It's a question I never asked myself. It's something that happens to you. And the question of why you do it comes later. First it happens, and then you begin to wonder, "Why am I doing this?" Composition is not a decision that you coldly arrive at and say to yourself, "Oh, gee, I'd like to write something for a large symphony to play and everybody's going to love it and they'll all applaud and I'll see my name in the papers." It's all very nice, but it doesn't really work that way. I suppose you compose because you hope to put down in terms of musical notes and sounds and colors the most profound theorems about being alive that you find inside yourself. I guess that's about as broad as one can make it. I suppose that every creative artist hopes that through his art, he will get down in some permanent form the very essential and most basic feelings about life that he or she had, in a form that is going to last and will tell future generations what it seemed like to be alive in our time. And the reason you hope to do that is you know that is what the great masters of the past have accomplished. So that it's not just a matter of expressing your personal feelings; the plan is much larger than that. I think that's basically the reason why one wants to compose. I know it was with me.

RD Could we go back a little bit in time and hear about your affiliation with Roger Sessions and the group that you represented in the '20s and '30s . . . that group and its relation or non-relation with the group headed by Edgard Varèse? What was the polemic at that time between the two of you?

AC I remember coming back from the three years that I spent in Paris and wanting to know what other Americans on the scene were also trying to write the new music of the '20s. I had met some of them during my student years in Paris; but there appeared on the scene in New York very colorful figures like Roy Harris, a fellow from the West who had lived on ranches out in California, who thought of himself as an odd type because he wanted to write symphonies instead of ride horseback. Roger Sessions had also been born in Brooklyn. I don't quite remember when we first met, but obviously he was a very serious and dedicated man. Walter Piston appeared on the scene up in Boston. Chicago was rather active at that time. At any rate, there were enough people around that the idea occurred to us to get together and give some concerts of American music. There were also, however, two other organizations. One was the League of Composers, an active group who introduced the new music from Europe. The new Bartók, Stravinsky, and Schoenberg scores were heard, many of them for the first time, at their concerts. The other group was headed by Edgard Varèse, who had a very powerful personality, and striking looks. He looked like a genius—an advantage sometimes—and he was definitely a leader of men and women and was in general a very strong character. You were either his pal and did his bidding, or you were out. So there was room enough in a city the size of New York for numerous groups. Those I would say were the three principal ones who occupied themselves with the playing of works written at that time by local composers or composers from all over America. And it made for a lively scene from the '20s to the '30s.

RD We played your Clarinet Concerto last week, and that was written for Benny Goodman. Is there any connection to the fact that Woody Herman got Stravinsky to write *Ebony Concerto* for him?

AC Well, perhaps that was in Benny Goodman's mind.

RD How did you come to write the concerto for Benny Goodman?

AC Why not? I had written some music based on jazz elements, and here was a big shot in the field who wanted to play in the concert hall, and he chose me to write a piece for him that would be comfortable for him and not too "far out" for the Carnegie Hall audience, and so I wrote the Clarinet Concerto.

RD Then, about the world of dance . . . what kindled your interest about the world of dance? I mean, you have your marvelous ballets, *Appalachian Spring, Billy the Kid, Rodeo,* and *Dance Panels.* Can you tell us a bit about that?

AC It's very nice to be asked to write music because somebody really needs it. Most dancers really need music, and there's a limit to how many beautiful Chopin melodies you can put to the dance. Finally, if you want to create your own world in the dance, you have to have a music that will reflect that world, especially if dancers like Martha Graham or Agnes de Mille decide that they want to dance American subject matter. This was rather unusual, because in the '20s and '30s dance was the Russian ballet, of course, and always had its origin in some foreign country. We had our popular dances that all the world danced, but no one had translated that into serious terms until Martha and Agnes and others of a similar stripe came along. So that when they in turn asked me to write something because they had an idea for a ballet which was American in subject matter, naturally I felt very attracted to the idea of providing the music. It was as simple as that.

RD How did Hollywood enter your life?

AC Oh, it entered at a very nice moment. Things were pretty bad in the '30s. I think if the truth were told, I guess I pushed my way in. I had a friend in Hollywood who worked at the studios. He said, "If you really want to write for the films, you have to come out here and meet some of the people who are in power who make these decisions." And so I went out, but my first visit produced no result at all. Then I stayed home and waited until I was called out. To be asked to take subject matter and pictures and a whole artisitic creation (we hope, if you apply yourself only to those pictures that you can respect) is a very interesting and challenging project. It's very interesting to be asked to add to that particular manifestation of the artistic feeling of gifted people, and to either help or worsen the result by the kind of music you write. And naturally the composer has the rather unique advantage of seeing these various scenes without the music, and then hearing the music put to the film. And in taking it away and putting it back several times during the editing process, you get a very good idea of the power of music to make things more humane and more alive and somehow more living than they are without the addition of this sound that builds along in certain scenes. That became a fascinating activity of mine. I only did five or six films in all, and each of them was rather different in subject matter. Two of them had to do with life on a ranch; but each one of them interested me, and I enjoyed the experience of being out there.

RD The awards that you have received are numerous, but there is one in particular that is curious. How did you feel about receiving the 1964 Presidential Medal of Freedom, "The highest civil honor conferred by the President of the United States for service in peace time?"

AC Well, I felt very good. It's not usual for Washington to pay attention to the serious composers of this country. We're not spoiled with attention, at any rate. The honor came absolutely out of the blue, through a telegram, and I went down to Washington to receive it from the hands of President Johnson.

RD You're written several works that have certain political overtones. I could mention a few like *Fanfare for the Common Man, Preamble for a Solemn Occasion,* and the *Lincoln Portrait.* About the last, what was your approach to dramatizing Lincoln and his words? Was it out of patriotism? How was the text chosen?

AC The work was written on commission from André Kostelanetz, who, during the Second World War, wanted to stimulate composers to write patriotically inspired music. I thought of doing a portrait of some great American. I first thought of doing a portrait of Walt Whitman; and why I ended up with Lincoln, I no longer remember.

RD How much did you consider the words of the text? I'm just curious because I've just been through the American Statement Seminar of Mortimer Adler, and we've been going over texts word by word.

AC The text of my piece?

RD No, the texts of various Lincoln statements . . . particularly the Gettysburg Address.

AC I remember reading a very interesting biography of Lincoln by an Englishman whose name I've forgotten. It was an interesting history, particularly because it was written by a foreigner, who wouldn't have known from his own experience what it was like to be an American in the era of Lincoln. In that book I found a number of quotations of Lincoln's speeches and letters that attracted me, and I simply excerpted them and combined them with a couple of sentences from the Gettysburg Address that followed easily. It wasn't a hard piece to write, and the first narrator who spoke the lines was Carl Sandburg. That was very lucky, because everyone connects Sandburg with his great biography of Lincoln. I've also conducted it in foreign countries with the text translated into Japanese and other languages. I'm delighted to say it seems to have universal appeal. There's something about Lincoln's personality that everyone can feel empathy with.

RD The last question, before we open to questions from the floor . . . just in general, Mr. Copland, what do you feel about the attitude of the government as it was in the '20s and '30s with the WPA program, and how do you feel about government subsidy for the arts today?

AC Well, I think America has a long way to go in relation to subsidy for the arts if we take Europe for a model. In European countries the governments are very aware of the artistic life. I don't know if the artists think they get sufficient support, but there is much more of a tradition of government support for the arts than here in the US. I think that during the WPA days, we artists began to think that we were being left out. If everybody else needs help, the artist needs help also. It gets tough for the artist just as it does for the people of the country in general. I went through a period when the idea of the government being involved in helping the arts to live and the artists to work became an accepted idea in our own country, and I'm all for it. Obviously, some money is going to be wasted; but on the whole, I would say there's more good to be gained from the government's helping than from the government keeping out of it.

***From the Audience:***

What is really done to encourage and to help a composer to write and also to get performed? This would be a most interesting subject, especially for someone coming from Germany.

AC Well, I think one might begin by saying that most Americans are brought up to think that the government really shouldn't get mixed up with the arts, and that somehow or other the arts ought to find a way to support themselves, either through private subsidies and well-to-do people, or heaven knows how; but the idea that the government is somehow responsible for helping the artist to create his art work has no long background of history in America.

RD Even in the New Deal, nobody assured an artist a job. If you go down the list of Roosevelt's recommendations for jobs, you'll see that nobody talks about artists. I can only quote here Harry Hopkins, the ex-social worker and WPA overlord. He said, "Hell, they've got to eat just like us!" There is an unnaturalness about our legislature looking at the arts, and I think people in general find the arts rather queer. And the issue about whether the government would control our aesthetic is another thing that bothers a lot of people.

***From the Audience:***

I was in Washington, head of two commissions, one in the humanities and one in the arts in 1962, and proposed that there be a foundation for the arts and a foundation for the humanities and was called up on Capitol Hill and told by several senators and legislators, congressmen, that that was a very un-American idea!

RD I have an article here that quotes Gerald Ford as still preferring football to art. It says here, when he was a minority leader in the House, he confessed that although he had opposed the National Endowment for the Arts and Humanities at its conception, he had since been converted by an Alexander Calder sculpture, which had really helped to "generate" his hometown of Grand Rapids, Michigan.

AC I ought to get a little credit for that. I wrote a fanfare to help establish that . . .

RD We have a very eminent person in the audience, a member of the Supreme Court, Justice Blackmun. Do you have a question or comment, sir?

JB I'll presume to ask this of Mr. Copland if you'll permit me, and I presume only because I think he and I are of approximately the same vintage. When you speak of the '20s, I of course can remember them vividly, as so many here tonight cannot. But the thing that has always impressed me about Aaron Copland is the perpetual, consistent, it seems to me, youthful approach to music and what he's trying to interpret. And if I may, I would like to ask . . . assume that today in 1975 in this rather mixed-up world that we have, that you were twenty again. And perhaps in Brooklyn. Where would you go; and would you view the next fifty years, which takes us into the '20s of the 21st century, with as much enthusiasm and excitement as obviously the past fifty years have meant to you?

AC It's very touching to be singled out as a representative figure of a certain period. . . .

JB I merely wanted company, Mr. Copland.

AC I would think it's almost hopeless to try and imagine what the world would be like 100 years later from when one was twenty. There certainly are going to be some things that are going to be the same; but on the other hand, there are so many intangibles. Certainly the situation in music will have been so differet from what we went through. The idea of an American composer existing and writing works that would interest the world at large was a very original thought in the 1920s. And I would hope that 100 years from that time it would be a perfectly natural thought. Everybody would expect America to produce such world-important figures in music. I'd be very disappointed if that turned out to be untrue.

***From the Audience:***

I'd like to address a question in general about the relationship between the current diversity of styles in music and how the composer achieves communication with the audience.

AC I think that you'll find that every composer would answer that problem in an individual way. Some of us are much more concerned about communicating than other composers. Some composers feel that their job is to write the very best music they know how to write and then it must take its chances. If it's good enough then everybody will love it. Others are more aware of what the audience capacity is, and might take that even subconsciously into account rather than think about it consciously. A little depends on the individual temperament. If you love your fellow man, perhaps you're more concerned that what you do would appeal to a fairly wide audience. If you're absolutely buried in the technical and artistic world that you live in, and can't be concerned about who likes it or doesn't like it, that's another point of view. You take your chances. I think it largely depends on temperament.

***From the Audience:***

Mr. Copland, I can hardly think of any well-known, successful artist, at least of your generation, who hasn't been, during their education, at least once for an extended term in Europe. Today, would you advise the young composer still to go to Europe for his education?

AC Well, I think the kind of dependence on Europe that was apparent in the early 1900s doesn't exist anymore. But that certainly doesn't imply that we can't learn things from other countries or that we shouldn't learn and be influenced by things from another country. I think of the creative artist as a person of an open mind, open to any possibility that might cross his path. It seems to me it would enlarge anybody's horizons to go to Europe and see how the other half of the world lives, and find out what they think is great. It's the job of a cultivated person to know what goes on in the world.

RD Well, I have just one last thing that I think is proper to say ... *Happy Birthday, Aaron Copland!*

# Narrative X

My experience with and curiosity about the music of living composers broadened significantly in the 1970s. Frequent invitations each year to conduct European orchestras such as the Berlin Philharmonic, the London Symphony, the Concertgebouw Orchestra, and many radio orchestras, not only increased my standard repertoire, but added to an ever growing list of premieres and performances of new music from the "other side" of the Atlantic.

While at home, two appointments appeared almost as enigmatically as they ultimately vanished. One was the artistic directorship of the Aspen Music Festival's Conference on Contemporary Music, which began in 1970 and ended in the disastrous final year of 1985. Calamity reigned in that last, 16th summer, an occasion that should have marked a new beginning due to the decision of the Fromm Foundation to use Aspen as a summer home. I had enthusiastically endorsed the co-existence of my sixteen-year-old program with that of the Fromm Foundation; and I personally spoke with Paul Fromm and his new adviser, composer Earle Brown, with great excitement and anticipation.

However, mismanagement by the Aspen administration, which was in a state of general confusion and economic crisis, and the persistent indifference of the Music Director, produced an atmosphere of near-chaos. Some performances had to be canceled because rehearsal schedules were disorderly, personnel assignments were incomplete, and the buck was passed in a dizzying circle when responsibility for these conditions was queried.

The potential for chaos was always "around the corner" in Aspen. But in the last few years of my stay it became more manifest. The quality of musicians assigned became less than first rank; and those willing young students who were assigned were treated like recruits in a military boot camp. I had inherited this post from Darius Milhaud and, despite everything, managed to continue the tradition of inviting Composers-in-Residence on an international scale each year. Those composers included: Brown, Goehr,

Crumb, Wuorinen, Xenakis, Druckman, Rochberg, Sessions, Amy, Carter, Nabokov, Yun, Reimann, Copland, Del Tredici, Davies, Knussen, Wernick, Penderecki, Tippett, Bolcom, Schat, Schuman, Glass, Rorem, Lutoslawski, Subotnick, Berio, and Rands.

The irony of the 1985 summer was that my program was terminated and the Fromm project was retained. Ostensibly the reasons were financial. Although Gordon Hardy, dean and president, generally supported me through most of my tenure, he was not in a position to influence what he described to me as "a social board."

The other position which came my way was that of Director of the 20th Century Music Series at the Juilliard School in 1972. This lasted until 1979 when, one day, Peter Mennin, president, announced: "There will be no contract for you next year, because *I* need your program." Absolutely bewildered after seven years of work, I left Lincoln Center that day feeling like a victim in that murder-mystery family game "Clue": "Professor Plum did it in the Conservatory with a lead pipe."

In addition to the music of some thirty American composers, we had also presented new works from thirteen other countries; and all in all, we performed forty-four works that had not been heard in New York before! The young professional Juilliard musicians were praised by the press for their outstanding ability to perform and comprehend this extremely varied and complex presentation of contemporary works. Many went on to champion this repertoire; and in many cases, many incorporated the idea of new music into their concerts.

During the course of the 20th Century Music Series, Harold Schonberg of the *New York Times* wrote, "One of the nicest things about the evening was the expertise of the young Juilliard players." And on another occasion he said, "These Juilliard kids have grown up in the idiom, and there was never a false move or a feeling of insecurity . . . The playing would have done honor to any organization anywhere." Leighton Kerner of the *Village Voice* wrote, "The Juilliard series of contemporary music seems to have been more rewarding, both to the participating students and audiences, than any other public activities of the school." In 1984, Andrew Porter wrote in the *New Yorker,* "The demise five years ago of the Juilliard's Twentieth Century Music Concerts left a hole in New York's musical life . . . Those Juilliard concerts were very well played: looking over my reviews of them, I find name after name of performers who have gone on to fame."

Mennin's remark—"I need your program" will forever remain a mystery in my mind.

# *Friedrich Cerha*

BORN FEBRUARY 17, 1926, VIENNA

**November 26, 1986** **Vienna**

November 23, 1986—5:15 p.m. on the night train to Vienna from Amsterdam—raining—cross German border after Arnhem. German guards on board with German shepherd dogs. After devouring two pairs of bratwurst—to bed. 7:00 a.m. breakfast; 8:00 a.m. Vienna. Saw some friends, did some business and on November 26 spoke to Cerha. The next afternoon I boarded the Orient Express for Paris to speak with Xenakis and Amy.

RD I understand that you are working on a new opera.

FC I am finishing my new opera—in despair, almost, because of the deadline. I hope I'll be done in time. . . . This opera is based on Zuckmayer's last play, *Der Rattenfänger.*

RD The Pied Piper . . . What made this play so interesting to you?

FC The way it deals with acute, present-day problems, such as the phenomenon of power, and the relationship of the individual to society. Can one adapt to society? What happens if one despises this society and its traditions—can one simply go away? Can one so simply abandon one's sphere of living, one's culture? Does this possibility, tempting as it seems, exist at all? Don't we have to live with other people? Has running away ever solved the problems of co-existence with one's fellow men?

RD The Pied Piper is the main character in the opera. Is he a metaphor for something bigger?

FC He does not stand one-dimensionally for Hitler, as some have thought. Besides dealing with power in many ways, the play—this is particularly important—also throws some light on the *origins* of power. It shows up those features in people which enable them to become leaders or seducers, and it also shows to what extent projections of others enable them to play these roles and thus exert real power. Although we have some ideas about the genesis and the functioning of authoritarian structures, we have not yet sufficiently analyzed the mentality of the authoritarian victim: the regressive longing for a paradisical, childlike existence; for a parental figure which satisfies all desires and solves all problems. And these longings frequently go together with an obsessive craving to submit to authority. All this we have seen during the Nazi

*Universal Edition*

period—the consequences were devastating, things happened in broad daylight and for all to see, and yet we didn't understand that it all was generated by the unholy marriage of childhood longings to the need for submission. Insecurity is at the root of it all, and insecurity causes people to trust blindly in leaders and ideologies.

RD So the implication then is on a social level, on the economic level and our larger sense of security.

FC Yes, and on a personal level. That's very important.

RD Would you speak about this?

FC That is not easy—one would have to disentangle a labyrinth, and that is well-nigh impossible. Well—I'll try to illuminate a few points or aspects, running the danger of sounding simplistic. Not only the Pied Piper, the rats, too, are a metaphor. Life in the city is corrupted by social injustice. Utilizing to the full all their privileges and advantages, the rich and powerful undermine the existence of all the others. Deep down in the ground the rats are doing their work. Destruction fattens the rats as well as the rich and powerful "upstairs." And people sense subliminally what goes on underneath: they get restless and anxious. The Pied Piper comes as a stranger. A permanent fugitive, he has learned to be on the alert. He is different; people notice him, and they notice in particular how securely he moves in a world of insecurity. He mesmerizes the weak and irritates the powerful. Here is someone with whom one cannot bargain, who cannot be bought; he doesn't curry favor with authority, he does not care about the reprisals of the law, and he does not conform to the established *mores.* The alliance of (false) security to business does not function. To use the Pied Piper and then drop him—that is the only way out. Someone powerful merely on the strength of his nature and personality: this is too much of a danger for the sham edifice which upholds the authorities. One has to get rid of this stranger! His desire to join the "community" is not even considered. Power feeds the ego and compensates for failure in basic human relationships. An example for this is the *Stadtregent* (burgomaster), who has to resort to institutionalized means, to routine procedures of govenance once he senses danger. And the "little" executioner, who claims "to have done nothing but his duty," exemplifies the insecure individual, who feels secure as long as he can obey commands.

RD This reminds me of the arguments used by your Bundespräsident when he tried to explain his attitude during the Second World War.

FC Yes ... the "Lamento of the Little Executioner" is dedicated to him. But, in contrast to Waldheim, the little executioner changes his opportunistic conduct at this point. He acts as his conscience tells him although he risks death. Dubious, shady laws, order and commands: he no longer wants to be part of this. But the piece was conceived long before the Waldheim affair. I don't like to bring in actuality in such an obvious way. I am interested in basic human attitudes and their labyrinthine consequences.

**RD** Tell me something about the women in your piece.

**FC** What goes on between the Pied Piper and the female protagonists—relationships, interactions—illuminates many aspects of power. Power, indeed, infests the psychic background of the man-woman relationships. And at this point the tragic disparity of the ideal and the real also comes into play. The proud, beautiful, and also frustrated wife of the Stadtregent, Divana, has always tried to compensate for lack of love with possession. Her weak husband is unable to satisfy her. Confronted with the Pied Piper, she is not able at first to distinguish between loving and possessing. It is in contest and battle with him that she realizes that he is the master; that she wants to be overpowered by him; that she wants to give herself to him. Like her daughter she is deluded by a childish ideal which never has been replaced by a truly lived and experienced relationship. But now a genuine confrontation has taken place; now, despairingly, she senses what she has not had and what she will never attain; and thus she looks for surrender and death by means of an intoxicated dance, which her sick heart cannot sustain. For her children, too, intoxication takes on the dimension of a true solution. Maris, Divana, and Rikke, the second female protagonist, move in a field rich in nuance. Its span encompasses everything, from total idea fixation to living, creative human relationships. Rikke, the daughter of the executioner, grew up as a pariah like the Pied Piper, excluded from the ways and norms of the city. She also has her fantasies and her delusions of grandeur, but she is not completely imprisoned by them. She senses and understands quickly what her lover can bring about; to what extent he has been provoked by the perfidy of the city people and that now he is eager to turn everything around and on its head. She urges him on towards—what one might call—a political goal. She knows the poor of the lower city and she instigates rebellion. The poor, looking for salvation, follow blindly—and that is what the children want to do also. They are driven by disgust with the managed and rotten world of their parents. They don't know to what extent they themselves have been shaped by this world; all they know is that they want to be different. These children, in particular, exemplify the "wrong way out": they long to embrace a chimerical idea, and they dream that all problems will be solved, all desires fulfilled. Another "wrong way out" is the flight into intoxication, into drugs. Still another alternative: the guru, the ideal father or Big Brother; of whom one can dream, although he has no hold on reality. He is seen as guarantor and sponsor of a better world. We have seen what happens when a Führer exploits this kind of mentality. The Pied Piper doesn't do that; that is not the point the author wants to stress. The children look for a new beginning, but they are looking for it in Utopia.

The children's exodus, as I saw it before me in my mind, was for me the starting point of my composition.

But not ideas only; history, too, inspired me. In 1982 I traveled in Mexico, in Yucatán. There I was, looking fascinatedly at the remnants of old Indian cultures; highly developed cultures, which were not eliminated by wars or epidemics; rather, cities simply left behind, probably because people were no longer able to stand the oppression of the dominant theocracy. They didn't want to live as one was forced to live there—they left. Human beings who simply abandon their sphere of life, their society, everything familiar and everything abhorrent; this idea fascinates me. I saw, with my inner eye, this exodus from an old, sick culture; all of these people possessed by utopian ideas of a better world somewhere else—it seemed to be almost like a sacred rite . . .

RD We don't know where they went . . .

FC There were so many old cultures, with one taking over where the other had left; something of the old was always preserved in the new culture . . . Egypt, Greece, Rome, Babylon, Assyria . . . War, Conquest, Assimilation . . . In Yucatán this curious cut-off took place. And the Pied Piper, in the end, goes away and takes the children along.

RD You spoke of Baal, the main character of your first opera, as not finding a place, a place in the world . . .

FC It's interesting that you mention that now. Baal and the Pied Piper are indeed connected, but only in an elementary way. Both confront an alien world, both rebel, both fight the "mores" to which they are supposed to submit. Baal fights, so to say, on two fronts: he fights for his (and any individual's) right to long for happiness; but he also is averse to any natural adaptation to human relationships. In the end he perishes. He is Brecht's provokingly exaggerated image of the life energy per se, menaced more than ever by our managed world. The Pied Piper is in a different position: those who want to leave are the children, but the Pied Piper does not promise them anything. He destroys all their illusions; he makes it clear to them that with the new problems they still have to solve the old ones. His decision to take them along as they are and to lead them into a hard, uncertain future—this decision, I think, is meaningful to all of us. All of our young people here have to face a most difficult problem: they look at their parents, who during the Hilter era were under the spell of ideological-utopian gibberish, unrelated to the world in which they lived. They begin to realize now that much evil in the world is caused by the disparities of Illusion, Ideal, Utopia, and Reality.

The Pied Piper leaves, and only two human beings remain in the burned-out city: a wise old man and a cripple. These three—Pied Piper, wise man and cripple—have decided consciously to face the world realistically, to put up with all the difficulties of hopelessly entangled "realities," and to get the better of them. That means: any new beginning, inasmuch as it is possible at all, no matter where it takes place, is a burden which can be taken up only by those not dazed and befuddled by Illusion—rather, these people honestly have to face life as it is and they have to do step by step what they have found out to be vitally important.

**RD** The word *Netzwerk*—our world as a network of relationships—evidently plays a larger role in the *Rattenfänger*. *Netzwerk* is also the name of one of your earlier works for the stage. You speak of the concept of *Welttheater*. We have your *Spiegel*, *Netzwerk*, and *Baal*. You speak of *Spiegel* as the observation of the mass and not the individual. *Netzwerk* is observation from above, where society seems insect-like . . .

**FC** May I interrupt: *Spiegel* looks at the evolutionary processes of life-events from a faraway distance. *Netzwerk* changes the perspectives. This "look from above" exists there, too, but you find in this piece also those small intermedia—I called them "Regresses"—which suddenly illuminate typical behavior models as if they were under a microscope. This is a kind of interaction between distant and close view—between perspectives—and this is mirrored in the music. To say it better: heterogeneous elements invade basic structures and cause disturbances, which affect the environment, to be in the end integrated in the environment. A disturbance which is too far-reaching has to be balanced out by the complete system.

**RD** In *Baal* you speak about the individual, even in a provocative, concentrated way, as you mentioned. Now comes the *Piper*. Can you talk about these relationships?

**FC** You know the term *Welttheater?* It figures prominently in theater history and refers to stage plays which deal with basic questions of human existence, of our world altogether and its meaning. My pieces contain *Welttheater* aspects which, by the way, have changed—so I believe—in a significant manner. It seems to me that—now as in the future—works of the stage have to touch and mirror the deepest layers of our existence. Without this aim one cannot be truly creative. Opera as simple spectacle—beautiful music with beautiful voices—something merely culinary—I have difficulties thinking in such terms. This, of course, cannot prevent audiences from thinking otherwise: they might not be aware of our deeper goals and enjoy our works in their own light-hearted manner.

**RD** And the other, more abstract aspect: the thinking in systems—a topic to which you always return—isn't "system-thinking" deeply connected with your work, particulary with *Netzwerk?*

FC Yes, in *Netzwerk* these ideas appear indeed in a particularly pronounced manner. But they played their role already when I composed *Spiegel,* and they again became more prominent in the *Pied Piper* compared with *Baal.* When I composed *Exercises,* the piece that gave rise to *Netzwerk,* I was involved with observations of organic growth. For instance: plants exist which are related on the basis of organismal structure. But different environments cause them to differ in appearance! In visual terms, at least, the kinship has disappeared. And the opposite also holds true: a unified environment creates similarities and resemblances among plants of a differing structural build. Thus, for instance, cactus plants can look like succulents and vice versa. In this way I learned to conceive of the *visible* as reflecting processes of development and growth.

*The Pied Piper* is characterized by a multitude of procedures, all of them defined by clearly determined basic contents. These procedures, in the end, form an interlocked, dovetailing, multi-netted system, which determines the characters and their interactions. Each "procedure" stands for a complex of powerful human drives, which cause events to happen. One might speak of *Leittechnik* procedures. Thus, for instance, there are polymetric fields determined by discontent, unrest, and rebellion. These "fields" can be organized in very different ways, and the results would also be different, while the audience, all the time remains aware of the basic kinship. But this is only one aspect of all that's interwoven, interlocked in this network-texture. I think, indeed, that this kind of "system-thinking" is more evident, more easily sensed in *The Pied Piper* than in *Baal.*

RD This way of thinking is probably most evident in *Netzwerk*. . . .

FC Yes—because the musical-technical events correspond so clearly with the happenings on the stage. At the time when I composed *Netzwerk,* people generally were not ready—as they are by now—to conceive of the world as a many-netted, entangled system, or as a multiplicity of such systems.

RD You observe the world and the problems and situations on two levels: one, the so-called reality, and two, the musical. How are the connections between the reality-situations translated musically?

FC I am not translating anything. I am interested in forms of organization, such as they crystallize in communal life, or even within our bodies, our brains. We are better informed than at any time about the genesis of such organic systems, of their possibilities and limitations. Think of the connections between the person of the artist, his environment and his work. Contemporary science helps us to understand this "network." The creative process takes place in an arena in which the artist (a uniquely profiled individuality), his materials, and all the "givens" from without fight it out among themselves. To be sure, there are connections between my extra-musical interests and the way in which I organize my music. And as I journeyed from *Spiegel* to *Baal* these connections naturally have changed.

RD In terms of theater, in *Spiegel,* you speak about the idea of conjuring fear, relaxation, brutality, desperate effort, emphatic impetus, shock, and resignation—these emotional controls or considerations. And you say you are frustrated that this cannot be realized on stage . . . using the stage, the theater you know. What is missing? Because *Speigel* has not been done to your satisfaction yet. How would you describe a successful realization?

FC I don't believe that it is impossible to realize my ideas. We are lacking conditioning and experience. *Spiegel* is a theater of moving forms. The music is tied to motions of objects, light and, very important, human bodies. These motions produce mass events *(Massenereignisse)* which suggest certain static situations or changes, influence, contact, repulsion—basic forms of human behavior. I would want to avoid anything too obviously symbolic, too billboard-like. I have proposed a scenic solution, but I realized that for *Spiegel* many detailed solutions may be considered. The audience should always be able to recognize the very different basic concerns of the seven scenes and the formal connections within the organism of the whole piece. Aesthetically and dramatically the stage would have to suggest formal discipline; an attitude corresponding to the attitude of the music.

RD No narrative. The theater of Robert Wilson comes to my mind, with very slow-moving gestures that have no narrative. Is this similar to your idea?

FC That seems possible; I have been told about Wilson, but I haven't seen any of his works on the stage. Regarding body motions: I did attend some Buto-Theater events, which come close to my ideas of a theater of motion.

RD You say that this concept of "theater" is not collage, not music drama, not opera, not ballet . . . what is it?

FC This exactly is the difficulty. All the so-called performing arts are sorted out and relegated to cubicles, which then are entrusted to specialists. If it's Dance, then there are ballets and choreographers; and Musical Drama is at home in the opera theater with the stage directors, designers, conductors, and singers. And the Fine Arts also are tied to their routine ways of presentation—but whatever it is, it has to fit into a drawer. Anything not fit to be put into a drawer creates difficulties. Not fitting into a drawer for a work of art practically means: having no chance. *Spiegel* and *Netzwerk* don't belong anywhere.

RD Performances of *Baal* or *The Pied Piper* are less problematical?

FC Naturally. It's always possible, of course, that something is not understood, viewed in a one-sided manner or misinterpreted; but truly good pieces cannot be killed in that way. Modernistic clichés, of course, can be dangerous; they are being applied to everything, to Schiller's *Jungfrau von Orleans* no less than to Ionesco, and thus they have lost all meaning with regard to new pieces. It is irksome, to say the least, when pieces as different as *Baal* and *Le Grand Macabre* by Ligeti turn into look-alikes on the stage.

RD Now, your orchestral work, *Fasce*. How does *Fasce* fit in the history of Cerha? I conducted the world premiere of *Fasce* in 1975 in Graz and Vienna on a program with Nono's *Per Bastiana Tai-yang Cheng* and Ives's *Holidays Symphony*. Both Nono and Ives were receiving their Austrian premieres. Two weeks ago I was speaking with Ligeti and he told me the story of seeing your sketches of *Fasce* in 1959, and telling you that you were writing *his* piece! You completed *Fasce* in 1974. Can you speak about *Fasce*, about this Ligeti observation, and what are the parallels, etc?

FC The years 1959 and 1960 were of the greatest importance for my development as a composer. This was a period of feverish planning, conquering, and rejecting. Imagining new worlds of sound was immensely exciting for me. At an even earlier phase I had been dreaming of creating musical *situations* (as opposed to drama, dynamics, progression)—and Ligeti told me that he had had similar ideas. Think of the music of the great tradition: time is divided up into segments, and the time experience is enhanced by the thematic processes, which permit and provoke experiences of music remembered and music anticipated. I had other ideas—this goes back to the years around 1950—I was aiming for a kind of music which would strictly limit, if not eliminate the experience of time. Early attempts in this direction are my *Sonnengesang des Heiligen Franz von Assisi* and the composition of a segment taken from the Chinese I-Ching. In *Sonnengesang,* for instance, I tried to minimalize time experience by means of fixed chords of equal duration, avoiding at the same time recognizable musical periods . . . Later I found this technique used in a posthumous piano composition by Satie, the *Danses gothiques.* I used other methods in the I-Ching, but my intentions in both pieces were only incompletely realized, because I had not found, at that time, access to a thoroughly suitable, de-formulized material. I didn't know anything of Ligeti, he didn't know me; our situation, as it was then, could be compared with Schoenberg's situation at the time of his *Opus 16.* Schoenberg wrote at the time that his ideas of sound-motion in space would open paths for future composers to follow. It seems that Ligeti and I were such future composers: my *Mouvements, Fasce,* and *Spiegel* are cases in point. Interestingly, Schoenberg's ideas (his *Imaginationen*) have something in common with the thought of Varèse. To be sure, Varèse started out with ideas more closely connected to the world of the present; but during the 1920s he, too, had difficulties finding a really adequate musical language—music that would satisfy him. As we know, he was silent for decades. . . .

RD How would you describe the decisive impulse furthering your and Ligeti's development after this period?

FC I learned about the workings of serial techniques, and I met members of the young international avant-garde. . . . These techniques did away with all the formulas of tradition; they inaugurated a weaning process. Ligeti became conversant with them later; we both independently were fascinated by the new ideas. We met as late as 1958 in Vienna. The music which we wrote after the Darmstadt experience shows clearly that we were—each in his own way—critically aware of the dangers inherent to the serialism of those days. The same holds true for Boulez. I am thinking in particular of the danger of uniformity—the uniformity resulting from the pointillistic splintering of the musical material. This method deprives the listener of all points of orientation. If I had decided to turn into a puristic, strictly serial composer, I might have been able to realize my wish for the elimination of the old musical time experience. I reacted differently. In the end, "liberation" seemed less important to me than form and expression. Perhaps I called one of my early orchestra pieces *Espressioni fondamentali* for this very reason.

RD But the musical ideas which you mentioned—has your style changed?

FC Certainly . . . but I haven't completely changed. Of course, your thinking might get a "push" here and there, doors might open . . . but there are bridges: some point toward the future, others toward the past. In truth I adopted serial methods mainly to realize my own ideas which differed from those harbored by the serial purists. At that time this wasn't understood. In the *Relazioni fragili* for harpsichord and chamber ensemble, a work which I performed with my wife in the "Reihe" concerts, I kept the building cells smaller, which made for more variability and better mediation between the different types of texture. I produced "blocks" with static effects. I still enjoy the oscillation of static and structural elements.

In the works of 1956–57 you'll find a new element next to the pointilistic structures: a blurring of rhythmic, melodic, and harmonic contours by means of one-dimensional glissandi—particularly in the third movement of *Relazioni,* which strongly impressed Ligeti. Ignoring the trends of the period and without knowing Varèse, I succeeded in creating "moving sound curvatures" *(Klangkurvenbewegungen),* which later would play an important role in *Fasce* and *Spiegel.* Also the interaction of static with dynamic elements—characteristic of all my work, I believe—was continued. For instance, in the *Intersecazioni* of 1959 I amassed musical "blocks" moving in their own way individually, each one endowed with its own timing. I did more work using this kind of counterpoint later on, for instance in *Mouvements* II (which excited

Ligeti very much) and in *Spiegel* IV. Ligeti's work with "multi-timed" procedures differs from mine, but in order to "eternalize" our basic relationship he later dedicated the third movement of his Chamber Concerto to me. At that time interpretations of this kind of music proved to be immeasurably difficult. *Fasce* as well as *Intersecazioni* I finished years later; performances seemed out of the question. This piece *(Intersecazioni)* was performed for the first time in 1973 by Michael Gielen, one year before *Fasce,* which also dates back to 1959.

RD And what was the last essential step which led into the world of *Fasce* and *Spiegel*? In which way did it occur?

FC This happened spontaneously. The *Trois Mouvements* are three studies of movements, relatively simple, which I wrote within a few days. In each piece I limited myself to one single characteristic sound-situation. Motion here does not indicate a basic change of "situation"; it happens, very gradually, *within* the situation characteristic of the piece. But the three combined form a process: the extremes designating this process are on one hand a sound event which is dense, colorful and intense in various, shifting ways and consists only of short, high notes; on the other, a setting which renounces all short note-values and is based on a single, deep-pitched cluster. The slow emergence of areas and colors, their appearance and disappearance—that is the single event of the piece . . . I always was fascinated by the motion of subtle nuances of color, by slow, hardly noticeable coloristic mutations. At first I didn't dare to utilize the time necessary to make these changes noticeable; in fact, I didn't even have the technique required for such effects. Berg talked about his love of the "art of small transition." I share this love with him, but I realize it in a different way.

RD Compared with traditional music, Friedrich, the pieces you mentioned look very different in score; in outer appearance, your music looks different. . . .

FC Yes, a new way of notation had to be found, I found it for *Mouvements,* which also in *this* respect was important for my development. Think of my ideas for *Mouvements* II, where attacks of the wind instruments are tied to slow, continuous motion changes in the string instruments—a kind of glissando of the strings—such ideas cannot be expressed in the traditional notation. To get hold of these ideas with the help of graphic signs, signs which in their ways reproduce what the music tries to express—this was an unbelievable liberation for me; a spell was broken; floodgates were opened. I never composed with such excitement as during these years—not before, not after—and *Fasce* was the first major orchestra piece, the first large concept in this new world of sounds.

RD How do you see your special situation as expressed in these pieces—I am also thinking of comparison with other composers, who created "sound color compositions"?

FC Before we can compare I have to talk about myself. This became clear to me after *Mouvements.* I would have to accept far-reaching limitations if I were to renounce all the beautiful complexities in the music from Bach to Schoenberg. I would be confined to material of great purity, material unrelated to tradition. Reduced to this limitation, my statements also would be limited and would suffer. To forestall total impoverishment and to find a substitute for the lack of communication, I had to resort to thinking in systems. Even in earlier years I had become interested in "system-thinking"; I had read Norbert Wiener. Now, for the first time, I consciously looked at a piece of music as if it were such a system, a system containing elements which influence, impede, disturb, eliminate: elements which a superimposed system could catch and balance. This kind of thinking was a challenge to my imagination. Many things became possible: continuity, development, simultaneity of processes which differed in terms of duration and effectiveness with regard to the whole piece. The result was a new kind of complexity, a new wealth of content. The title *Fasce* indicates that the single voice, the single event is always tied to a group—consisting usually of sounds close to one another in terms of pitch and also in terms of color. These bundles or bunches relate to one another in various ways and find their place in large, spatial musical events of a different dimension. These ideas I pursued in various ways in the seven parts of *Spiegel.* But during the ten years I needed to complete them—in fact, already in 1962 in *Exercises*—changes in my outlook began to occur. I began to leave this puristic world behind, to invent (or accept) heterogeneous material and to integrate it, into a musical organism. *Exercises* then became the basis for *Netzwerk*—we talked about it before. But no matter how emphatically I invented new music, no matter how persistent I was in working it out, I was sufficiently critical to know that I had reached an extreme point with *Fasce* and *Spiegel.* Perhaps this was the reason that I was so conscientious of my work, so careful and scrupulous. I knew that it had to do with a singular point in time; with ideas which had to be handled with the greatest care because this would no longer be possible at a later date; because something was to be fulfilled in them which could never be repeated.

RD How do you see your and Ligeti's position in this situation?

FC We agree in realizing that at that time certain ideas were in the air, and that these ideas matured at the same time; thus, we (and others too!) pursued similar ideas independently. Whether someone did something similar a few months later or earlier—this doesn't matter at all. What matters are the results—and, as one can see today, they differ in terms of detail as well as in terms of principle. We are different people with different experiences, with different backgrounds—as human beings, as artists . . . Interestingly, it is mainly the static aspect in *Fasce* and *Spiegel* which Ligeti noticed—this reflected his main concerns at the time. There are such parts in *Spiegel,* for instance, III or V; but at the same time I thought "dynamically," in terms of processes and evolutions in contact with other musical phenomena. Even in 1959 I looked at the overall form of *Fasce* as a dynamic-organic happening. What, you may ask, is the main difference between my music and the music of other composers of the period—the music of Ligeti or Penderecki or Scelsi (of whom no one knew anything)— or Lutoslawski? It is my concentration on spatially vast, complex formal developments and relationships. I am not in agreement with Ligeti with regard to the importance of the static elements in my music as it was written in that period. Perhaps he was at that time more interested in cellular organization, in subtle, highly differentiated processes of working out minute details—not so much in complicated, large musical forms. And he could not have taken an interest in the dynamic organization of the complete time-stretch of a piece. . . . Yet there are resemblances and connections.

RD Friedrich, we talked about different experiences, different backgrounds. Could you describe or trace your musical genetic code?

FC I did it already, to a certain extent. It is not so easy to pin down . . . Many influences come into play as long as one is young and in the process of finding oneself. Works of art created simultaneously may mean totally different things in terms of art history . . . Here in Austria one was cut off from all development; developments which took place during the war, and also before the war, because of the Nazi concept of *entartete Kunst* (degenerated art)—one was cut off from what happened in one's own country, cut off from the roots. After the war it was particularly Neo-Classicism, the music of Hindemith and Stravinsky which became known. Stravinsky affected me much more than Hindemith; but Hindemith, of course, also had to be studied. The Viennese School—the music, the people from Schoenberg's circle who were still here—could be approached only in a round-about manner. I owe a great deal to Josef Polnauer, who analyzed Schoenberg's music with me and had many good suggestions to offer when I was rehearsing with my ensemble Die Reihe. The way to Webern I had to find on my own. During the 1950s I analyzed and thought about Webern's music a great deal, and this confrontation did much to form me and to make possible everything that was to follow. And it is true to say that later, after the direct influences had disappeared, I always fell back on my experiences with Webern's music—in fact, whenever important changes (or turns) in my development took place. One doesn't know too much about this. . . .

RD Thanks to you the third act of Berg's *Lulu* can now be performed. How do you see your relationship to Berg, the composer?

FC . . . Much has been said and written about the relation of my music to Berg's music since the world premiere of *Lulu* in 1979 and the performance of my opera *Baal* . . . I think this relationship is overrated. At the time when I wrote *Exercises* I was indeed fascinated by Berg: by his ability to build a complex musical organism with the help of very different components. I think that there are also connections between Mahler and Berg, although their languages and their ways of thinking and acting vary greatly. Nobody was aware of my working at *Lulu* when *Exercises* was premiered in 1968. What I did in *Exercises* and *Netzwerk*—my attempts at integrating different elements, having them interact in all possible ways, etc., might have been partially inspired by Berg's procedures. But my initial position was different, and so was my material.

RD And *Baal?*

FC *Baal* reflects the permanence of one of my main interests: the development of an abundance of speech elements (the ways and means of any kind of statement) and their integration in a complex musical organism. This first Brecht drama was influenced by Wedekind stylistically as well as thematically, and it seemed natural to me to point to this connection—not by quoting music, but by quoting a certain musical situation. Such "situation-quotations" in *Baal* refer not only to Berg, but, for instance, also to Schoenberg's *Erwartung*—which, again, no one noticed because of the way in which I am "labeled" at present. . . . There are also connections to Weill. . . . Much music appearing in *Baal* has nothing to do with Berg, but refers back to *Fasce* and *Spiegel*. . . . I believe that I found my own varied and also organic world of speech, and this speech-world I have developed further in *Rattenfänger*. . . . One simplifies things if one ascribes to Mahler or Berg all music which has a continuous flow and which is also expressive; many people do this nowadays. Time changes perspectives: today one sees Berg as a Romantic; a few decades ago one couldn't do that. Temporal distance furthers the ability to distinguish. I have time, I can wait. . . .

RD Can you specify other experiences which you think were essential and which have occupied you?

FC When I was involved with the music of Schoenberg I experienced a kind of rebellion against everything academic, frozen, formulaic. That was around 1950. It was at that time that I detached myself most emphatically from the "development-thinking" of the Viennese School. There was a whole group of young musicians here who formed a kind of counter-movement. Some liked the aesthetics and the provocative tone of *Les Six,* others, strangely enough, were influenced by Carl Orff, but even more by J. M. Hauer, who for a short time was my teacher also. You know of Hauer's twelve-tone games. Neither Orff nor Hauer helped us on. We were lured by "statism," musical "situations," which at that time I tried to master for the first time. The group around the famous "art-club" developed a kind of minimalist thinking—radical limitation, concentration on circumscribed, sparse basic conditions, and the testing and the playing with variations within this framework. This trend against Academia and Dynamism could also be found in some other countries such as Hungary: countries which, like Austria, had been cut off from the international scene. Our minimalism differed from what later was called minimalism in the United States, and it had no international echo. Nothing could come from Vienna: we were too isolated. This was a special development of some historic interest. And it helped me in developing my interest in "statism", and it helped me also to get ready to work creatively with limited possibilities—influences which can still be noted in *Fasce.*

Another influence to which Vienna in particular was exposed came from the East. There was, for instance, Josef Marx, a leading personality during the postwar years—later he plotted and schemed against everything new and also against the Viennese School—but he told me very early about the music of Scriabin and Stravinsky. He introduced me also to other Russian and Eastern music, and it seems to me now that at that time Hauer and Scriabin were of great importance for my musical way of thinking. Clearly serialism was also an important factor in my development. Besides, I should mention two composers—I studied them and I was fascinated by them: Varèse and Satie. Musically they have very little to do with one another. They were, as well, unknown in Vienna, and I made it my goal to perform them regularly. Varèse was interesting for me particularly because of his formal concepts. He is one of those who limit themselves to the use of a few elements only—a fact not generally noticed. Satie impressed me with his cool, factual manner and with his commitment to "drawing," that is, to the pencil rather than to the brush. I appreciated him so much because I had worked with large masses of sound before. Different as they are, Satie and Webern do touch in certain areas. To come back to Webern: his musical vocabulary was one thing, but besides I appreciated immensely his general artistic attitudes, his artistic integrity and his almost fanatic concentration, his fastidious care in dealing with his musical material. In these respects I still consider him to be my model, and he still is an actuality for me.

RD Well, we followed your whole development as a composer. Let us have a last look at "here and now." In one of your texts on music you speak about the fact that innovation today is found in the demand of a high degree of spiritual and emotional differentiation, and not in orthodox musical systems.

FC You are right—neither one-dimensional orthodox systems nor the invention of totally new ways of shifting musical material around—nothing of that would be innovative. Others have said this also; this is no longer controversial.

**RD** Yes, well, can you speak about the so-called higher spiritual and emotional differentiations and innovations?

**FC** The changes that occurred during the last decades were not as easily recognizable as the innovations of the 1950s and 1960s. At that time many of us concentrated puristically on certain ideas and possibilities and we attained extreme positions. But a creative artist reacts when he finds himself in a dead alley. He cannot repeat himself all the time, nor does he desire it: not in principle and not within a tightly circumscribed framework. He has to differentiate, and all our creative musicians chose that path. Ligeti, Boulez—each of them in his own manner. Once one ceases staring at one possibility, a second way opens up: one sees one's own position as one within a plurality of possibilities. Differentiations (which, incidentally, put many composers in touch again with aspects of musical tradition) are not the only changes which took place. It happened now that well-known and controlled principles of organization were seen in a wider context, and that they were tested with respect to their possible function within such contexts. This happens gradually, and if it is not generally noticed it is because the outer manifestations of this development differ greatly among the composers. As said before: it seems that we have, so to say, run out of new constellations of material; the possible contribution of computer technique I wouldn't want to judge. It happens now that our thirty-year-olds take up ideas of the 1950s: they create at best more or less differentiated remakes of positions which have become historic. They are "new" only for those who don't know well the art of our century. . . . It is, of course, still important to invent new possibilities, but these, it seems to me, are to be found on an altogether new plateau. . . . Differentiation—ingenuity in the differentiated mastery of complex problems—that, to me, is innovation. Regarding your question: I think that I was able to clarify a few points, but perhaps I still was not quite clear enough. Consider: monomania drives you on in one fixed direction; you can submit to this drive or else it gets the better of you, it crushes you like a steam engine—but there is no choice. The multiplicity of phenomena attract and tempt you, but it also has a nerve-wracking effect. The art of our century reacted in so many ways to our terrifyingly kaleidoscopic world.

One talked about the "protective primeval ground of atavistic forms"—this referred to the attraction of primitive art as manifested in the works of Arp and Brancusi . . . These powerful forms are not primitive. But it is as if they had led us to the primeval sources, to the core of all form and forming. These artists found their way back to the *Grundphänomene* earlier than we musicians—with the exception, perhaps, of some sections of *Mouvements* and *Spiegel.* Think also of Mondrian: the basic "order" as manifested in geometry, the whole trend towards "geometrizing," which also has its psychological aspects.

Trends towards making self-limitation of this kind creative, in terms of concentration, borderline-consciousness, sensitizing—such trends exist today. They do not lead into new territory, but a good composition contains effective moments of differentiation. Besides these are movements which express the multifariousness, the grotesque discords, and paradoxes, all that's unassimilable in our complex world. Such movements are surrealism, collage techniques, which work with elements of rupture and with alienation. This, too, continues today. It isn't new. In the way the world looks, one can understand that artists react in that manner, but the methods have become so widely known that they have lost their provocative sting. New works of this kind do not stimulate intense spiritual partisanship or a great deal of feeling. Why do I insist on lively reaction? Because so much in our world is likely to roll over us like steamrollers, and people who have been victimized in this manner or people who don't mind being manipulated are not what we need in the world to survive.

Big words, perhaps, but we are dealing with an important subject. . . . New musical ideas and differentiations enlarge our horizons and stimulate our thinking and feeling. Highly sophisticated music which mirrors the complexities of the world can claim for itself a high degree of relevance. To be sure, I would call such works "innovative," no matter what their chances are with the general public. It might be that composers who write music which is monomanic or primitive or orthodox and imprisoned in ideologies have an easier time. But these are questions of Life and Survival, and we have now gone back to the beginning of our conversation. . . .

*Universal Edition*

# *György Ligeti*

BORN MAY 28, 1923, DICSÖSZENTMARTIN, TRANSYLVANIA

**November 5, 1986** **New York**

Ligeti is in New York for a celebration of his winning the Grawmeyer Award for Music Composition 1986 and a concert of his Études for Solo Piano and Trio for Violin, Horn, and Piano. When I arrive at his suite in the Plaza Hotel he is reclining on the bed. We have very little time because he is due at the Academy of Arts and Letters in about an hour. It's three in the afternoon.

RD I'm very interested in what you consider to be your "musical genetic code." For example, in this country, Copland, Carter, Piston felt the need to go to France and worked with Boulanger. This, together with the influence of Koussevitzky who performed new music, produced a kind of musical genetic code. One of the obvious examples in Europe is the one that centers on Messiaen and his offsprings. You come from Hungary. Can you speak about your own musical genetic code?

GL I come from Hungary and I studied in Budapest; and the great Hungarian musician was and is Bartók. So, in the beginning my main influence was Bartók. But very soon, in Budapest in the early fifties, I felt I had to do something totally different. Hungary as a Communist country was totally isolated ... today it's not so isolated.... I had, during those years, absolutely no knowledge about my generation in the West, about Boulez, Stockhausen, Cage, and so on. Also, it was the time of Hitler, and the time of Stalin ... modern music, modern art were absolutely prohibited. So the Viennese School ... we knew the names of Schoenberg, Berg, Webern, but not the music. I knew a little bit of Stravinsky, *Firebird* and *Petrouchka.* The *Sacre* maybe I heard once on the radio, very much blurred, but it was not vivid in my mind, and it was not possible to see the score. It was a very isolated situation. Also, Bartók was prohibited ... his main pieces. His first period was allowed, until the First String Quartet, and then the last pieces, the Third Piano Concerto, *Concerto for Orchestra;* however, the *Music for Strings, Percussion, and Celeste,* while everybody knew it from before, was not allowed.

RD There was a decision that it was not official.

GL Yes, it was too modernistic. And you know Shostakovich's troubles in the Soviet Union.

RD Does any aspect of the political concept of Central Europe that seems to be re-emerging now through the writings of Václav Havel (Czech), György Konrád (Hungarian), and Adam Michnik (Polish) interest you? Or perhaps the Czech writer Milan Kundera in his essay, "The Tragedy of Central Europe"?

GL I like Kundera, but I never met him. I like his writings about all these problems of people who are émigrés, and I feel a very deep solidarity with people like Kundera. But the concept of "Mitteleuropa" you know . . . it was the old Austro-Hungarian monarchy; and it doesn't exist anymore, even though there are still many traces of it. You can call it "Mitteleuropa" or *not* . . . it's merely nostalgia. Hitler and Stalin destroyed it. But to come back to my genetic code, it's very clearly in the beginning, Bartók. And then I developed what I wanted to do, this kind of "static" music; also my conceptions . . . I changed a lot during these three decades from 1956 to now.

RD May I guide you through those?

GL Yes. So first Bartók and a little bit of Stravinsky, nothing more.

RD Can you reflect on Bartók as a man?

GL I never met him personally. When I was a student I knew his music. You should know that Bartók lived in Hungary until the year 1940, when he emigrated to the United States.

RD Yes, and he died there in 1945.

GL But it was in 1940 that he was very much against the Nazis. He was not forced to emigrate. The Jews were forced to emigrate, but not Bartók. He was not Jewish. But he was absolutely against Hitler. He was so against any dictatorship he couldn't stay anymore. Now what is my feeling today about the music of Bartók? I don't think that he is the most important composer of this century, which is what I was thinking when I was young and in Hungary. Bartók was the first and Kodály was the second, and everyone else came after that! Today, in my perception, I think someone like Stravinsky is much more important.

RD I remember reading somewhere about your emigration from Hungary. It was rather dramatically described.

GL It was not an emigration! It was a flight, by foot!!

RD Which you ended in Cologne.

GL I came to Vienna first in December 1956, after the revolution was suppressed, and then I had an invitation from Herbert Eimert to come to Cologne. I had to wait for a German visa; and it was in the beginning of February 1957 that I went to Cologne.

RD And you met not only Eimert but also Stockhausen.

GL Yes. It was very important, it was my second phase, coming to Cologne . . . In Budapest I didn't know electronic music, but I heard about it.

Eventually, I also listened to some on the radio; but it was during the revolution. Before that, everything from the outside was jammed. During the revolution there was no more jamming. On November 7, 1956, there was a terrible fight in the city between the Soviets and the Hungarians, but it was possible to listen to the radio. I heard the first radio broadcast of *Gesang der Jünglinge.* Stockhausen and I had already exchanged letters. He wrote to me, and I had already written him a letter, but this was before and he did not know that I had heard the piece. And then two months later I was in Cologne. In the moment when I decided to leave Hungary, I knew I wanted to go to Cologne, because it was the center for electronic music. More for electronic music than the concept of serialism, which I also didn't know, just read about it. But I thought electronic music was really the medium of the future. And I wanted to go totally into this medium. So for two and a half years I learned it, I made some pieces.

RD You realized *Artikulation,* there.

GL *Artikulation* and the piece before it called *Glissandi.* And a third piece . . . in those days I liked French titles, so it was called, *Pièce électronique Nr. 3.* There is a score . . . then I had problems with the limitations of the studio and I realized that "the electronic" was not for me.

RD Could you speak about the musical activity that was present at that time, particularly in Europe, and the attitudes for instance, toward order, toward form, toward these concepts that, I suppose through serial thinking, were investigated in a very thorough way. What do you feel you took from these various concepts that were in the air?

GL A musical thinking that was very clear, very intricate, with complex construction. Even before, in Hungary, I had this predilection for clear construction, but I didn't know the serial method. I never wrote serial music. I analyzed Boulez's *Structure Ia,* because it was my interest. So I wrote the article for *Die Reihe.* It was the first thing I did when I came to Cologne in 1957. I wanted to really know the technique, both of the electronic studio and of serial music. Therefore, I analyzed this Boulez piece. At first I wanted to take on *Le Marteau sans maître,* but I realized very soon that it was too complex. I could not analyze it totally. I wanted a piece that I could completely see through. *Structure Ia* of Boulez is his only piece where this is possible. Everything is rationalized. I took from serial music a certain kind of thinking in construction, but not more. Not the method, maybe the discipline. And I think I maintained it until today. I am very constructive, but also emotive and poetical. I hope so. But I never used construction like this old serialism, the Boulez, Stockhausen pieces. I say old serialism, because later both Stockhausen and Boulez became less involved with orthodox serialism. And never in the way of, for instance, Iannis Xenakis, who builds musical form on certain algorithms and mathematical principles. I am very interested in mathematics and I have been interested in Benoit Mandelbrot's fascinating "fractals." I want to, not only want, because I already did, develop certain musical forms which have nothing to do directly with mathematics, but which are a little bit influenced by this "organic growth" which is possible to simulate with the algorithms of Mandelbrot, for example. Anyway, I am interested in mathematics, but in my music I don't use mathematics in a direct or strict way. It's always ... thinking in music—first, construction—second. First is imagination and if you want, emotion. I have always had an interest in picture puzzles, paradoxes of perception, for certain aspects of growth and transformation of structures and you will find these constructive aspects in my music, but they are always dominated by emotive aspects.

RD There are many younger composers that are disillusioned by the idea of serial thought, both in Europe and in America. They see that, although it once was thought of as *the* way of the century, it isn't any more.

GL It never was.

RD Boulez made a statement in an article recently, saying "I was only a serialist for two months." I don't know whether it's actual fact or not.

GL Even if he didn't say that, it's true.

RD But this disillusionment does not prevent composers from still using what they find interesting from the whole study. Would you agree that this research into serialism was an important 20th-century musical achievement; and would you give some credit also to Messiaen, especially in his *Modes de valeurs et d'intensités*?

GL Yes, yes. It was an important step. Even so, today, I feel very far away from this thinking and also very far away from the whole experiment. I make musical experiments, but that particular experimental attitude, also the attitude of everything that is connected with John Cage ... something I was very much interested in at the beginning of the '60s ... Now, I feel very, very far away from it all.

RD Some of the issues at that time were determinacy and non-determinacy.

GL For me, determinacy and non-determinacy ... both are obsolete. Not only today, but in the last twenty years, maybe.

RD Well, for you, a compositional turning point, in a very positive way, was in your orchestral pieces. I've done, for example, *Atmosphères.*

GL The first one was *Apparitions.* Do you know that piece?

RD Yes. Perhaps you could speak about that as a departure, because that was close to the time we are talking about.

GL Yes. In fact, this idea of a "static" music, and music that is constructed with very complex webs, polyphonic webs, this was an idea which I had in Budapest, and I even remember the exact time that it first came to me, although at the time I didn't write anything down.... It was in 1950. I had imagined it, but I didn't know how to write it down. The first piece that was purely "static" was *Visions,* which I wrote in the summer of '56 in Budapest, and *Apparitions,* the first movement, is a third variant of this piece. There was a first variant in '56 titled *Visions,* and then the second one in '57 in Cologne and Vienna, with the title *Apparitions,* where the first movement was the same, but more sophisticated after the influence of Boulez, Stockhausen, Koenig, Kagel, and so on, in Cologne. And finally, the third version is *Apparitions.* I say the title in French, because in English it has a different connotation, like ghosts, while in French it means appearances. The first movement, I finished in '58, and the second movement in '59, and the performance was in '60. Then it was absolutely finished at that time. By the way, it's very interesting that, independently from me, Friedrich Cerha . . .

RD Yes . . . I know.

GL So you know it . . . before *Spiegel,* before *Mouvements,* his first piece in a "static" style in fact, as I remember, was *Fasce.* He finished it later, but he began it already in the late '50s.

RD I meant to bring this up, and I'm glad you did; because I wondered, when I did the world premiere of *Fasce* in 1975, what the two of you were talking about. It had to do with the fact that there were some sketches or something in the beginning, previously, and then he finished it later. What were those parallels in terms of you and Cerha?

GL As I told you, in '56 and '57, I had already written the preliminary *Apparitions.* But the final version of the first movement was in '58, which was the first kind of completely "static" music. While I composed it, I knew Cerha and I knew his music, some of the earlier pieces like the pieces for violin and piano and then *Relazioni fragili* in 1956–57, for harpsichord and chamber ensemble plus "vocalise." And we were very good friends; but I didn't know what he was doing during that time. One day I went to him and he showed me the first sketch of *Fasce.* My immediate reaction was, "You are writing my piece!" So in fact it must have been something happening at the same time. By the way, I know now today, but I had no idea before, that in Rome, Giacinto Scelsi was doing similar things at the same time his pieces, *Quattro pezzi su una nota,* from 1959. So from a certain point of view he did something which I later did . . . having one note, like in *Lontano* or the Cello Concerto, starting from one note. . . . And Scelsi did it in the late '50s. So there must have been something of a general idea in the air.

RD Well, there were many things. You speak of the one note, and then the expansion from it; but also there was debate at one time about who was the first to compose a "glissandi-cluster."

GL It was Xenakis!

RD I think it was Xenakis. Stockhausen wants to feel that it was he, but I think it was Xenakis.

GL May I absolutely, clearly answer it was Xenakis, in *Metastasis,* in 1953–54. Stockhausen did it also, but later. And I also did it later. We should give the priority to Xenakis. And also, when I wrote *Apparitions,* I thought that I had invented the complete division of all strings, you know, it was a score of 63 systems. I didn't know *Metastasis* in those times. I knew it about 1960, 1961. It was also something in the air. If we choose the man who did it first, the "glissandi" and also the concept, not of "static music," but of "sound masses," it's Xenakis.

RD But I can go one better. Consider the work of Charles Ives. The context is different, of course, But the "glissandi-cluster" appears in the *Holidays Symphony,* particular in *The Fourth of July,* and also in the Fourth Symphony. These were written between 1909 and 1916!

GL I have to tell you that Ives made a very, very big impression on me; but I didn't hear the music of Ives before 1962, 1963, I knew the name. It was when the first record of the Fourth Symphony came out with Stokowski . . . it was the first Ives record I heard, in the early '60s. Then very soon many other pieces, *Fourth of July, Decoration Day,* and *Three Places in New England.* I had no influence from Ives, but then the priority is Ives, or if you want, Cowell, or clusters of Cage, maybe.

RD Well, yes, and Cage and Varèse. The concept of cluster as a musical "noise."

GL My music of this time, *Apparitions* and *Atmosphères,* end of the '50s, had nothing to do with the conception of noise. If you go back to the question of a musical genetic code, I like Varèse, I like many composers. But I never had a direct influence from Varèse. Nor from Cage. And my conception was not the conception of cluster. I don't even like to use the word. My conception was of a very complex polyphonic web . . . then I called it micro-polyphony, because it's the changing of this web, and the inner structures of the texture which was important as a musical idea, beginning with *Apparitions.* And from ths point of view it's not Ives and not Cage and also not Xenakis. It's something very different, but very close to Cerha.

RD That's what we're getting at . . . very close to Cerha. It's quite remarkable.

GL And close to Scelsi, from what I discovered in the '80s. But I think the priority questions have absolutely no importance. I laugh a lot when I read about Hauer and Schoenberg arguing who invented the twelve-tone system, the dodecaphony . . . it's stupid. It was in the air. Or when Schoenberg wanted to sue Thomas Mann because he didn't mention Schoenberg's name in *Dr. Faustus.* It's ridiculous.

RD Well, what we're getting at of course is to try to recapture some of that "air," and then gather your point of view.

GL I have to add something to my genetic code. I have to say that when I went to Cologne, the impact of Stockhausen and Boulez, and the concept of order was important to the composers who were around the studio in Cologne and Darmstadt. But the man who was especially important for me was Gottfried Michael Koenig, who is unfortunately much less known. I learned the technique of electronic music . . . how to work in the studio, from him. He has an electronic piece, *Essai,* I don't know if you know it . . .

RD I have only seen the name, but I have never heard any of his music.

GL There is a score, but you cannot read the score while you listen to the music, because the score is just numbers and prescriptions of how to make the piece. He composed this piece in 1957. I was there during the whole time he composed the piece. And in the studio I was a little like an apprentice. His thinking was very orthodox-serial in a certain special way. Then I learned directly very much from Stockhausen. He was very nice to me, and the first six weeks that I was in Cologne, I lived in his apartment. So every day we had conversations. He was less "Great Man" at that time. Today I wouldn't be so interested in living with him. I have to say though, that then he was extremely nice and extremely helpful. In music I learned a lot from him, because he was finishing his *Gruppen* in 1957 when I lived in his apartment and I could follow that. Do you know him well?

RD I met him for the first time in 1963, when he came to Buffalo to do the American premiere of his *Momente.* Then later, in 1971, I did *Carré* with the Residentie Orchestra in The Hague and in Paris, where Stockhausen came; and also in 1972 in London with the BBC. I was also very impressed by talking with him during and after the Paris performance.

GL You did *Carré,* wonderful piece. I was at the first performance in Hamburg in 1960 with Kagel, Markowski, Gielen, and Stockhausen. So you know . . . he can be very nice, but he can be very "important."

RD And formidable.

GL And a guru, and I don't like gurus. I have to say, I am very ambivalent even now, today, with Stockhausen as a person. I like and respect his music, but I have certain reservations. But in those times, it was very important for me to know his method, how he worked, and it had a lot of influence on me.

RD One of the words he keeps using in his method is to "mediate" something. He seems to come to the conclusion that there is nothing on earth you cannot "mediate" if you put it in a kind of polarity and understand the differences between the two extremes.

GL It's well known in old Persian religion, the antagonism between "light and darkness," all this dualistic religious thought. It's just *one* way to think. You can also think that everything is unified, in a holistic way. You can put it in a reductionistic way, and whatever . . . It doesn't help. All these ideological schematic patterns of thinking are empty. But I don't want to discuss Stockhausen's philosophy, only to say that what I learned from him was not on this level. It was on a very direct level of musical apprenticeship, a practical level. I was there when he finished composing his last couple of notes of *Gruppen,* which is maybe his *main* composition. And then I was there for many rehearsals, preparing the first performance of *Gruppen* in 1958. So the first performance of *Gruppen* and later the first performance of *Carré* and many things in connection with Stockhausen were very important for me. Today, I feel much closer to Boulez, for instance; but in those days they were very significant to me . . . Stockhausen, Koenig, Kagel, also Evangelisti, who was less known . . . It was a whole society. Stockhausen, in the clouds somewhere as a half-god, before he became god, and then we were the civilians.

RD You speak of Boulez. It has been said that there might be a thread in him, in terms of the synthesis of Webern . . .

GL And Debussy . . .

RD And Debussy, and Messiaen. Does the idea of synthesis interest you?

GL No. No duality, no synthesis, no analysis. I hate all these pseudo-philosophical over-simplifications. I hate all ideologies. I have certain musical imaginations and ideas. I don't write music naïvely. But I imagine music as it sounds, very concretely. I listen to it in my inner ear. Then I look for a certain system, for a certain construction. It's important for me, the construction. But I always know it's a second thing, it's not a primary factor. And I never think in philosophical terms, or never in extra-musical terms.

RD Perhaps you could speak about your music today.

GL I can tell you some things which influenced me, some aspects. I went through a gradual change during the late '70s and maybe from the beginning of the '80s, I began something really quite new. I have many pieces which are a lot less well-known than my earlier pieces . . . I am the same composer, so what it left is the conception of musical form; "closed" musical form as against "open" musical form. I was always a partisan of "closed" form. Music is not everyday life. Art is artificial, it's an artificial product. It's *closed.* It has to be well constructed, also poetic, but construction is important . . . complex, sophisticated, not minimal but "maximal"! And this is what I always did and I do today. But something changed, because I don't want to always repeat myself. And I had a very deep crisis at the end of the '70s, beginning of the '80s. The crisis was partly personal, because I was ill; but partly stylistic. Also, all this discussion about Modernism, post-Modernism and so on. You find it here in New York between "uptown" and "downtown" . . . I have in myself an "uptown" and a "downtown." And finally, what I am doing now is neither "modern" nor "post-modern" but something else. The Trio for Horn, Violin, and Piano (1982), as well as the Études for Piano (1983), which will be performed here next week; and a Piano Concerto (1985–86), which was performed two weeks ago in Graz, and a couple of chorus cycles, represent my new musical thinking.

RD You speak of a crisis. How did it manifest itself?

GL I felt that this kind of modern music we spoke about which began in Cologne and Darmstadt is, for me, obsolete. And also what I did myself . . . I had to change. At the same moment, I don't like post-Modernism, neither in architecture nor in music. I don't want to go back to tonality or to expressionism or all the "neo" and retrograde movements, which exist everywhere. I wanted to find my own way and I finally found it. I don't do "static" music anymore . . . no. I have found certain complex possibilities in rhythm and new possibilities in harmony which are neither tonal nor atonal. Maybe the Six Études for piano are the most typical.

RD May I list the titles? 1. *Désordre* (Disorder); 2. *Cordes vides* (Open Strings); 3. *Touches bloquées* (Blocked Keys); 4. *Fanfares* (Fanfares); 5. *Arc-en-ciel* (Rainbow); 6. *Automne à Varsovie* (Autumn in Warsaw). You say that in these études lies a new conception of rhythmic articulation. One that began with the idea of superimposing rhythmic grids of various densities, in *Poème symphonique* for 100 metronomes (1962); and later you developed what you call an illusionary rhythm. Can you explain this?

GL Yes. In *Continuum* for harpsichord (1968), the interpreter plays a very fast, even succession of notes, but as a result of the frequency with which certain notes occur, what we primarily perceive are slower rhythmic configurations, which are not played but are heard. This idea is further developed in *Monument* for two pianos (1976), where both pianists play similar musical phrases, one in duple and the other in triple meter. A new kind of overall rhythmic pattern emerges from the homogeneous fusion of sound and from the complex contorted polyphony which is achieved by superimposing two relatively simple phrases. These are examples of what I call an illusionary rhythm. Then in the '80s, when I first heard the music of Conlon Nancarrow, I was very intrigued. I was struck by the rhythmic, metric complexity of his polyrhythmic music. Obviously, Nancarrow comes from blues and jazz, and from an American musical tradition, which is situated far away from my musical background. (If you want, I am "Mitteleuropa"!) However, I felt myself being very close to Nancarrow concerning the essence of musical ideas. My interest in polyrhythmics and different simultaneous speeds, which I heard in Nancarrow's marvelous music for mechanical pianos, gave me the impulse to consider looking for ways and means by which *living* interpreters could perform a similarly complex music. I asked myself, how can I find the possibility to have one performer, for instance, a pianist, play in different speeds at the same time? On the other hand, I was also influenced by Charles Ives and his thinking in different musical layers.

RD Collage, perhaps?

GL Not quite collage. In the '70s, in my opera *Le Grand Macabre,* there are collage aspects. My newer music should not be understood as an "Ivesian" heterogeneous layering. On the contrary, it is the homogeneus fusion of sound which is of utmost importance. More in the Nancarrow direction, but naturally it's not Nancarrow.

RD Are there any other European precedents that you can add to this rather large area of rhythmic investigation and layering?

GL Yes. The meter-dependent *hemiola* as used by Schumann and Chopin. The *hemiola* arises from the metric ambiguity posed by a measure of *six beats,* which can either be divided in three groups of *two* or in two groups of *three,* and stems from the mensural notation of late medieval music. To this whole distinct domain I add a second, separate musical thought process . . . the additive pulsation principle of sub-Saharan African music.

RD Exactly what African musical influences are you speaking of?

GL Especially the polyphonic music of Central Africa, for instance the Banda polyphonic music, with orchestras of twenty or twenty-five people. I listened to the recordings made by the Israeli musicologist, Simha Arom, and was struck by the complexity of the metric ambiguity. In this music, of course, there are no measures in the European sense of the word, but instead one finds two separate rhythmic levels. A ground layer consisting of fast, even pulsations which are, however, not counted as such but rather felt; and an upper layer of occasionally symmetrical but more often asymmetrical rhythmic patterns of varying length, though always multiples of the basic pulse. Consequently, as there is no "meter," we are not actually dealing with an "ambiguity." There are no accents, only an evenly flowing pulse. You will not hear quotations of African music in my music, but in my thinking of rhythmic and metric patterns, the knowledge of the African principles are decisive.

RD And the result?

GL One often arrives at something qualitatively new by unifying two already known, but separate domains. In this case, I have combined two distinct musical thought processes, the European hemiolas and the African pulsation principle. By combining the two ideas, I developed a method that enables the pianist to play several different speeds at the same time, having a very fast common pulse as a common denominator. You can follow the different speeds, but you have to listen carefully. It is very different from any music which exists until now.

RD Apart from combining these two musical aspects which have influenced your thinking, you spoke earlier of your desire to develop certain musical forms which reflect the concept of "organic growth."

GL Yes. Aspects of growth, of the generation of musical form departing from a conceptual "genetic code" are at the center of my interest. However, in my music one finds neither that which one might call "scientific" nor the "mathematical." But rather a unification of construction with poetic, emotional imagination. There are two non-music influences I can mention. First, I am very much interested in the ideas of artificial intelligence, for instance, in the different layers of language. Concerning this domain, I very much like the writings of Douglas Hofstadter. My second non-musical influence is closely related to the concept of organic growth and comes from the new field of fractal geometry, mainly from Benoit Mandelbrot—we already spoke of it. I always have been interested in complex ornaments like the Islamic ornaments in the Alhambra or in medieval Irish art, the Book of Kells. The fractals are the most complex ornaments ever. They are the ideal model of what I want to develop in my own music.

RD I know you must leave soon for the Academy of Arts and Letters, but one last question. As a youngster, what led you to music?

GL Ohhh . . . nothing. In fact I don't know how it came. My parents, they didn't want me to . . . especially my father. . . . I was not allowed to study an instrument, and at fourteen I said, "Now I want to study piano!" We had no piano, so I had to go somewhere everyday to practice. I wanted to be a scientist, wanted to study physics, but then I turned to music. It was not a decision, it was the result of a chain of accidents.

# Narrative XI

By the mid-'70s, I had performed most of the large works of the 20th century, from Stravinsky, Berg, Bartók, to Varèse (including *Arcana* and *Amériques*) to Schoenberg (including *Gurrelieder*). I also continued to promote American music in Austria with the ensemble Die Reihe, the ORF Symphony in the Steirische Herbst in Graz, as well as in the Holland festival with the Concertgebouw Orchestra, and also in the Zagreb Music Biennale. My debut with the Berlin Philharmonic in February 1976 was an all-American program: Ruggles's *Sun Treader,* the German premiere of Druckman's *Lamia,* and Ives's *Holidays Symphony.*

Promoting American music around the time of the Bicentennial Year was of course a bit easier to do. I remember Aaron Copland saying to me, "This is a rich time for American music, but they'll soon forget us."

However, it was my Dutch connection that was and remains particularly interesting. I have performed a lot of Dutch music and have admired the attitude of the Dutch toward music in general. I have conducted most of their orchestras, from the Concertgebouw, Rotterdam, and Residentie to the various radio orchestras, to orchestras in the provinces in Arnhem, Groningen, and Leeuwarden. And it has been a spectacular example of a devotion to music that is absolutely ingrained in their lives. Small as a nation—but big in musical heart.

Over the years, I have developed very strong ideas about how to program contemporary music, especially in festival circumstances. I feel the need to "hang" 20th-century music, much like a museum curator "hangs" 20th-century paintings, with an eye toward aesthetic, stylistic, and perhaps historic connections, so that a listener has a chance to grasp existing musical inter-relationships.

The word "festival" is often used too loosely. In many cases, it turns into just a gathering of odds and ends. A festival should have strong underlying thematic or connective elements. Music festivals, in particular, should reach out to the other artistic disciplines for enhancement. There is a great opportunity to bind and bundle our century, with regard to periods, movements, circumstance, and philosophy. A festival should become a total show, so that the audience gets more than a glimpse or a scattered presentation.

By now, there are composers who deserve one-man shows. These are our 20th-century classicists. They represent our tradition, *and we don't have any other.*

*Jill Richards*

# *Krzysztof Penderecki*

BORN NOVEMBER 23, 1933, DEBICA, POLAND

**February 14, 1987** **New York**

Penderecki was conducting subscription concerts with the New York Philharmonic. I visited him at the Hotel Mayflower.

RD I must first plunge into an issue that is from the past; it is one that, through these interviews, has emerged as a source of crisis in some composers' lives, a source of disapproval or approval: and that is the musical initiatives of postwar Darmstadt in the '50s. What is your particular view of that atmosphere now, looking back at it?

KP I was not really connected with Darmstadt. I wanted very much to go to Darmstadt, but at that time, I couldn't get a passport to go abroad from Poland. It was impossible to travel. The first time I went there was after waiting about five years. Then suddenly I got permission to go . . . it was either 1960 or '61. I spent a very short time there, only one week, and they played some of my music. First of all, I was disappointed, because of the isolation that was forced upon all of us in Poland in the late '50s. A political change opened up in '56, but until '56 it was not possible to play any new music. It was forbidden. The first time I heard *The Rite of Spring* in '56 or '57, it was a shock for me! So from the isolation I started really to develop my own ideas, my own music. Of course, I was rebelling against my professors in school, who were very traditional, and I had to study many years of counterpoint and orchestration, etc. At that time we had no technology, no tape equipment to listen to music.

RD Some of the composers mentioned in your biography were Malawski, Skolyszewski, and Wiechowicz.

KP Those were professors who were very conservative; but, nevertheless, I think they gave me a very good background. After I finished school in 1958, I started to go my own way. I wanted to forget about the past. This was also the first time that I discovered, for and by myself, Schoenberg, Webern, and Boulez. It was already very late, '57 or '58. I was fascinated by this music for a short period of time between '57 and '59, and I listened to it sporadically. But I still couldn't travel.

**RD** I spoke recently with Ligeti and he has a similar tale in one way, because he fled from Hungary in 1956.

**KP** I think it was exactly the same situation.

**RD** He also speaks of the isolation in Hungary. But in your case . . .

**KP** I stayed, of course. He was at least ten years older than I. So he fled from Hungary while I remained in Poland. But there was a big change at that time.

**RD** What was the atmosphere then, politically? What was changed?

**KP** Oh, politically there was absolutely a big change. It was the end of the Stalin era. My older colleagues started a festival in Warsaw, called Warsaw Autumn, which was established in 1956, the year of the change in Poland and the revolution in Hungary. So I began to study music which was previously forbidden. I already mentioned *The Rite of Spring.* Ridiculous, no? But it was true. This kind of music had not even been played on the radio.

**RD** Nor Schoenberg, Berg, and Webern?

**KP** No. Later I discovered this music in a very short time, and in the beginning it influenced my music very much. However, in 1959 I started a piece, *Anaklasis,* which was commissioned by the Donaueschingen Festival. I would say it was the first step towards *my* music. I introduced the cluster technique which I developed later, and I developed a new kind of notation. Then I wrote some pieces like the First String Quartet, *Threnody for the Victims of Hiroshima, Polymorphia.* . . .

**RD** *Fluorescences,* later . . .

**KP** That was in 1962.

**RD** Then, of course, *Emanationen* from 1959.

**KP** Oh yes, that was performed in Darmstadt. But when I went to Darmstadt, I already had my own ideas about music. I didn't want to copy Darmstadt. I didn't find Darmstadt so fascinating.

**RD** Yet there are interesting parallels one can draw. Who cares who was really first, but I've had certain discussions, for instance, with Xenakis about who was the first to use the glissandi cluster . . .

**KP** He is obsessed with the idea that he was always the first! I didn't know of Xenakis at the time I was writing my music. Maybe Paganini invented the glissando, who knows?

**RD** You didn't know of these things?

**KP** No, of course not! It's a very funny thing, because if you ask Stockhausen, he would tell you that *he* invented everything. But it is not true. There were many composers discovering a very similar kind of music at the same time. Ligeti was one of them, with *Atmosphères,* or my *Threnody,* for example. It's a different music, but it has a similar approach.

**RD** But, of course, *Metastasis* is from 1954.

KP Yes, but it was not played! We didn't know it!

RD Well, that particular approach, let's call it the "S-S approach," the "sound-search," what led you to searching for those sounds, besides rebellion?

KP The electronic studio was established in Warsaw in 1957. So I went there to work in the studio. I was a student at that time and used to go every month for a couple of days at a time. Eventually, I worked at the studio for two or three years, writing music for short films and theater. I think I learned a lot from that electronic studio. I think that what I mostly learned was discovering something absolutely new and, I thought at the time, with unlimited possibilities in music. In the late '50s, beginning '60s, I was the first to introduce a new way of playing the string instrument. And I think *my* ideas, using the cluster, using "noise," discovering new ways of playing an instrument, came from working in that studio, not from studying others.

RD Not from Varèse?

KP No.

RD Did you know his music?

KP At that time, yes.

RD Because "noise" is a factor in his music, as it is in Cage.

KP They played Varèse early on in the Festival in Warsaw. Maybe in 1957 or 1958.

RD So the sound sources came then as an investigation of yours?

KP I was a violinist at that time. I even wanted to be a virtuoso violinist, not a composer, and I was very interested in finding new possibilities for using the instrument, for using string instruments. So I experimented in the studio, playing the violin and making clusters from one note, or playing behind the bridge, on the tail, on the bridge, or simply the highest pitch, different kinds of vibrato, slow or very fast, quarter-steps . . . of course, quarter-steps were not my invention. This had been done before, but I was interested in all of these techniques.

RD The search for sound sources was a big thing, whether it was in Darmstadt or Warsaw. It was all over the world. The other big thing, however, in terms of Darmstadt, was the solution to formal conceptions. And there, historically, we have this battle between determinism and indeterminism. It's curious that many of the composers from behind the "Iron Curtain," as we called it then, used indeterminist technique more than they did serial technique. Is that a fair observation?

KP Yes.

RD Why is that, I wonder?

KP It's hard to explain why. It is curious, perhaps. Maybe it's a kind of improvisation, a freedom. Maybe it comes from folk music, but I am not really sure. I don't know. Indeterminism began in Poland and then influenced all the composers behind the Iron Curtain. The Warsaw Autumn was the only festival where they could come and listen to forbidden music.

**RD** It had no political overtones?

**KP** Maybe. I didn't think that, for me, this strict serialism and post-serial pointillism was a technique which I could use to write my own music and say something that I wanted to say, because it was so impersonal. Of course, I wrote some pieces using this technique. For example *Emanationen* was written strictly in this technique.

**RD** But you had an initial feeling that it wasn't a technique for your expression.

**KP** Yes. I remember the first time I heard this music and saw the scores. It was when Luigi Nono came to Poland for the first time in 1958. He brought with him not only his own music but also the music of Berg, Schoenberg, and maybe some Stockhausen. And Boulez I remember for sure, because for a short time I was very much influenced by his *Improvisations sur Mallarmé.* I don't really remember all the pieces now, because it's such a long time ago. But I did see scores and some of these pieces influenced me. The piece *Strophes,* which I wrote, is certainly influenced by the *Mallarmé.* But this was before I started my *own* music.

**RD** *Strophes* was one of the first pieces I saw of yours.

**KP** That was 1959.

**RD** So that in using the more indeterminate approach, you also had to solve larger, formative questions. And, of course, we had at one extreme, Xenakis with his probability theories and the law of rare events and Markov chains, etc., and Stockhausen on his path of super-organization/total-organization. How was it that you became satisfied with your approach to the larger, formal aspects of your pieces?

**KP** I was not interested in using mathematical formulas for my music, because I didn't think it was the way for music. In pieces like *Threnody, Polymorphia, Fluorescences,* the form became free, it was an evolutionary form. Maybe it was a reaction to another form of political dictatorship and totalitarian domination, which my colleagues and I disagreed with in postwar Poland.

**RD** Would you call it open-form, rather than closed?

**KP** No, it's closed. But it does not have any background of something strictly mathematical.

**RD** There's an interesting aspect to that period of your music which I found very fascinating. I found it in other composers too; and that is how a piece was executed, by a conductor. For instance, the piece you conducted the other night at the New York Philharmonic, *De natura sonoris, no. 2.*

**KP** It very much depends on the conductor, because when I write for a small section, for example, that the tempo is only the seconds which I indicated, 15 or 7 seconds . . . even I make it up as I go along, while conducting. It depends on the acoustic of the moment, it depends maybe on the audience. Perhaps for the old ladies I do it a little faster, or I let the brass play louder to scare them a little bit.

RD And then the psychological atmosphere of the musicians . . .

KP It's strange . . . I mean this piece is still fresh in my mind, it was written in the '70s. In this kind of music the conductor's role is very creative. But even with the pieces which I wrote in the late '50s and '60s, there is still a kind of resistance. The musicians don't understand it, they don't like such a music and they don't want to play it. Really, that's a problem. I understand why conductors don't want to do such a music, because they have to fight against the orchestra.

RD Do you think it has to do with the tasks that you give them as players? They think it is beneath their virtuosity?

KP They are not willing to make the effort to learn something new. They want to have everything written out for them so they won't have to do any creative thinking! Just to play! If I would have written this piece in the conventional 3 and 4 meters, there would be no problem. But because it's a free kind of notation . . .

RD Do you think it embarrasses them?

KP Oh yes. For example, I saw in the performance, some of the musicians, while the brass were playing, they were holding their ears! It's ridiculous, but this is typical of New York.

RD I personally feel that many conductors don't perform this kind of music because they are unable or unwilling to have their "creative" thinking challenged. I have two observations. One, I heard your performance and then last night I listened to your recording of the same piece. What struck me very much about the recording was the atmosphere of the piece, as opposed to your performance, which was a good performance, but the atmosphere was missing.

KP Of course! During the making of this recording, I had unlimited time to work with the musicians. The Polish musicians were familiar with my music. Here, I think I had only half an hour to put it together. Because, also on the program was the Shostakovich Fourteenth and, as you know, it is a very hard piece. And it's a long symphony, fifty minutes. So I really had to concentrate much more on Shostakovich then on my own music.

RD That was only one aspect. The other observation was the fact that in that time, when your music and the music of others was just beginning to come out, there was a hope that the musician was changing, that he was sort of curious about the new techniques. In that period there were at least some musicians who felt that way.

KP Yes. I'll tell you my observation, because I'm conducting a lot and I always try to do such a piece of mine on a program. The older musicians who, at that time, played as younger musicians in an orchestra, are more open. They are really trying to understand and are trying to play. The young musicians today don't even care. I'm not only talking about the New York Philharmonic. Around the world, it's the same problem. I don't know if this is your feeling also, that with the older musicians, they try to do something that the young ones don't.

**RD** I think you are right. I think the professionalism of the older ones, having gone through all that, comes through. They feel that they must be professional and try to play it. While the young ones still have a rebellion of some sort to go through.

**KP** There is something else. At that time, and we are talking about the late '50s and early '60s, I didn't write any long pieces. The longest piece I wrote was maybe twelve minutes. The form is very easy for a piece that lasts ten minutes or so.

**RD** Yes, there are a lot of twelve- or fifteen-minute pieces in the world.

**KP** Particularly from this period, yes. The first really long piece in which I had to deal with the large form was the *St. Luke Passion.* And as a matter of fact, in this work I went back to an old, traditional form, which was the form of the Baroque Passion.

**RD** So, as a composer, that particular shape was like an "old friend" for you to use. It did break a certain pattern in your composition. You changed from this shorter piece idea, but also the sonorities changed.

**KP** Yes. In 1962, I wrote *Fluorescences,* which I consider a very decadent piece.

**RD** Why decadent?

**KP** Because in this piece I tried to avoid the "normal" sound of an orchestra by only using "noise" instead of "normal" sound. There is no form, it's very open. It's really written *against* everything. It's a very rebellious piece.

**RD** And you were conscious of being rebellious?

**KP** Yes, of course! And then at the same time, in the same month, I wrote *Stabat Mater.* This was my first step, in 1962, when I still really didn't know where to go. I wrote five, six, maybe ten pieces in a very short time, and I was no longer interested in going the same way and in producing another *Polymorphia* or another *Fluorescences.* I was looking for something different. I studied 16th-century, late Renaissance counterpoint, and the Dutch and Flemish Schools.

**RD** Ockeghem, Obrecht?

**KP** Yes. This kind of music fascinated me at that moment. And as a result of this study, I became very interested. I listened to this music and studied scores for maybe a month or two. So I wrote *Stabat Mater* in a very clear, contrapuntal style. It goes back to the counterpoint of the Renaissance. Also, this re-discovery was a very important moment for me because, from that moment, I really knew how to write the big form.

How to approach the big form. In the *St. Luke Passion,* these two paths merged; this re-discovery and the "discovering" of sound and all my experience with instrumental and choral music of the '60s. So I wrote my first long piece.

RD I would like to speak further about your very long list of pieces that reflect a kind of religiosity, anywhere from the *Passion* to *Utrenja* to *Dies irae, Magnificat, Lacrymosa, Te Deum,* and the *Polish Requiem.* First though, I must tell you that there's another composer who comes to mind whom I have also interviewed, and that's Max Davies, who also continues to have a fascination with Renaissance music . . .

KP Maybe more English Renaissance . . .

RD Not necessarily. What I'm getting at is, aside from his fascination with certain compositional techniques which he uses for himself, he also is fascinated as a human being with the survival of the people in the time of the Renaissance, how they survived religious persecution or intolerance. To this day he feels it was a motivation for him. In fact, he's writing an opera called *Resurrection,* based largely on what he feels are false or hypocritical religious beliefs of today. Now, in two cases at least, you have made reference to atrocity. One was *Threnody for the Victims of Hiroshima,* and isn't necessarily religious; and then I believe, another one about Auschwitz.

KP *Dies irae.*

RD *Dies irae.* Let's use both examples of your religious reflection and your sense of humanity in terms of motivating these large works. Can you speak about that?

KP To understand this, you have to understand what the situation in Poland was like in postwar time, when we were forced into another system, which we didn't even want, which came about by force from the Russian army. They came and installed the government; the Church was persecuted. At the time that I wrote the *St. Luke Passion,* it was impossible to perform religious music. It was forbidden. All the orchestras and choirs were state orchestras and choirs. They were not allowed to play or sing this music. I remember a problem . . . they printed a poster for the *St. Luke Passion,* and they were not allowed to write the word "saint." So it was the *Luke Passion.*

RD Really? That's pretty strong.

KP Oh yes. Since that time, it has changed a lot, and I think things have changed too for my music. Since it was such a success in the West, they had to recognize this piece. So they started to play this piece and I was able to force them to play it in a church. The Church in Poland played a positive role in history, not like in Western Europe, where the Church was always the opposition. The Church in Poland was with the people, played a political role . . . to be free, to be a Pole, to be able to speak Polish, which was not allowed after the partition in the late 18th century. Poland really disappeared from the map, because it was swallowed up by Prussia, Austria, and Russia. Poland did not exist as an independent state until the First World War, after which it became independent until World War II. So the Church, during the time that Poland did not exist, was preserving Polish culture and tradition. And it stayed so, and was even stronger, after the Second World War, when the Communist government was installed. We were all anti-Communist and we had to be together with the Church to survive. I just wanted to write religious music to show my position. It was a political approach.

RD So you were strongly motivated by political, anti-Communist feelings.

KP Yes. Even now in the '80s, the Church plays a very important role, because Solidarity is very connected to the Church. In Poland, 95 percent of the population belongs to the Catholic Church.

RD Yes, that's a very strong thing, going beyond nationalism.

KP So maybe you can understand why I became so interested in writing religious music.

RD There is one area today that seems to be burgeoning, and that is the relationship that we are reading about between Poland and America. Apparently, now there are some new advances between the governments.

KP I hope very much that this is true.

RD At this moment, Krzysztof, the whole political atmosphere is becoming more free?

KP Yes. It has become more open, because they understand that isolation is not a way . . . there is such a difference between East and West. We were always behind about twenty years, but now it's fifty years. They were not open to exchange. These Communist countries, they were close to collapse. They did collapse already. They are not able to produce anything compared with Western standards.

RD You mean artistically, as well?

KP No, no, not artistically. Artistically we are free. Maybe in the future you will even see American television in Russia. There is no way to block, as they used to. Jamming no longer works. There are *some* changes going on. We can only hope.

RD You know that my great experience with your music was in doing the North American premiere of *Utrenja, Part II,* in Aspen in 1977. Not only did I enjoy doing the music, because it is on such a wonderfully large scale, but it also touched on my specific background from my grandparents, who came from Central Europe, from what is now the eastern tip of Czechoslovakia. I told you at that time, that I'm a Slav through and through, a sub-Carpathian Ruthenian. In other works you use Latin, primarily; but in *Utrenja* (The Resurrection) you used the Old Church Slavonic language. . . . It is my understanding that this language is a precursor to Russian.

KP Yes. Some call it the old Bulgarian language. It was introduced by the brothers St. Cyril and St. Methodius; and they were from Salonike, from Byzantium, Macedonia (Greece). Church Slavonic is really Macedonian, an old Macedonian language. And St. Cyril, previously known as Constantine, and St. Methodius were the ones who translated the Bible in the 9th century from Greek into Slavonic. They translated into the language they spoke, which was a Macedonian language. And this became, like Latin became in the Western Church, a sacred language. Until now.

RD Let's move now to the *Polish Requiem,* which I guess was the last big work that you wrote using a religious text. . . . Did Pope John Paul have any influence on your writing a *Polish Requiem?*

KP Not really. I wrote one piece which is dedicated to him, and it was written right after he became Pope. This is the *Te Deum,* written in 1979. The *Requiem* is more connected with what is going on in Poland in the '80s. I started this piece with *Lacrymosa* . . . in fact, it was requested by Walesa for the unveiling of the Solidarity monument in Gdansk. And it was first performed (from tape of course, because it was open-air) in December 1980. At that moment I didn't know I was going to write a whole Requiem. I was just writing a short piece for soprano, chorus, and orchestra. After the death of Cardinal Wyszyński, I wrote *Agnus Dei,* a cappella, which was performed at his funeral. So I started to think about writing a Requiem. Of course, I had always wanted to write a Requiem. After that I started to write it, in late 1981. And in 1983, half of the piece was first performed in Washington, D.C., by Slava Rostropovitch. And in the following year I finished the work. So it took a long time.

RD You told me once that you knew the Pope when he was a priest in Cracow.

KP Yes. He used to come to my concerts, whenever I or somebody was doing my religious music. *St. Luke Passion* or *Utrenja* . . . he came. He was very interested and he *is* very interested in art. He writes poetry himself. He is a very cultured man.

RD Do you still maintain a relationship with him now?

KP Yes. I conducted the second act of *Paradise Lost* with La Scala for him in Rome. He invited us with La Scala to come to Rome, to the Salla Nervi in the Vatican. A terrible acoustic! For 10,000 people! And I also conducted for him in the Castel Gandolfo.

RD Ah yes, the Pope's beautiful summer residence. Is there a place to perform there?

KP Not really. There is an audience hall, which is not really a concert hall, but it is big. That was only with an a cappella choir. I have a concert soon in Rome in March with the *Te Deum.* So I hope I can perform in his presence.

RD Was he the priest of your parish in Cracow?

KP No.

RD Well now, just to include another side of you. . . . We spoke about the search for sound, and a number of pieces that showed this. We've talked a little bit about the chorus and orchestra pieces that have a certain religious connotation. Now, your virtuoso concerti of course, are another whole branch. The string pieces particularly are so flashy! I remember a performance I heard you do with the Rotterdam Philharmonic, not too long ago, with Christiane Edinger. The Violin Concerto. I thought she played brilliantly!

KP Ah yes, of course, you were there! You were conducting in Amsterdam.

RD You had written that piece for Isaac Stern.

KP Yes, in 1976. But this belongs to another epoch. I should say from '74 maybe, until the Viola Concerto in '83—I fell in love with "Romanticism." Obviously, it is unlike the 19th century, but there is some common element, especially in the kind of orchestration. In my earlier music, the orchestration was very specific. In the First Symphony, I use only blocks of instruments, like woodwinds, then brass, the pizzicato strings, or percussion alone, etc. Maybe it was the moment when I started to conduct that changed my feeling for orchestration. I was searching for a rich sound of the orchestra.

RD And you feel you got that through conducting?

KP Yes. I think that was what developed it for me. I remember rehearsing with the London Symphony. I did the first performance of the First Symphony, where there are no normal sounds for the strings. They play sul ponticello, they play pizzicati, but there is nothing they really can *play.* I remember what a terrible time it was for me. They didn't want to play! They *did,* but they absolutely resisted! Of course, it was not only that. After I wrote those pieces, I became interested in another kind of expression. It was the same time that I began working on *Paradise Lost,* and this also probably changed my ideas. Writing this big oratorio . . . it's not really opera, it's oratorio. I became more interested in a big, rich sound from the choir and orchestra. I discovered "Romanticism" again for myself and I wrote several pieces. Even the Viola Concerto is like this. It also has elements of what I would call Neo-Baroque music. It has elements of both in it.

RD And melodic fragments . . .

KP Yes, of course. But there are a lot of melodies in the *Requiem.*

RD And then another major concerto . . . recently, I saw your performance with Rostropovitch of the Cello Concerto on Public Television. A very excellent performance.

KP That must have been from the Aldeburgh Festival.

RD It sounded like a very good performance; and, of course, he's a fascinating performer, just incredible. You just told me that you have written a new piece for him.

KP It's for choir. I wrote it as a surprise for his sxtieth birthday on March 27th. It is part of the Mass which I might write later.

RD Which part?

KP The *Song of the Cherubim.* It is a wonderful, mysterious text. I will conduct it for his birthday in Washington. And next week we do the Cello Concerto with the New York Philharmonic.

RD We were talking about that aspect of your composition which has to do with the virtuoso.

KP As I said before, as a young boy and for many, many years, I wanted to be a virtuoso. It was my dream. In the '60s, it was impossible to write virtuoso pieces. This was another time. But I started to write them anyway, at a time when other composers were not really interested in such a music. For instance, I wrote a Capriccio for Oboe in 1964, then a Capriccio for Violin in '67, which is a very virtuosic piece.

RD Which you also conducted in Aspen.

KP Yes. And then I wrote my First Cello Concerto and then the Violin Concerto for Isaac Stern. Then the Second Cello Concerto and then the Viola Concerto. Now I would very much like to write a concerto for trumpet.

RD Trumpet?

KP Yes, for Wynton Marsalis. He's a wonderful performer. And he also comes from jazz, so it's fascinating. He is also an interesting performer and I would like to write a concerto for him, because of his background and experience in improvisation.

RD In addition to the surprise for Rostropovitch, do you have other composition projects that are in the works now?

KP I'm becoming very much interested and involved in opera.

RD In the past you have written *The Devils of Loudon, Paradise Lost. . . .*

KP *Black Mask,* now in Salzburg. So now I'm even more interested in opera, because I have written so much religious music. After the *Requiem,* what can I write anymore? Maybe I will write another Mass, a cappella.

RD But voice and chorus and orchestra and the drama are things that you are very attracted to.

KP I think so, yes.

RD Do you have a particular project?

KP I have the theme, which is the Dreyfus Affair. We are approaching the centenary of the Dreyfus trial in 1994.

RD Well that's been dramatized a lot.

KP But there was never a piece. There is no opera.

RD What fascinates you about that subject?

KP The problem of intolerance, and it's a very political piece, too. I think it's a very dramatic subject.

RD As an example of the larger aspect of human behavior.

KP Yes. And I have another subject, which maybe I will do for the Munich

Opera, which is something I found in a book by a Polish-Jewish author. I am very interested in Hasidic music, which was something that started in Poland in the 18th century.

RD Did it really?

KP Oh yes. I like this music and this atmosphere. And no one has really written a piece about it, using this music. I'm very much interested in this culture, which almost disappeared completely. There are still some Hasidim who live in New York and in Israel; but it was part of our culture when I was growing up in Poland. And I would like to do it.

RD So those are two possible dramatic areas.

KP Oh, I have more. *Antigone,* which also interests me. Strindberg's *Miss Julie,* which I think is a wonderful text for an opera; and I have an interest also in the Russian author Mikhail Bulgakov, his *The Master and Margarita.* It's a wonderful book. There are more subjects that I am interested in than I am able to write.

RD And how will you handle the text? Will you write the libretto or will you have to engage someone?

KP I will do it.

RD You write your own libretti?

KP Usually, yes.

RD So you feel no crisis in your compositional life?

KP Do you know what helped me to overcome the crisis which many people have in their life as an artist? Two things. The first was, I started to conduct. That's part of my second profession. In the beginning I didn't take it very seriously. I just wanted to help support my music, but I got more and more involved. So when I have had enough of my own music and I am tired of writing, like this year . . . I finished an opera which took me two years . . . then I just conduct for one year. Writing small pieces, yes, but not trying to start a big piece; and this gives me new ideas.

RD And the second thing?

KP The second . . . I am terribly interested in botany. I have my own arboretum in Poland. I bought the land, and I am planting the trees, and taking care of it when I am there. I am very involved in it. I think it's very important to have something else. Because if I were only to write music, I couldn't survive! It would be impossible for me to do only one thing. It gives me another dimension.

# *Sir Michael Tippett*

BORN JANUARY 2, 1905, LONDON

**November 13, 1986** **Chippenham, Nocketts, England**

Took the 10:30 a.m. train from London to Chippenham. I was met by "All the Fours" taxi (654444). This was my third visit to Sir Michael. The first was to invite him to Aspen. The second included a car trip to Bath. I asked the driver where the name Nocketts comes from. He didn't know. Nobody knows.

Sir Michael's house is a private enclave within a much larger expanse of farmland. It's a rather isolated frame house that looks out over the countryside, and he has been living there for over fifteen years. He greeted me warmly, and after lunch, and after scraping our plates off on the terrace for the animals, we went to his work room, where he reclined barefoot on his favorite couch and proceeded to talk.

MT There's a fresh work on my mind. In fact, I'm fully absorbed in it. Two or three years ago I found it very difficult to say exactly what it was. Because I didn't want to say I was writing another opera . . . but in fact, it is like that. I'm concerned with the fact that the use of the term "opera" itself is confusing, and especially as there are sort of standard, stereotypic views of what it is. I'm very interested, as I have been for some time, in the degree to which the theater as a whole, the mixture between gesture and spoken word and everything else, and singing of course, has to do with that word "opera." Therefore, I've simply said a work for theater, for voices and stage. Now I'm quite clear it *is* a kind of opera. All right. My normal practice is a well established one now, in that I have many years of consideration as to what the dramatic material would be that I would work on. I've only lately discovered that it's very much the way Wagner did it. In other words, you gradually make a scenario, and that is discussed with whomever you wish to discuss it with. And then eventually from that you make some kind of text. The text always has to be concerned with music, to my mind, because you don't need as many words for a coloratura aria as you do for another kind of an aria, and so forth.

RD And is this text in a tradition of your own?

MT Yes. There is no other way to do it, if you are able to do it. You have

*Franz Berko*

to learn how to do it. You have to be good enough to make it. And also, I take a lot of trouble and effort to make clear that I'm not imagining that I'm writing literature, in the sense of something that you read.

RD Well, are you writing a stage work?

MT A stage work, but also a work that is going to be sung. And there are portions in which there may be no words at all. As Wagner showed, you can fill a stage with music, and he did for *The Journey to the Rhine;* and whatever happens is, somehow, in that extraordinary music he produces. As T. S. Eliot once said to me, in a literary sense of course, "The words come last." I'm in that tradition of writers and creators or what ever it is, in which the concepts come first, and then a lot of work and imaginative processes, until eventually, when you're ready, *finally* ready, you look for the actual notes. They are not entirely spontaneous, they then have to be found. The *Anfall,* as the Germans would say, then has to happen. That is more difficult. Starting off the notes themselves is always a little more difficult. Once there, by whatever means one uses, then as far as I'm concerned, I can proceed from the start to the end. I can even write it down in full score, and I can let that full score go to be made into a vocal score and be printed act by act before anything further is done at all. I never have to go back.

RD Well now, just to retrace. You get to the scenario by concept and then, when do you begin to apply the notes?

MT In between the scenario and the notes is some kind of text. If you go to a librettist and say, "Make me an opera text or libretto" out of a well-known story that is already written there, then in a way, the characters are already given ... to a limited extent. If you are inventing, then you are inventing characters. That is a very different thing ... to make characters who will hold the stage in their own right, if you are writing a character opera. May I give you an example of a very fine opera which has just come out of England, Harry Birtwistle's *Mask of Orpheus.* There are really no characters. They are all mythological people, hardly characters at all. It's a different kind of opera. If I took something that came out of European history, like Shakespeare did or Moussorgsky did, that's *another* kind of opera. If you are writing an opera, I think you need to know, first of all, the genre that you are proposing to use. So far, the genres I've used are very distinctive. *Midsummer Marriage* is really out of *A Midsummer Night's Dream* and out of a certain tradition of fantasy and imagination, etc. Well now I'm in a new world, becaue I've been fascinated with what the "sci-fi" world *really* is, what it's doing, why people want it. And then, these accidents that take place ... I'll give you an example. I was, I've always been very keen on the nature of the Russian literature ... I don't speak Russian and I unfortunately can't read it; therefore I have to read it in translation. But I've always been fascinated in many ways, read a great deal of the literature, and read quite a lot concerning it. There was a book published recently by somebody whose name I've totally forgotten, who had been concerned in literature and the whole political world, also of Russia, for some long period of time, at least 100 years. And a set of essays came out. I went off to a bookshop in Bath to get it but unfortunately I got it wrong. The shopkeeper produced instead one called *From Pushkin to Pasternak.* In that book was a fascinating account, finally, just before we reached Pasternak, and *Zhivago,* of two utopias that were made in Russia in the early 1920s. One was where the Communists go and find, as it were, a Capitalist utopia, which is bloody hell. The other was where they arrive in a Communist utopia, which is marvelous! It's a critique. I then considered the whole tradition of the utopia, which is "some other place," I mean literally, from the Greek meaning "nowhere." The only other very early one that I knew was William Morris's *News from Nowhere* ... a very small and curious example of his own world and what the future might be or should be. And then came one from Samuel Butler called *Erehwon,* which of course is "nowhere" spelled backwards. But more interesting to me and right from my twenties was a scientific utopia from H. G. Wells called *Men Like Gods* . . . fascinating! So by accident, as it were, I learned that this tradition of putting a critical concept ahead, or back, might return you to Eden, or you might go forward into some either good or bad world. For instance, *1984* is an example of a concentrated utopia going wrong, as it were, and if you go that way you will not reach the goal.

RD How does all of this apply to your present interest in the world of "sci-fi"?

MT The point that interested me was that because of "sci-fi," possible beliefs that we really could have space travel, it looks as though you really could go "there," wherever "there" is. . . . The other deeper thing is our desire for messages, you see, that we might "find out." I guess all my operas have the one messenger who comes from the gods or whatever it may be, but perhaps comes from another world. So that was one side of it. Another side of it was an actual television program called, "The Flip Side of Dominick Hyde," a very beautiful program which takes place about three generations ahead in the same country, in England. In it, a young man comes back and engenders himself. The rest of the story is what happens to the young woman he finds. It's very curious and beautiful. . . . At the back of it, you get the kind of dream utopias we might have . . . those islands where everything is okay. And then comes the payoff line, "You'd better believe it, baby, because if you don't, it'll be much worse." That's all because in this mad man's universe dreams are cheap. So I'm interested in, have been all my life but more specifically now, in whether these dreams knock you out, turn you over, or give you strength. On that basis, I found the characters.

The woman, a loner . . . that's rare in theater. Men are usually the loners, but there were women loners, too. They interest me very much. And the people who've always been the closest to me . . . the professional healers, in a world that needs it. So already there are two orphans, if you like . . . a boy and a girl. But the girl is going to come through, and the boy is going to go under. All right. You get as far as that. That may be the initial concept. Then you have to structure, and the structure would be the scenario. That's a very long process indeed, because you have to try to put the jigsaw into position with your own imagination. At the same time, if you were writing a novel, you can describe the background of your characters, where they come from, in whatever technique you use. You can describe even their childhood, or their parentage or whatever you like. In an opera or the theater, you can't. It's immediate. The man or woman steps on the stage and at once has to be who he or she is. There's no background. So it's much sharper . . . you have to know what you are doing.

RD And you're at the stage of writing the music.

MT Yes. The music began about last January. I'm now nearly through with one act. One act takes about a year.

RD After the scenario and after the libretto, do you write the music very quickly?

MT Well, "quickly" is a difficult word. I write at my own speed. *The Mask of Time,* which was about ninety minutes of music, took me three years. This work will involve a similar time-span.

RD You seem to have some question about whether it might be *this* form of opera or *that* form of opera . . .

MT No.

RD Are you presenting a music drama?

MT I don't know. Part of the problem would simply be that the names we give like symphony, for example, but certainly opera, they have a long history. They change in many ways, and yet they are used over a wide range. If I want to specify, of course, it's a work for the theater; and it's certainly a work of musical imagination. That's as far as I can get. The real question you should ask me, is whether I am ever interested in producing it or directing it. Some composers are. I am not. I know quite a bit about the theater, quite a lot about how it is done. I shall have some idea in my mind as to what it could or might look like. But I haven't got the imagination, theatrically, or what is it, scenically, to match my own musical imagination. That is what I would seek in a director. That *he* would have that imagination, and enough sympathy to produce it. That's why I never go to any rehearsals.

RD You never go?

MT No. Unless they want me.

RD You have in the past, dealt with certain ideas of rebirth in some of your texts and certain aspects of intolerance. Are these sort of generic ideas in this opera, on a different tack?

MT I have never wanted at any point to go back into the same material, not so much the same musical material but the same dramatic material. Or the same genre of opera. I want a much wider world. So once I'd done the *Midsummer Marriage* and its extravagant lyricism and its abundance of sound and its illuminations . . . I went straight from there to try to recover an heroic world. A tragic, heroic world. And for *King Priam* I had to learn another technique. So you see, by the time I've reached *The Knot Garden,* I'm in the world of Chekov. All right. If you ask me where I am now, I will only say that the last big piece was a piece, not for the theater, but for the concert hall . . . a more contemplative piece . . . *Mask of Time,* a very large scale work, indeed. Very serious. All right. I can't go back into that world. I won't go back into that world. In works of these dimensions, in a sense, you move from one to the other. So *now* there's a lighter-heartedness. But with me, behind that lighter-heartedness must lie the problems. The problems that a lone dreamer, a woman, might have in trying to face up to what she calls the terror town. Once you're there . . . it appears that you are where you were before, dealing with the past material and ideas, but you're not, because I've never dealt with problems of urban society or whatever. The other concern is where do her dreams or anybody's dreams come from? And can you produce, for instance, characters, apparently real characters you see, which come from another world? All right. Therefore, from that I have to produce a scenario which is quite different, because there are three characters from somewhere and today, and three characters from nowhere and tomorrow. Now that's already a new genre. That's all I can say. Once I get drawn into that world, I begin to live in it. And begin to imagine; and it has endless problems, of what interrelates. The real *thing* for me, Richard, is the pre-affair that goes on for three, four, five years before I actually reach this point. Talking to people . . . I have a certain very limited number of people close to me, to whom I will show that scenario, and talk about it a bit; and listen to what is said. Until *at last,* I see the interrelations of the characters I'm inventing.

RD I understand that. I'd like to bring up something I think you said, about the fact that music shouldn't mirror the times we're in. Did you say that?

MT Something approaching it, yes.

RD But yet this particular project has to do with mirroring something.

MT I don't know . . . "mirror" is a difficult word. I mean, I would use "mirror" as Shakespeare used it. You put the mirror up to nature. Okay? I can never not be aware of the times, yes, but the technique is for me generally surrealistic rather than real. Therefore, the other imaginative elements take over. And that is, to my mind, where the theater is. I go to talk to people. . . . I went and talked to Peter Hall, I said, "Look, supposing I came right out of the opera house and into the theater. What would happen?" He said, "Well, the modern theater. It's clear that what the audience wants is a great deal of imagination and music in the theater. The realist theater is really gone." He made it clear to me that, as there is no subsidized theater in America it would be useless to do it as a musical. He could do it for me at the National Theater in London, but that was it. But what he said to me was quite real. And that is, that at the present time, we are moving more towards a theater of imagination. Take a figure like Berio, who's younger than I am. He's simply steeped in it. All right. The amount of relation you have specifically to the problems, or the notion of "our time" is more difficult. Because after all, if you are an eighty-year-old person like I am, your whole attitude toward your time might be something different than it is for the young twenty-year-old. It's not as simple as all that. But my biographer would, I am sure, make clear to you, as he has done in his own books, that I can't ever be *disrelated* to it and can't entirely depart into a world of total imagination. But then comes my craft, and I have to control it.

RD Of course, that's another thing. Your craft. I don't think you were ever noted for espousing a system, were you?

MT No. I'm not systematic.

RD System versus intuition. Could you say a little bit about that in your work?

MT Well, not a system of musical notes, no. Of course you have a semantic problem, what do you mean by a system?

RD How do you produce order then?

MT That's the point. When I was a younger person, I looked for it always. For instance, strangely enough . . . I'm not very much drawn to the length of Wagner's enormous librettos; but he did it in almost the same way. They are scene by scene by scene. Now, they can be very short, but if you read, *Oper und Drama,* there is a long discussion about how fast you can move in big stories, what you can do and so forth. I had to learn where I could. Stravinsky had the same problem, when he came to it. He wanted to go with Auden into a pastiche opera . . . a pure "number" opera. But I already knew that all operas are "number" operas, in a sense. All right. This already means that there *are* musical numbers. I mean, you see *Porgy and Bess,* it's all in a set of "numbers," believe me! So that is, in a sense, a system, but not quite. You have to begin to order the goings on of who is on the stage and the number of people. Very early on, you know, Richard, I used for myself the method which is used in French drama, published drama, and that is that everything is in small scenes, all the way from Racine and Corneille and the lot of them. And it always gives you before each scene exactly who the people are who are in it. "Scene II," and then comes a list of characters. Very interesting. I then did that for myself, so I would never misunderstand myself. I needn't do it now . . . All right. The next part of the craft, you see, is that if you are going to write the libretto as well, you have to enter into this world of theater in a different way. All right. I go to Shakespeare, who is god for me. And I notice that in almost all the plays, you are *in* the material *instantly.* There is no beating about the bush at all! *There you are.* In the first scene of *Lear,* Lear is about to distribute his stuff to his three daughters; and before you know where you are, the whole thing is put before you . . . In capsule for you, before you even begin. Now, I look at a work like *Don Giovanni* and see that Mozart did the same, or his librettist Da Ponte did the same . . . You've got Leporello outside, then Don comes out, chasing, someone is murdered, and it takes exactly, I've timed it, it takes three and a half minutes. What happened after that was more difficult, you see, because then they had to make a whole opera. It takes a long time before you get that speed again. Wagner's method is quite different. Take the second act of *Tristan.* I don't know how the hell long the big scene is where the two of them are making, not real love, but some type of imaginative love, God knows! It goes on and on and on. *And then,* when Mark comes it's with such rapidity and it's all over and Tristan has been wounded . . . Wagner knew what he was doing exactly. He knew *precisely* that timing, because he used that technique. I had to learn timing; so I *learned* timing. But that timing also concerns the musical gestures, you see. So that already you are in a somewhat different position from what happens if you go and have to talk with a librettist. A librettist has got to understand what you want. It's up to you.

RD Well, it's got to be convenient for you! Yes, of course.

MT I remember going again to Eliot, who is my master; and Eliot said, "If you are very slow in your material and you don't know quite what all your music is going to be, then be very careful if you go to a librettist as such, because they have their own ideas. And you need to know the nature of your own music very clearly before you ask them." Because, he then said, Auden played a "tour de passe passe" on Stravinsky. And now for me, who is not Stravinsky and who has to fight my own way to my own things, he then said, "Don't therefore, go to the librettist. Go to talk to the theater people." So I went to talk to Peter Brook, I went to talk to Rennert in Germany, and tried to learn from them. But I had to teach it to myself.

RD I can see that when dealing with a librettist, there are two different clocks working, aren't there?

MT Absolutely. And if I may put in English operas which were really crazy . . . one was by Arthur Bliss, a very good composer. He said to me, "I am going to write an opera. And it's obvious that I take the best dramatist of the time, J. B. Priestley. . . ." What they produced was an opera very much like *Midsummer Marriage, The Olympians* which came out before *Midsummer Marriage.* But the whole point is that the Priestley libretto was, in fact, set by Bliss word for word. Priestley's notion was an idea of the spoken word on stage. The result was impossibly overlong. They didn't do the elementary thing . . .

RD Preliminary editing . . .

MT Well, I mean, the musician must force, as Verdi and Strauss did, he has to make *his* work of art. That was ingrained in me. Totally. Those were my aesthetics. They were Stravinsky's aesthetics and they were Eliot's and they were quite a clear aesthetic. That even in a mixed art, *one* is operative totally.

RD How do you feel about the success, as long as you mentioned Stravinsky, of *The Rake's Progress?*

MT Well, that's the opera he wanted to do, and he does it. Okay. I find it troublesome, but that's neither here nor there. I find the whole scene with the bread machine a sort of excrescence on it, but I'm not going to argue the toss. It's a lovely work.

RD Some lovely music.

MT Lovely music! You can learn negatively as well as positively. What I learned from that is that you have to be careful . . . but after all, Stravinsky had an absolutely *superb* gift for pastiche of a certain kind. It became Stravinsky as he did it. I'm not that; therefore I couldn't use that kind of libretto, because I couldn't go back to the 18th century. Auden offered me one virtually, you see, he really wanted it. He believed he was a natural librettist for English opera writers. But his whole style was not mine. His verbal style was too ornate for me. But also, he was drawn so back into the 18th-century techniques. I wasn't. That's all. Okay. Okay.

RD There you are. That's his time clock.

MT That's no criticism particularly of Auden or of any of the operas he did. It's simply, you have to find what *you* have to do. If I can't do it for myself without Auden and I don't find the right librettist, then it's my own fault.

RD Well let's go back to your new work. It's called *New Year.*

MT Yes. In other words, it's about the winter solstice, when we think we're going to go over from the bad to the good, only after the new year. Just as in *Midsummer Marriage,* where it's a summer solstice. In a sense, it will be very intriguing.

RD Could you describe it as something that has a resolve or is it a question mark, or both?

MT What do you mean by resolve?

RD I mean in its scenario, in its philosophical development . . .

MT Resolve what?

RD I don't know!

MT Wait a minute. Yes, there is . . . there's a heroine, all right. And who does in the end come through.

RD Well you said, she went up and he went down.

MT Yes. She in other words is strengthened to come out of her capsule into the outer world; although she's a loner. All right. She is going to, we think, she will be able to stand as a loner and a helper and a healer within what she calls the terror town. All right. She's had to find this. And others are not going to do that. And the general public . . . there's a middle act in which everybody's on the stage . . . you might be in the middle of a Berlioz opera or something . . . when that happens, then her dream knocks all the collective over. They just go over the edge. And then it becomes a little frightening. But it doesn't want to be frightening.

RD What about this lighter-hearted aspect that you mentioned?

MT I remember Colin Davis said something to me about it. I said I'd like to write . . . I think I use the term "comic opera," or the Italian term "opera buffa." His face fell about a mile, because he couldn't imagine . . . all right, but then I was meaning a technical thing. It's lighter-hearted, it has lots of tunes. But then, that is a problem in itself. I have real musical numbers. I have to see how Gershwin did it, how Cole Porter did it, how anybody did it. Because I have always been fascinated also by techniques from the musical. But it isn't a musical, because a musical, as Peter Hall quite rightly said, "If you go into the theater and produce a musical, in England, you have to have full houses for three months to get back the cost of the musical itself. In New York it might be double that." That forces you, you see, to write a certain kind of music. I can't. So I have to learn how to do something other, which is avoiding a sort of gala performance opera. But I don't know . . . I'll get there. Also, I want television! And also, the magical world is there very strongly. The whole business about "sci-fi" is a magical world, to put it mildly.

RD Are there any other sources or figures you might draw upon?

MT The figure, of the great masters, which I am closest to is *Ibsen!* Ibsen never produced a stage character until he had found it in the papers. He used the newspapers. That isn't quite true. . . . I find what I want in all sorts of ways. But because you find a lot, you have to sift it out.

RD Let me ask you a question. What about musical thievery? Stealing! I've put that question to some composers, and they often refer to Stravinsky and how "He made it his own!"

RD Oh, I don't ever do that to any degree at all.

RD But your awareness of musical history . . .

MT Doesn't worry me. I don't give a damn. I'm far too strong my own self.

I could take what I like if I wanted to take it. The oddity about Stravinsky really is that he didn't write any long pieces. *The Rake* is the longest. But he does what he's such a master at. He came out of Russian tradition in which the work of art is independent of the time. He flees away, as it were, from whatever happens in Europe and finally gets to California, where he's happy. That's a Russian tradition that comes from Chekov and Turgenev. He couldn't read the others. He only read these works. There the idea is that the work of art is somehow in a magical world which is polarized against the wicked world. I've got that, but I've got the other very strongly. This also explains, you see, the kind of "neo-classicism" which happened to him, after the tremendous influences of the Russian tradition, of Rimsky and the rest of them. You end up with a very sharp "neo-classicism" which came out of the French world and God knows what. Okay . . . I can't do that. I never could do that. I'm never frightened about any of it, nor have I ever been concerned, in all seriousness. . . . After all, if you think, I came into music in 1923 or 1924. . . . That's a long time ago. The fashions that have happened since then, I've been quite independent of them. I'm not ever really worried or moved by that.

RD Well, there was this mid-century domination of a push toward total organization, and you've sailed right through that, I can see.

MT All right. You have to do that with a certain strength when you have to fend for yourself; but I would never try to make that into an objective for everybody . . . that is who I am.

RD It's curious, this great disillusionment with that whole business today.

MT Yes, for the younger ones, I think it's very hard. They have their own problems.

RD In speaking with them, it seems now that the doors or pens are open! They are running like hell in every direction. And maybe that is healthy.

MT I'm sure it is. But I can't answer that for any young person except to say that they should be ready to go into that open world. How they hold themselves, going back to what you said before, how you retain your craft and your structure, that is for you. You see, another thing . . . you take Sibelius, a fascinating character. He wasn't somebody that went on renewing himself, or had to do it. Unfortunately, or fortunately, I don't know what it is, I am like that. I was asked in Germany lately, "Have you ever been particularly influenced by one writer?" And I sort of thought and said, "Leider nicht." And it was quite deliberate. I was quite serious! Unfortunately, I had to find my own way. It seems to me, at the risk of a generalized statement, that you are not born with every gift. You are born with a limited gift. I was born with a very limited gift in the music world. If you want to do other things, you have to add to that gift by hard work. I think the only gift I ever had was the very good gift of timbre and ensemble. The others I've had to learn all the time.

RD So much has been written about you and your life and personal things, everything about you; and you've been a free, free spirit about saying what you mean and what you've done. Is there anything else left to say?

MT I haven't the slightest idea and I have no interest in it. That's for the others to do. I have not even read Ian Kemp's book about myself. I refused. I simply said, "Look, Ian, I will talk to you, because I'm not going to talk to everybody, believe it or not." What he then said was, "What you'll really do then is try and help me get a biography, at least; a chronology, which is quite difficult enough." All right. After that he said, "I would like to come to you and talk about what I think are critical works." And he also wanted to know whether I could agree with him when he had tried to find out what, say, Jung meant to me, whether I had gotten it right or not. The rest of it I said, "You've got to do it yourself. It's yours." I don't really look upon myself as a guru. I am now unfortunately put in that position with younger people. But the world is too wide, too huge. I'm fascinated, for example, about what is going to happen in the world. But I'll tell you, I was once asked publicly in Germany, "Did you ever go to that Mecca of modern music, Darmstadt?" So I said, "Für mich, MECCA ist in Arabia." I was being naughty. I can't take this Mecca business! I mean, it doesn't matter. People went if they wanted to.

RD What were your feelings about Schoenberg and his system?

MT To me, that was always, "*Ah!* And now we have it! . . ." It was phenomenological nonsense. It was alphabetic. I mean, I could trick anybody with that system if I really wanted to. You see, I was by nature, as the Germans would say, "von aus heraus." I really belonged to a polyphonic world which was possibly English in temperament. And that means you make all sorts of sounds by different means, and they cannot be categorized. Purcell was the same. From this you can get B *flats* and B *naturals* at the same time, for example. Which comes right out of Weelkes. When I first heard that in the world of blues and all the rest of it, you see, these things all seemed to be joined together. Now, when I had to find harmony for myself, none of the well-known disciplines, as such, made any sense to me. When I looked at *Harmonielehre* and saw all these lists and lists of all the possible chords in fourths or whatever, I mean I shut the door and said, "That's that!" I've never heard a chord *on its own.* I hear a chord laid out for several instruments; and as my own teacher, who was very clever, said, "And you've only got to put it in a different way; and that chord would sound quite different if you put five trumpets on the seventh or none on the bass." I mean, you can play any trick you like. And it wasn't until I read Vincent d'Indy, who was a very odd figure, who said, "There ain't no chords." I mean really! That it's all *movement.* All right. After reading *Harmonielehre,* I also read the Hindemith book, and in that you take about forty pages to reach *Note 1!* I reached that *1* and said, "Oh, Christ! Shut it up!" Lutoslawski once said a very interesting thing, I heard it on the radio, he said, "The only things that are going to survive are the greater works of art. And most of those, in my opinion, have been written *despite* the systems." Now, I wrote him a postcard saying it was nice to hear that spoken. But I don't want to feel that I know the answers. I don't.

RD The worm turns all the time. But there was such a dominance, I must say, and a surety about all this. And *bang!* It was *the way,* you know. What about someone like Luigi Dallapiccola, who evidently felt that serialism was "Our language?"

MT Well, of course in the end, he knocked it all to six in his own way. What the *hell!* If I go to hear his *Il Prigioniero,* it doesn't occur to me that it's any system at all. I hear it as I do anything else. And therefore, I hear other things in it. I hear the tradition from which it has come, the Italian tradition. Nono is no different in a way. But then some of them could be killed by it. They really shut themselves in. Well, so it seems to me . . . Come on. What happens now is a nice question. All sorts of things will happen now. I would like to live a little longer and see what happens.

RD What is your impression of some of the younger composers?

MT Some of the younger ones, I'd like to see what happens to them a bit. They turn up and they're very young and they are immensely gifted . . . a lot of very, very gifted youngsters. Then comes the problem of character, or whatever. What I want, feel I love, when the young ones come, is both the intellectual strength and some other strength in themselves that is ready for rigor of some kind, and that is individual. It is the first thing I see. I very often see it in the face itself. I'm not accurate at all in that; but I'm longing to hope, and hoping that the real things will happen. My dream, you see, is that a really big figure will come out of America, because I can't see where else it's going to come from.

RD Maybe it's a valid dream, I suppose. But what discourages me from imagining it is this vastness of possibility.

MT Oh, that is right. I meant it quite seriously because I'm not the least bit sure. I remember Barbara Hepworth saying to me, "We are the last generation in which it could happen." I don't believe it for one minute. I think there are other figures. But still, I know what she was trying to say, that she felt the "break" had come. The problem I have with any discussion, say, on Boulez, is . . . where does this bloody "break" take place? Some people thought it took place in Vienna on such and such a date . . . and when does the modern world, when does the avant-garde . . . begin?

**RD** 1945 was supposedly a break. That was the death of Webern and Bartók and the end of the Second World War! What is your view?

**MT** I'm much closer to a figure like Ahkmatova. As Hepworth said, the real beginning of the 20th century was 1914. These figures deeply move me, they lived *in* the bloody thing, in a way which we bland English hardly knew. We did know it, but not in the way that *they've* known it. And they are intensely real to me.

**RD** These meaning . . .?

**MT** Meaning the figures that were of her generation, Ahkmatova, Pasternak, Shostakovich . . . *all* of them . . . not Solzhenytsyn, whom I don't like at all. And therefore . . . I can't think . . . what does a young composer have when he's locked up in say, Czechoslovakia? I *cannot* but not know they're there, Richard! . . . That is what I mean by not being of the world. They are close to me . . . and therefore, they are looking over my shoulder, and I must understand that they have lived in a catastrophic world. Now, one hopes all the time that it is slightly over!

**RD** But is it?

**MT** I don't know . . . I'm always regarded as an optimist. It isn't that. As I said in *The Mask of Time,* what can we affirm? What can we praise? We do it as best we can. If we can do it *at all.* See, a figure that, in my generation, would be very difficult for me, would be Francis Bacon. I shudder when I see these things.

**RD** Why Bacon?

**MT** Well, he's lived in the wrong world. He's lived really in the horrible world, and put it on the canvas. I cannot, I would not . . . Please, this is no objective criticism of Bacon, because he's a great figure; but of my exact generation, okay, I have to go the way I can. And that is all I meant by the original thing . . . I don't put down the actual horrors in that way. *Child of Our Time* is the oddest of them all. That was written, almost, before anything happened. It's now to be performed at Yale in memory of Martin Luther King. Because it's *real.* But it's real because there's something inside it, beside or beyond the obvious. There are some marvelous films which the Library of Congress has done, especially one with an old black man, with his white hair sitting at the piano, saying, "We began in the brothel. And the first thing was, before the customers came, there were the girls and they were behind you and you had to warm everybody up, you see. So you had to learn a different technique, you see. You sat sideways and played like that just to warm up the girls." And he did it! These things are lovely to me because they show you somehow how real it is. Can you see it? And these things go into operas and into the stage and into the theater . . . that's why I love the theater, and performances. Really . . . in the end, that's why I don't go into tape. Because I want performance. As a conductor, you should be delighted.

# *Peter Schat*

BORN JUNE 5, 1935, UTRECHT

**April 18, 1987** **Amsterdam**

We are in the garden of the Hotel Jan Luyken in Amsterdam and it is the day after Good Friday—Easter time. Peter and I have spent many wonderful evenings together over dinner at his 17th-century house (next to a canal) on Oudezijds Voorburgwal in the "red light" district of Amsterdam. I first got to know him on the occasion of the North American premiere of his opera *Houdini,* which I conducted in Aspen in 1979. I had invited him to be composer-in-residence along with his English librettist, Adrian Mitchell, who was poet-in-residence. This is the second of two formal talks. The first occurred in his kitchen on February 15, 1986.

RD There was an important thing that happened in Europe in music—and that was post-World War II Darmstadt. During the summer courses for new music held there, the thoughts of Schoenberg seemed to be carried further. It seemed to be something that composers had to reckon with in one way or another. Since you are someone who actually was there, I feel I can ask you some questions about it.

PS Darmstadt was a turning point in European music. I was prepared to go there and prepared for the whole concept. My teacher, Kees van Baaren, whom I went to in 1952, gave us the musical inheritance of Schoenberg, Berg, and Webern. He was the only Dutch composer who taught these things. So we knew about them. I became familiar with the concept of the series, the concept of the row, etc. These were things that struck me immediately when I had my first lessons with van Baaren.

RD So the idea of organizing pitches obviously attracted you.

PS Certainly. However, that is due not only to my connection with van Baaren, but also because of the Flemish-Dutch schools, the late medieval schools, the pre-Renaissance schools which had the same idea. Schoenberg derived the series, the concept of the row from the "Niederlandische Schule," as he himself said. So from an historical and cultural background, it's not alien to me.

RD What went on in Darmstadt as far as you are concerned? What did you experience there?

PS The newness. The shock of the newness. The new sounds. I think that

PEDDLERS
AGENTS

is also the weakness of it, seen from today. But when one first heard *Le Marteau sans maître* or the *Improvisations sur Mallarmé*—it was an incredible shock! And to hear *Carré* for the first time a little later was like hearing the devastated cities of Germany "in sound." That is how it sounded to me, but this is not an interpretation of mine alone. Boulez has also often referred to the fact that because of this cataclysmic event—Europe in ruins—the post-World War II situation was very important for the development of serialism.

RD You lived through the war as a child, and you once said, "I remember when it started. I remember every day that passed, and I remember when it ended." Spiritually it must have been very significant for you.

PS Yes, of course it was! Yet, it is something whose meaning no one can really quite grasp . . . the Second World War, the Holocaust, Hitler. I think Hitler's 1000-Year Reich is a success. He will be studied for 1000 years. They will study and look into Hell, into Treblinka and Dachau, if only because of the fascination. It cannot really be absorbed in our lifetime. But these are historical events, and they will continue to fascinate.

RD But of course the Schoenbergian manner or organization wasn't necessarily attached to that kind of development. . . .

PS Connect it with the traumatic experience of Fascism. Fascism was understood to a large degree as irrationality. Thomas Mann wrote his book *Das Gesetz* (The Law) against the fascists. And I think that's significant. Fascism is the breaking of the law. It is against the law and Schoenberg *was* the law, the old Judaic idea of the law. He wrote *Moses and Aaron,* and he had this constitution of twelve tones. That was also against the Fascists. So in that sense, serialism was also an answer to Fascism.

RD But as it turned out, it may have developed a new form of Fascism!

PS That has been said before. I think it was Lukács who said, "The new Fascism will look like anti-Fascism." But it was true that after the War and in Darmstadt, the relationship with history was traumatic. There was a revolutionary cult which suggested that we should start all over again.

RD And that sounded new to you?

PS That was the newness. Today I find this to be the weakness of it.

RD After your experience in Darmstadt, you studied with Pierre Boulez in Basel, Switzerland. There you furthered your technique of using serial organization.

PS Yes, I did then work two years with him, and he was incredible! An incredible man. He knew *everything!* Impossible! I composed *Enteleche* I and II in 1960 with Boulez.

RD And *On Escalation* came later?

PS *On Escalation* was in 1967 and it represented a kind of résumé. It was as far as I could go with serialism. It was a mobile score . . . well, you know it, because you conducted it!

**RD** Were you organizing just pitches, or were there other parameters?

**PS** Other parameters also: time, proportions, timbres, rhythms. I can think of about seven parameters. But I think now that this is the big mistake of Darmstadt. To equalize the parameters, to say that all parameters have an equal importance is not musically true. I think there is a hierarchy of musical events, and on the top is pitch, and immediately after is rhythm. They are very connected. And then timbre and then etc. There is a hierarchy, and pitch is the main thing for the definition of music.

**RD** What about intuition? Inspiration? Expression? And so forth. Where do they come in the hierarchy?

**PS** They are not parameters!

**RD** What are they then?

**PS** They are the *most important* things, and they are not "parameterizable"! So you can see that the whole concept of parameters, or organizing parameters is *shit!* Because the most important things you cannot organize.

**RD** Well, OK. There we are. If you were brought to a summary point with *On Escalation,* it obviously meant that there were things that displeased you. What kind of music did you go to next in order to express what you wanted?

**PS** First, let me say that *On Escalation* was dedicated to Che Guevara. He was the hero of those days. And he represented for us a revolutionary attitude towards life in the extreme—giving one's life for one's ideas.

**RD** But that's extra-musical, isn't it?

**PS** That's extra-musical. But it was the *revolution,* as such. And Darmstadt was a revolution.

**RD** A musical one, though.

**PS** A musical revolution and a breach with the past. A new start—all these things. But in those days, you saw that also in other scores, like *Sinfonia* of Berio. After *On Escalation,* together with a collective team of composers Andriessen, deLeeuw, Mengelberg, and van Vlijmen, we made an opera called *Reconstructie* (Reconstruction). This was part of a concept to de-individualize composition. In the opera, there was a very neurotic relationship with history. For example, you could only bring back tonality, in the classical C major sense, in an ironic way. You see this also in Berio's *Sinfonia.* History came back this way, in quotations. . . . We saw in *Reconstructie* Guevara as the Stone Guest, the Commandante, in a roccoco uniform. This was not done with a Socialist hero, the Cubans protested.

**RD** So in that sense you used history.

**PS** Yes. And Don Giovanni was a capitalist. An imperialist. And we used the countries of South America for the female roles. It was a very clear concept. We had a lot of fun with it. We went to a monastery somewhere in the east of Holland and worked there for weeks. Lived with the monks. We had our fun and the monks had their prayers! The relationship with history was the problem. Our experience was that you could not merge serialism–Darmstadt with history. And that is logical. Because it was a revolution. It was a breach.

RD In specific instances, whether because of Stockhausen or others, the idea of melody was cut, the idea of harmonic reference as we had known it was cut. The equalization of most things was ever present. This bothered you then?

PS It was a dead end. *On Escalation* was my limit. I could go no further than that. It made no sense to further the language itself, in complexity or mobility of the score. You had a taboo on melody, because melody was too personal and romantic. Romanticism was connected with fascism. Romanticism was seen as the source of fascism, political fascism. Political romanticism is fascism!

RD Who conjured that? Strauss? Wagner?

PS Strauss and Wagner and all the German thought. And the effect was that Romanticism was blamed ... one of the most beautiful and fruitful musical periods in history. Romanticism was blamed for Hitler! And that is a big shame. In that sense, we succumbed to Hitler and gave him power over Romanticism.

RD After the war, people didn't want to hear any more grand gestures, because there had been too many coming from Mussolini and Hitler. That was translated as being overindulgent, ergo Romantic.

PS ... and dangerous. It was the same thing that the intellectuals did while Hitler lived ... they succumbed to him. They gave up. A lot of them, the majority of them, were very enthusiastic. Heidegger, etc. German intellectuals were very enthusiastic about fascism. We must not forget it. And in most postwar intellectualism in music, they did the same thing. They equated Hitler with the "boss" of Romanticism. But Romanticism was one of the victims of Hitler.

RD That's a fascinating thought. I mean in the sense of Romanticism being a victim, because it is something that comes from the soul, and it does not need a Hitler to dominate it at all!

PS Hitler tried to destroy the soul. It is really a cowardice. It is really the defeat of the intellectual in the face of the horror of fascism. It is defeat to betray Romanticism. I see it now like this.

RD You spoke a bit about Boulez. What about Stockhausen? Did you ever have any personal contact with him outside of Darmstadt? Did he come to Holland in the '60s?

PS He came to Holland in 1969. That was the year of *Reconstructie,* the collective opera. He came to the Holland Festival with *Stimmung* and a piece about Mao Tse-tung with a lot of electronics. His concert was in the Concertgebouw, and it was disrupted by some idiotic imitation activists from the Amsterdam Conservatory.

RD Why do you think they were interrupting? What did they have against Stockhausen or Mao Tse-tung?

PS They were weak in the mind. I was furious when this happened during the concert. It happened two or three times. Stockhausen had to stop the concert and start again; and then again they interrupted *Stimmung*. Not the Mao Tse-tung piece, because it was loud and electronic and they could not interrupt it. But in *Stimmung* they started to sing with it a little bit, making the same kinds of sounds. That was very disturbing. Stockhausen was furious, and he disappeared under the rostrum. Louis Andriessen and I called for a discussion: why this? I went to the rostrum and the microphone and I defended Stockhausen. And he heard this and he knew this. I said these people were *idiots* to disrupt a progressive concert, because it was something new. I started to attack these activists and ask them questions. There was no answer, because they knew they had been wrong to behave as they did. Stockhausen later told the papers in Holland and Germany and the television in Germany that *I* had disrupted the concert!

RD So that was a misunderstanding.

PS No. He knew exactly what it was all about. When he came back recently to The Hague, I let him know, "You have to correct this situation." He said no, he was not interested. He lied about it. He always poses like the Pope, embracing everything, but just like the Pope, he is very controversial in the sense that he pits people against each other.

RD I think there was a lot of envy and jealousy that ran through this period. The competitiveness was on a high level, wasn't it?

PS Yes! But the official ideology was that it was all equal.

RD There are now two things I want to bring up. One: in spite of all your displeasure with organization, or super-organization as it became, you still maintain that it is necessary to have something "outside" a composer for him to use.

PS Not to use, but to have a relationship with.

RD The second thing you recently said was that it's not only important to realize what is thinkable, but what is viable. Let's consider these two points: the idea of having something "outside" yourself as a composer with which to have a relationship, be it a system or an invention of your own musical genetic code. And then this idea of discretion, or making what you conjure not just thinkable but viable.

PS Yes, that's it. The whole problem of Darmstadt, the whole problem of Stockhausen, of serialism, is that you can develop incredible thoughts, complex developments in parametric development; but it has no musical meaning. That can happen. It has happened before in history. It's not the first time.

RD Can you give me an example?

PS At the end of the Dutch schools, you have an over-development of counterpoint and an incredibly complex score. It's a very good comparison with our postwar situation. You had complex riddle scores in those days. How to make riddles . . . you didn't write a score, you wrote the parts; and you can read them upside down, for example, all these fantastic tricks. But this kind of complexity was ended by *Orfeo* of Monteverdi. It had its horizontal complexity; but what was happening at any given moment was just a chord and a melody. Whereas the complexity of the late Flemish schools with all those voices, sixteen complex voices, counterpoint, came to an end. Then you got the regime of the triad. There is a difference between modal thinking, which is more like serial thinking, like thinking in rows and counterpoints; and tonal thinking . . . tonicality I call it, which is the regime of the triad of Rameau. Which is dominant, subdominant, tonic . . . this incredible thing, these triads of chords which dominated music for three centuries.

RD What about the viability of Schoenberg's thinking?

PS Schoenberg had the same dilemma when he presented his first twelve-tone pieces in 1924. He declared to his pupils, his disciples, the concept of the row. If you are free to do everything, you can do nothing. That is what he discovered. So in order to have something viable to relate to, he declared *dodecaphony*. He said, "Thou shalt count from 1 to 12." Which is of course a law. But is it a musical law? You don't count notes when you listen to music! You register distances between tones. You don't count!

RD These things became more and more apparent to you, and you voiced your artistic displeasure. What did you do about it?

PS The first thing I did was to start taking singing lessons. I went to singing classes a few times a week for a while. I had such an urge to sing, and there is such an incredible difference between instrumental compositional development and the possibilities of singing! I remember that Luigi Nono was astonished when I said that I had taken up singing lessons. He was a bit worried!

RD Did he think you were deserting compositional life?

PS I don't know what he thought! But I remember his shock when I said that I had taken up singing lessons. Maybe I was too old for it? But I did it. We can do everything we like, and we can think of incredible developments in instrumental electronic complexities, but we cannot change the human voice. We have used the same cords and the same ears and the same mouth for 100,000 years. History comes with singing.

RD That is a spiritual development also.

PS Certainly. And it's ecological. Musical-ecological. The voice became a musical-ecological problem.

RD What did you get from that? How did that experience manifest itself in your music?

PS That came first in 1969 with *Anathema* for piano, in which I had serial principles and tonical ideas, an E minor melody. It did not become a unity—it remained a breach. You hear the breach between serialism and history. I want a language that is a unity, that does not have this breach. Because otherwise, you cannot be understood. You become unintelligible, mannered, if you have this language with breaches. And spiritually, I feel that serial music became unintelligible, socially speaking. The breach between the audience—the music-loving audience—and new music became bigger than ever. The entire musical dynamic was taken over by pop music. You can't sing away the troubles of the world. It is moving, in a way, the fact that pop music basically uses the regime of the triad as declared by Rameau in 1722 in his *Traité de l'harmonie.* The whole system is old, but it sounds as though it has always been there. There has been a development which was formulated in 1722 into a constitution, a tonal constitution, which pop music has spread over the whole world. If you go to Indonesia, where they have a different musical tuning, slendro and pelog, or to India or China, they have the same transistor radios and the same pop music. It is like an oil slick over the world now.

RD Are you approving it? Are you disapproving it?

PS I do not mind some of this music. Some of it is lovely and some of it is terrible. It can be very good music to move and to dance to and to remember . . .

RD And to be intimate on a very quick basis, I suppose . . . kind of like a drug.

PS Well, you can dance to it. I don't dance to Stockhausen! I'm not looking down on it at all, because that would be silly. If you see some of the influences—for example, the political influence, in terms of the liberation of the masses—pop music has helped. It is incredible what this has meant for people living under dictatorships. And through New Orleans, in the harmonic development of the "blues," it merged into something which is now very important in the world. I would rather see people be enthusiastic about some chords in G minor played by four musicians on a stage, than by the concept of racism or what have you. That doesn't bother me. People like to be together and to go out of their minds. And I don't blame them!

RD I would like to pursue your notion that a composer needs to have a relationship with something outside of himself.

PS There is a need to have something. For instance, look at the fruitfulness of a thing like a tonical C major law: tonic, dominant, subdominant. When you say, "I wrote a Symphony in C major," you thank Mr. C major for the cooperation. You try to do something with C major. You have a relationship with it. You do something nobody has done before with it and you say, "Thank-you, Mr. C major." You need that kind of relationship with music. Schoenberg understood this need. That is why he came up with dodecaphony. But it was not the right thing to do. It was not musically viable. And it was not a law that is inside the music. It was a law that was inside the numbers. You count. Thou shalt count! That is the law of Schoenberg. That is not a musical law, but it is a law. And the whole of serialism came from that law: thou shalt count. The tones, the rhythms, the durations, the timbres, what have you. Thou shalt count. And do it in the computer, also. That's not musical. And that's the whole mistake of serialism. Schoenberg's idea of dodecaphony—that all tones have equal rights, that they must have equal opportunities—*that is a stupid idea! Can you believe it?* All tones must have equal rights, so they must come in an equal number of times. Incredible! This is Socialism, and we know Socialism does not work. This whole stupid thing of saying that notes should have equal rights has dominated musical thinking for 50 years! Sometimes the whole world gets the measles, and doesn't know it!

RD Peter, how did you begin to solve these problems in your . . .

PS Sometime after starting singing lessons, and after thinking about the problems, I experienced, in a split second, a harmonic solution. I can hear it still.

RD Did it have to do with your piece *Canto General?*

PS Yes, I can trace it. I can demonstrate it in the chords in the central section of *Canto.* I maintain the relationship and shift the chord. Then you get a family of sounds. You get a tonality—but a chromatic tonality. What I started to develop was a chromatic tonality. In Darmstadt what one had was a discrimination against the consonant.

RD Unless, as you've pointed out, consonance happened ironically.

PS Yes, but that came only later in the '60s and not in the beginning. The core of Schoenberg, the emancipation of the dissonance, was translated into a discrimination against the consonant.

RD That's an interesting point.

PS So, you have another kind of discrimination! What I wanted was another musical language, in which all the intervals could be used; because I could not stand the idea that I could not listen to a fifth without thinking of Stravinsky. Or the idea that a fifth was something not used in modern music because it was too consonant. In singing, for instance, it is a very handy interval! I hated the idea that thirds and fifths could not become part of a language, because, after all, they are there!

And everything that is there msut be considered if one is to create a language in which all possible intervals are accessible and emotionally acceptable.

RD How did your opera *Houdini* figure in this development?

PS The interval is basic for *Houdini.* Later came the triad—the chromatic triad.

Rd A triad not in just that old-fashioned sense of major-minor ...

PS No, any triad or rather, tri-chord ...

RD Any three notes put together is a triad? Or a tri-chord?

PS Yes, in the chromatic scale. That is important because all triads are used and all intervals are used. I don't like the word system, but how could I unify it? The chromatic scale is so chaotic. There are billions of possibilities between all the twelve notes and all the triads. How to organize it? Where to start? I discovered that there are only twelve possible triads and they can be steered through the chromatic scale, using the entire chromatic scale, which is very important. This became a method of having access to this incredibly difficult chromatic scale. If you consider that you have seven tones in the diatonic scale, the composer has to juggle these tones, to keep them in the air. Composing is not some bureaucratic action—it is trying to play with these seven tones and keep them in the air, using some method of making them sound well. But if you have twelve tones, then it becomes very difficult! The juggler had seven balls, and now he has twelve. In order to keep them alive, he must keep them in the air. It is not a bureaucratic thing. Serialism gets out of this problem by calculating ...

RD ... administrating ...

PS ... administrating the balls. But you have to play with them. That is more difficult. To that end, I declared the regime of the twelve triads over the chromatic scale, and I call it the tone-clock. There is no other tone-system that has as many triads as it has tones. For instance, in a seven-tone scale, you have four possible triads; and in a nineteen-tone scale, you have thirty-one triads. Only the twelve-tone scale has twelve triads. This exists outside of me.

RD The fact is that not two notes (the interval) but *three* notes together defines the triad. And you feel that when you have three notes of any relationship, it constitutes a new harmony. Is there anything more to say about the fact that there are just three? What is this three? Without getting *too* spooky!

PS Ah, three! It is in nearly everything you can think of, in theology or in biochemistry ... the "triplets" ... the "threeness" of mother, father, and child ... the "trias politica" ... with two you wobble, but with three you stand firm. And in the chromatic scale, the triad is steered, or modulated, four times. $3 \times 4 = 12$. Four times these triads and you have the universe of the twelve tones. I sound like a salesman!

RD Yes! I might buy five of them!

PS I'll give you five! But that's another story!

RD You began to develop through *Canto General* and through *Houdini* your method which you now call "tone-clock." Let's examine those two words . . . tone and clock. What do they mean?

PS The word is made of the raw material of music: tone and time. It is not in the dictionary; but in Holland they now use "tone-clock" as though it were in the dictionary. It is a word I made up. There are twelve hours in a clock, and there are twelve triads, so there is a relationship. If you depict the twelve notes on a clockface—for instance, at one o'clock the C, at two o'clock the C sharp, at three o'clock the D, etc.—you have the twelve notes of the chromatic scale. If you put these twelve triads on the clock face, you can make a clock and read the time in music. You can name it and memorize it. That is what I am working on now.

RD Well, there are two things here. One is the principle of the tone-clock being used in your composition for any piece, including the opera *Symposion* that you're working on right now. And two is the actual, physical clock that would stand somewhere.

PS Yes. An actual clock is going to be built in a new "polder." And so that it can be heard *every* hour, without being repetitious, I have made a piece for it, with light and sound. It is a further development of the cycle of fifths, as in Rameau and the *Well-Tempered Clavier* of Bach. I am convinced that no music exists without a tone-system.

RD A tone-system?

PS There is no music without a tone-system. Whatsoever. The Indonesian system, slendro and pelog, or the Chinese pentatonic system, or our chromatic, dodecaphonic system. There is always a tone-system first. So I think that composers who think or who say they have only to close their eyes and they hear an orchestra playing . . . how could this orchestra know in what tone-system to play? Why, in Indonesia, do they have an entirely different orchestra than we do behind their closed eyes? There is no music without a tone-system first.

RD Of course, there are movements today which are trying to amalgamate tone-systems that are not exclusively Western. They are trying to bring Eastern elements into their music as well. I'm not sure, but that may be an abortive gesture.

PS Yes, we are going towards a cosmopolitan music, based on the chromatic scale. Because this scale has so many advantages. It comprises pentatonic, heptatonic, and chromatic things. Whereas other tuning-sytems do not have the same possibility of modulation. It is more malleable. It is the most malleable compromise in tuning.

RD While each ethnic order you mention has its own tonal-system, our system, as it has developed in Western thought, is the most malleable, according to you. Not the only one, but the most malleable.

**PS** Yes.

**RD** And what you are trying to do with that realization is to carry it further with your idea of using the tone-clock while you compose.

**PS** Yes. The problem lies with this incredible, difficult scale—the chromatic scale. There are more possibilities in these twelve tones than there are cells in your body. More possibilities than there are suns in the Milky Way. The possibilities of these twelve tones are so incredible that you are placed in front of something chaotic—it is too much for the mind! You cannot compose with one tone. Someone has said that after twelve-tone music, only one-tone music is possible. And I think there is some truth to that. Indeed, serial music often sounds like a heap of gravel. The composer knows where all the stones are, but no one can see any shape or meaning in it. The chromatic scale was developed out of the diatonic scale. Schoenberg wanted to see the chromatic scale as an independent entity. For me, the regime of the triad is very helpful in having access to the chromatic scale. Every day I enjoy how it works for me. Many composers work for a system; but this system works for me!

**RD** Do you find that this is something that will remain unique to you, or have other composers shown an interest in using your tone-clock theories?

**PS** It is too early to say, because I can appreciate that they have to find their own solutions. And they are right to do that. So it will take some time. I don't mind. With my pupils, I see that it is really very helpful. And that's all I want. I don't want a style or a school. Not at all! That's the last thing I want! Just a helpful tool, which is musically based, so that you can learn to hear the different triads. There are no other triads, after all. If you can master that, I feel you are better off. The chromatic scale presents very difficult problems. That is why there is such confusion in 20th-century music.

**RD** Do you feel that this confusion was brought about by the traumatic event of the loss of "classical" tonality, the demise of "classical" tonality?

**PS** YES! And where are we now? We only use "tonality" to explain Rameau, and for the "classical" C Major thing. In my conception, I call it "tonicality," because it is based on the tonic. C major is *tonical.* Whereas the chromatic scale (Schoenberg's) is in essence *atonical.* Not atonal, because that is a stupid expression. It means "without tones." That means nothing! So you say tonical and atonical. The latter avoids the tonic.

**RD** Peter, do you have any thoughts about the meaning of music?

**PS** Music means my life, for whatever that is worth. I am not a very successful living being, but to me music is life. I live for it. I can't do anything else. In many areas, I am a mistake, but I think in music I can do something. For Bach, music was harmony. The "living-together" of the tones was so important for him, because it was a mirror of peace, to the Kingdom of God. There are some people who say that we owe the entire 20th-century mess to the loss of tonality. I am not so naïve as to say that, because that is magical thinking. But there is importance in music. If music cannot be harmonious, if the tones cannot live together, how can people live together?

RD That's a very big life-principle!

PS Well, I have no mission about it. The one thing I hope is that I am not a bore. That is the worst thing you can be in the world. You might be a murderer; but to be a bore is even worse, I think. Thou shalt not be a bore! That's the main command. Especially for an artist.

*Charles Abbott*

# *William Schuman*

BORN AUGUST 4, 1910, NEW YORK

**May 22, 1986** **New York**

RD You were the president of Lincoln Center when Lincoln Center was brand new and the New York Philharmonic sponsored the French-American Festival in 1965 and the Stravinsky Festival in 1966.

WS That's right. Philharmonic Hall hadn't been open very long. It opened in September 1962. Avery Fisher Hall, né Philharmonic Hall.

RD Yes, I remember being at one of the opening concerts. The hall was done in that dark-blue, concrete material. It was during that time I met you for the first time.

WS I thought the hall was beautiful in those days. I'm not talking about sound, because that's a controversial subject, but I thought it looked very handsome. It hasn't look as well since then.

RD Prior to that, of course, you headed the Juilliard School.

WS For seventeen years . . .

RD Isn't it quite unusual for a composer to take on such administrative responsibilities?

WS I'm often asked that question, and people always have felt sorry for me. And they shouldn't have; I did it because I wanted to. I loved administration and the only reason I stopped was because I reached a time in my life where I wanted to spend the better part of my energies on composition. I didn't take administrative jobs because I had to. I could have taught; or in more recent years, my composition could have supported my family. But, I really enjoyed it. I wouldn't enjoy running Macy and Co., nor would I have the ability to do so. But if you run a school or the Center, you're always involved in the arts. And that is what made it interesting.

RD In the course of Juilliard, for instance, how much emphasis did you put on producing a musician who could cope with, or handle, the music of his time?

WS I tried to do it by getting composers on the faculty. When I started there in 1945, the first thing I wanted to do was to revise the curriculum drastically. I didn't believe then, as I don't now, that the way to study music is to study all the bones of the body and not take care of the body. In other words, I don't think you do it by just studying sight-singing and studying harmony and studying rhythm, and studying counterpoint and studying orchestration and studying history. I think you start with a piece of music. My desire was that the students have a knowledge of music beginning with today and going back into antiquity. And it seemed to me that the best way to do that was to hire composers. So I assembled an extraordinary faculty of composers. You know many of them, of course.

RD Who were they?

WS Well, for example, Vincent Persichetti,[1] Hugo Weisgal, Norman Lloyd, Peter Mennin, William Bergsma, etc. We brought in about twelve or fifteen composers, and got rid of all those theoreticians who were teaching harmony, that is, the bass goes up *four* one week, and if you're lucky enough it goes down *five* the next week! Whatever it is. We started something which was named Literature and Materials of Music, which still goes on. It's sometimes very successful and sometimes isn't. It depends on the teacher. Persichetti is a good example of someone who's brilliant at it.

RD I've known many Juilliard graduates who have said that about him. He is an exceptional teacher. Were you teaching composition?

WS No, when I left Sarah Lawrence, I gave up teaching, because I wanted to teach teachers. And I wanted to teach by revising the curriculum, which is what I did.

RD In terms of your compositional background, whom did you study with?

WS I didn't really have any teacher. You have to remember to place this in history. I'm ten years younger than Aaron Copland. When Aaron was thirty-five and I was twenty-five, he was an established elder statesman and I was a youngster just coming on the scene. This is just an anecdote, but in those days, the difference of ten years was tremendous! About five years ago when Aaron celebrated his eightieth birthday and I my seventieth, there was a program of his works at Symphony Space, a whole-day program, and some of his colleagues were asked to perform some of their music. I put on some songs that hadn't been done before; and when the songs were over, Aaron walked down the aisle, patted me on the back and said, "Bravo, my boy!"! This is an eighty-year-old to a seventy-year-old. And in his mind it hadn't changed from thirty-five to twenty-five! I digress from your question . . . you asked me about my compositional background. In the early '30s, I heard Roy Harris's '33 Symphony, which is not his Third Symphony nor even his Second Symphony, but actually his first. There had been a Symphony No. 1, which is sort of an Andante which he eventually withdrew. But the '33 Symphony had, for me, the most exciting music I'd ever heard. It was abso-

1. Vincent Persichetti died August 14, 1987.

lutely new and fresh. I still think, in fact, that Harris hasn't been given his due in the whole history of the development of American music. He's absolutely an original. In his best scores I still think he's quite exceptional. And his early music just set me on fire. I had graduated from Columbia and I was asked to teach at Columbia that summer. I saw that he was teaching across the street at Juilliard, this must have been 1933 or 1934. And I went across and registered for three of his courses. We got on famously, so I asked him if I could bring him some of my scores. He said sure, so I would trot down every few weeks to Princeton, where he was teaching at Westminster Choir School, and show him some scores, and he would make comments about them. In that sense I was a pupil of his, though not formally. I learned an enormous amount from him, not so much about new music, because he was not as enthusiastic about Stravinsky as I was. However, I learned a lot about older music, especially 16th-century music, which I hadn't known before. He introduced me to that. I was very enthusiastic about Stravinsky, as most people at that age were. It was either Stravinsky or Schoenberg, that kind of silly thing, kind of like Brahms and Wagner. It seems so ridiculous in retrospect.

RD Well, it had its consequences, who went which way.

WS Yes, certainly. Well, he wasn't anti-Stravinsky. He just didn't have my enthusiasm. We got along famously and he was an *enormous* influence on me. But it was Aaron Copland who really launched me. It wasn't Roy. Aaron, as you know, has done more for composers than any other single figure that I can think of.

RD I've heard some composers speak of that fact recently. How did he launch you?

WS I'll come to that . . . Aaron was remarkable because unlike Hindemith, who wanted to turn out little "Hindemiths," he always tried to understand what a composer was trying to do and try to help him reach his own language in his own way. I was never a formal student of Aaron's, but I started to take him scores. To answer your question specifically, I was in a contest in New York, and this was before I met Harris; the judges were Harris, Copland, Wallingford Riegger, Roger Sessions, and perhaps someone else . . . Maybe Bernard Wagenaar, who was on the Juilliard faculty . . . a most distinguished jury. I submitted my Symphony No. 2 and I was given the prize, and this brought my music to the attention of Copland. The only person who voted against me was Roger Sessions and I heard about this. So I wrote Mr. Sessions, saying it would mean a great deal to me if I could show him the score and go over it with him. He invited me to spend a day with him in Princeton, which I did. By the end of the day he was convinced that I shouldn't withdraw the Symphony, and I was convinced that I should! We had a marvelous time. I liked him. He was analytical, objective, and helpful. Then, the

Symphony was played by a WPA orchestra, the first support for music in the United States during the Roosevelt era. But it wasn't to support music, it was to give employment to unemployed musicians. The orchestra consisted mostly of people who had only played in pit bands and could hardly read. A few had symphonic experience. Edgar Schenkman was the WPA conductor. Aaron came and listened and was very impressed. At that time Howard Barlow conducted an orchestra every Sunday on CBS (just to show you how different those times were). The program was called Everybody's Music. CBS put a note in the newspaper that Mr. Barlow wanted to see the scores of new American symphonies. That was the first and last time I ever submitted anything in my life. I submitted a prize-winning symphony and he selected it for performance. When it was performed it got the most terrible letters of protest and CBS showed me those letters because I asked to see them. I was embarrassed and I said to Barlow, "What are you going to do?" And he said, "What am I going to do? I'm going to play it again in two weeks!" What they wanted was audience reaction.

Aaron sent me a penny postcard, which again shows you how long ago it was, saying, "Dear Mr. Schuman, please send a score of your symphony to Serge Koussevitzky, 88 Druw Street, Brookline, Mass." I sent the score and thought that was that! And about six weeks later I got a letter from John Burk, the program editor of the BSO, saying that Koussevitzky had scheduled the Symphony for February and would like to meet me; there would be a pair of tickets at the box office for the next concert, and would I please come backstage afterwards. Frankie and I went there and were taken back by Harris and Copland to meet the great man . . . this was all through Copland. Harris heard about it from Aaron, who invited him to come with us. All I remember about it is that Koussevitzky looked me up and down (you've heard descriptions of Koussevitzky . . . this is an accurate story), and said, "I will play the Symphony, but you must change your name. We cannot have two Schumanns!" I remember just looking at him and smiling and thinking, should I change my name to Koussevitzky or just let it go? Well, I let it go! Then Aaron asked to see other scores and he wrote about my music in *Modern Music* and brought it to the attention of Paul Rosenfeld, who was a leading critic of the time. Then Rosenfeld wrote an essay in the *Musical Quarterly* entitled *"Copland, Harris and Schuman"* which was a great boost to me as a youngster. Aaron was enormously helpful. I remember taking him the score of my Fourth Symphony which, parenthetically, he heard a few years ago at Tanglewood. I wasn't there. He called to tell me it was wonderful. He said, "I never heard this before!" And I said, "You certainly did. I took it to you at Tanglewood thirty or more years ago and you made a marvelous comment about the second movement which caused me to rewrite the ending of it!"

RD That certainly is an example of Aaron's concern about American composers. He is one of the rare ones.

WS If he believed in you he was wonderful. He could be very, very strong and a wonderful critic, but always in a balanced way. You didn't have to agree with him but he might say, "Well, that second movement sounds to me as though it's in another piece." And then he'd give you very good reasons why. Obviously, he was a person with his own aesthetic predilections but he was special in that he tried to understand yours. That's what made him unusual.

RD That's been borne out again and again by even younger composers, who say when they took their music to him that he seemed to have that quality of being able to stand aside from himself.

WS Absolutely. And very few composers could do that. I think in Roy Harris's case . . . if, by accident a dominant seventh chord was slipped into one of your scores, with Roy it was like blasphemy from a bishop. It was out. It was not permitted! He had very strong convictions. Of course, he had a very strong personal language. But he wasn't an intellectual in the sense that Aaron is.

RD I remember meeting him on several occasions. Before we go into some of your works, can you reflect on the atmosphere for music during World War II and afterward?

WS It's very difficult. Everybody writes as though the war itself had an enormous influence on composition at that time. I don't think it had the same kind of influence that it had after the First World War, when Stravinsky wrote *L'Histoire* because he couldn't have any musicians or whatever the practical reason was. In the first place, I had the personal sorrow of not being accepted when I was all set to be a Captain in the Specialists' Corps. I couldn't get in because of some physical disabilities, and Carl Engel, who was the head of Schirmer's, said, "Well, you're a composer, write something." So I wrote something and called it *Prayer, 1943.* I later changed it to *Prayer in Time of War.* All that was changed was the title, and Nicolas Slonimsky wrote a book in which he compared Schuman's two prayers. One felt very much aware, as any citizen would be, that a war was going on. I never felt that composers were especially affected by that war. Of course, the events of the day cannot help but affect everybody, and naturally it comes out in music. But to try to give it a specificity I think is on thin ice.

RD So you think, just "business as usual" as far as composition went?

WS I would think so, even though obviously, your mind was not "as usual," nothing was "as usual" because of the war. But it was a very different feeling, if you compare that with the Vietnam War, when very few thinking people wanted it. World War II was a very popular war, a great common cause, that was the great difference. It was a very optimistic time in American life.

RD Now, to some of the works that we did in Aspen on the occasion of your seventieth birthday in 1980. I think we designed them to have a chronology, at least a cross section of a lot of things. I programmed and conducted many of the works and if I can remind you of a few of them . . . the String Quartet No. 3 (1939), Symphony for Strings (1943), *The Young Dead Soldiers,* that's 1975; we had the *Variations on America* which has that connection with Ives (1962), we had *In Sweet Music* which is 1961 . . .

WS No, no. *In Sweet Music* is much later than 1961. It was in the '70s.

RD Didn't it have some earlier origins? There was a transformation to it.

WS That's right. It was based on the song, "Orpheus with His Lute" and *The Song of Orpheus,* which was a cello piece. That may have been in the '60s.

RD And also, I think there was some reference to Billy Rose?

WS That's right. Rose had commissioned the original song.

RD Then, other pieces that were done at the same time in your presence: *The American Festival Overture,* the Piano Concerto and the Violin Concerto, *New England Tryptich,* and we showed the film *Night Journey.* With those works as a kind of sketch of things, how do you perceive your style, in terms of how it changed? Or did it change?

WS Well, you know the wonderful thing is that I can't analyze my own music, and I don't try. I let others do that, it's better for me. I'm not good at that. I don't like to talk about my own music in that sense. It's all of a piece as far as I can see; but obviously, I'm sufficiently objective to know that harmonic language changes and develops over the years; lots of things change, we change. But I wouldn't know how to be more introspective than that about it.

RD There was one emphasis by composers of your generation, and that was the symphony itself. What was that all about?

WS It never occurred to me to write anything except symphonies. Before I could write anything I wanted to write a symphony. That seemed to me to be the most logical thing in the world.

RD Was it sort of the traditional thing to do?

WS No, I never thought of whether it was traditional or it wasn't traditional . . . I wanted to write symphonies. I never thought about it. Of course, I was very naïve. I came to music self-taught.

RD Do you have a remembrance of your experience as a performer?

WS Yes, I had a dance band. I was the leader of the band, I played the violin and the banjo; and I was the vocal soloist. I wasn't very good at the banjo; I was even less good at the violin. But I was a marvelous vocal soloist! Our big number was a song called, "My Sin" which, if you're not nice to me, I'll sing.

RD And so with that you did . . .

WS *Dates!* Burton Lane, of *Finian's Rainbow* fame, played piano with us for a while. We had very good musicians and we just played dates, whether college proms or weddings or bar mitzvahs or confirmations . . . you name it. It's a wonderful experience for any serious musician, because you really had to read. I did a lot of arrangements for the band and I got paid to do them later on.

RD But the banjo?

WS I loved the sound of the banjo. There used to be a commercial for Cliquot Club Ginger Ale and they had a wonderful banjoist. I loved the sound of it, so I bought a banjo in a hock shop and learned to play it.

RD Do you think this dance band experience filtered into your thinking when you composed?

WS Not the dance band so much as the fact that I was brought up on popular music. It made American speech a very natural source of musical expression for me; and I think the melodic turns of my music are based very much on American speech patterns. (I was nineteen before I heard serious music for the first time.) For example, in the String Symphony, the last movement certainly affects that idea . . . (singing) Ba-da Ba-da bum/ta-*Dee-dum*/ta-*Dee*-dum/ta-*Dee*-dum/ba-da-da-*Bee*-da bum . . . I suppose they're really jazz figures. They are difficult figures, but I think they stem from jazz in some way.

RD Well, the way you sing it, I mean, it's almost like "scat sound."

WS I wrote "scat" music. One of my first published choral pieces had really "scat" sounds and I did it again in *In Sweet Music*. I love the idea of making syllables that sound the way the music sounds. For example, Randall Thompson had a Second Symphony that was widely played at one time . . . (singing) *Dum* da-*Dum* da-*Dum* da-*Deee* dum/*Dum* da-*Dum* da-*Dum* . . . And Koussevitzky used to sing it . . . (singing) *Hul*-la/*Hon*-da/*Heen*-da/*Hoy*-da . . . A kind of Russian thing. He could never quite get it. I think there is a natural sort of syllabilization that goes with music, when you sing.

RD Perhaps a little catalogue of all the "ta-da-das" and the "loy-loy-loys" and the "hoy-hoy-hoys" should be made! So much for syllables and American speech. You've set a lot of words to music. Do you have a particular system?

WS No, but when I've set words to music, it all stems from my setting of popular song lyrics, because I like the naturalistic setting of words. Britten, for instance, makes an art of not setting them naturalistically.

RD How do you mean, naturalistically?

WS Well, if I would sing . . . "Where am I go-ing to-day," one note per syllable, I would say that would be a perfectly literal setting of the word. Whereas, you could set it in a more distorted way.

RD You mean a melisma is added, in some way.

WS Yes, or a purposeful non-literal setting of the word. I like to set words naturally.

**RD** And not repeat a word?

**WS** No, I don't mind repeating words, although Ned Rorem has taken me to task for that . . . He said you shouldn't repeat words.

**RD** That's his idea, that a poem should be about the same in a piece as it is without music. Other composers don't think so.

**WS** There's a lot of merit to his point of view. But I think when you set a poem to music, you're trying to add an element that is particular to music. That's what you're trying to do. You're not trying to improve the poem in any way; you're trying to give a different *something*, a different reason for being. I think generally speaking, he's correct. But I wouldn't be a slave to his theory.

**RD** I've heard composers say that they "rape" a poem. I've heard composers say that they really don't think that a poet would object to a fragment taken or rearranging the order of poetry. Do you feel that way?

**WS** It depends on the poet. If you're using Edna St. Vincent Millay, she would be very, very strict. You couldn't change any punctuation; everything had to be just so. But if you take a Walt Whitman, who is a poet of thought and line rather than form, I think he can withstand much more handling. Because you can't start setting all these verbose things of Whitman . . . there isn't that much time in the world. It depends on the poet.

**RD** And what is your view about the subservience or domination of either music over poetry or poetry over music?

**WS** I think that if you set a poem, your obligation or your goal is obviously to produce a different result; otherwise, why set it? I don't believe in the business of "raping" a poem. I think I choose a poem because it means something to me. I feel so strongly about it that I always put in the score, "Please print these words in the program book." I think it's very important that the words be heard and understood in a performance.

**RD** You've dealt with poets no longer living. What was your experience with *live* poets?

**WS** I just had a recent one. I'm doing a work on a consortium commission from six or seven orchestras, and it will be done for the first time in the fall. It's a theme derived from the Statue of Liberty celebration; and I asked Richard Wilbur to work on it with me. He's written an absolutely marvelous text. We did it as a collaborative effort, by discussing the kinds of thoughts we wanted in this particular piece. I find that a very natural way to work. Poetry is filled with rhymes, "eye-rhymes," that don't sound like rhymes, they only read like rhymes. And so for music, you obviously have to have something that sounds like a rhyme or it doesn't come off. You don't have the element of sight. That's the first thing you have to explain to a poet, the element of sight is missing. Sight is very important in the writing of poetry; but in music, it doesn't mean anything, because you don't have anything to look at . . . you only hear it.

RD That's one form of your imagery.

WS That's right, you have to have that. And the other is, I don't believe that you can have very great subtlety. In lieder, for example, you can be as subtle as you want. But if you're writing a piece for chorus, I think you have to be very, very clear and the text has to be clear. It isn't the place to probe the depths of philosophic thought.

RD You mentioned lieder. Are there any examples of composers who have written lieder in the past who have served as a model for you?

WS I think all the great lieder writers had the words uppermost in their minds, when they set them. When I think of Schubert and Schumann and Brahms and Wolf and Debussy and Duparc and Fauré . . . I think they set the words in a most telling manner. I think Ned Rorem does, too. The words create the atmosphere. I think Sam Barber did that very successfully, too. There are not too many successful American writers of art song.

RD What about your own dramatic experiences, in terms of the stage?

WS I haven't written much theater music. Mostly what I have done is ballet music. I've done four ballets with Martha Graham and one with Anthony Tudor called *Undertow,* which created quite a stir because of his choreography. I don't know why I didn't do more in the theater, because I am very fond of it. Given a choice, I would always rather write a symphony, I guess.

RD Which brings me to the next question. How difficult was it to get a performance of your symphonies?

WS I was very spoiled having Koussevitzky's interest. You can appreciate a conductor being in a position to say, "Everything you write, I will play. All you have to do is write it and I will play it." How extraordinary! So I grew up knowing the Boston Symphony would play whatever I wrote. I just took a virtuoso symphony orchestra for granted.

RD What was your association with the poets or painters of your day?

WS Really none. I grew up in a completely bourgeois atmosphere and background, there was no serious music in my family. My father played the *William Tell Overture* on the pianola every morning before going off to work, and we sat around the piano on Sunday nights and sang. I was in my twenties before I became part of the "musical scene."

RD What about Varèse? Did you have any contact with him?

WS Only that I played the lion's roar in the first reading of *Ionization.* I was not at Juilliard, but I was walking down the hall, and Nicolas Slonimsky was there and he said, "Come in, I'm doing a reading of a new piece of Varèse and I need you to play the lion's roar. I had read about it in an orchestration book, but I didn't know it. So he said he would show me how to play it . . . I'd met Varèse socially but I didn't know him at all.

RD Do you have any urge to conduct your music?

WS Not at all. I was a choral conductor. I did that for many, many years. The reason that I don't like to conduct my own orchestral music is, that I am not good at it. I couldn't rehearse it. I wouldn't know how to handle the complexities. I wouldn't know where to go, I wouldn't know what to do. I've worked with all the great conductors of my time with the exception of Toscanini, I was too young for that, and there's nothing like a wonderful conductor to work with.

RD You mentioned Koussevitzky several times, what other conductors do you consider champions of your music?

WS Bernstein, of course, was the leading one, and Fritz Reiner, believe it or not, when he was alive, and Monteux, and of course in the present day there are a number of them. Leonard Slatkin would be one. He has a very good place in St. Louis, he's developed a first class orchestra and does lots of new music there.

RD During your career as a composer, did you have a view of what was happening in Europe?

WS Yes, I was aware of it in terms of what would almost sound quaint now. I remember my excitement in seeing the *Mathis der Mahler* score for the first time. And so, of course one was aware of what was happening in Europe. But I never felt the need to go there to study myself. I've been there very rarely.

RD You mentioned the originality of Roy Harris. What about that word, originality? Who's original?

WS Well, I think that's a wonderful question you ask because originality only applies to the first time you've heard it. The next day it's no longer original. I don't think originality has much to do with quality; it has to do with novelty. Ives is obviously an original voice, but his place in history will not be because he was original; it will be because of the quality of his music. And what that quality will be one doesn't know yet. I'm not as enthusiastic about many Ives works as other people are.

RD What about the notion of pluralism and the broad approaches today. Do you have a word about that?

WS *Yes!* I think that one of the worst things that's happened today is the idea of equating the worth of all different *musics* . . . "musics" now has become a plural word . . . in the last twenty years. When you look at some of the evaluators in the press, and you read the reviews, there will be a concert review of the most revered master of our times, and the next review will be of some pop artist. Everything is equated and everything is supposed to have equal value. I don't believe in that. I think that journalism, for example, by design and intent, is meant to be ephemeral. It's not the same as literature. And every once in a while you get a journalist who is so extraordinary that he's more than a journalist; there are various examples that one could cite. I think to say that pop music is the equivalent of a Beethoven symphony is absurd: one storms the heavens of intellectual and emotional depth; and the other is musical entertainment. Now they are both perfectly worthy, but you don't say that *Life* magazine is the equivalent of Tolstoy. It has a very different function in the world. I think all these things have different functions and not to honor the special function is to be a snob; but all these functions are needed and useful. But to try to equate them I think is *sick!*

*Charles Abbott*

# *Witold Lutoslawski*

BORN JANUARY 25, 1913, WARSAW

**March 21, 1987** **Philadelphia**

I drove down from Connecticut to Philadelphia to hear the Philadelphia Orchestra with Witold Lutoslawski conducting an entire concert of his music. The next morning, a Saturday, we spoke in his suite at the Barclay Hotel.

RD I would like to set the stage, by asking you briefly about the Warsaw Autumn Festival.

WL It started in 1956, and then the next was in 1958. And then every year with one exception, 1982.

RD Because of martial law?

WL Yes.

RD I choose this departure because historically what was going on at the same time were the musical initiatives at Darmstadt. But first, the circumstances that brought this wonderful festival to light. It is said, in your material, that you were instrumental in organizing it, or helpful in organizing it.

WL Yes. In fact, the originators of the Warsaw Autumn Festival were two composers who died prematurely: Kazimierz Sirocki and Tadeusz Baird. They asked me to join, and I of course did. But it was their initiative. It started earlier as a festival of Polish music, but it only took place once. Then in 1956 it was an entirely different kind of artistic manifestation . . . an international festival of the music of the 20th century.

RD In speaking to your colleague Penderecki, he describes it as a wonderful moment. He was much younger; but he spoke about the change of atmosphere that allowed such a thing to happen in 1956. That is to say, the freedom to invite international figures and in fact bring international contemporary music to Poland. He described a certain isolation before that, as did Ligeti in Hungary the same year. The musical isolation before 1956—could you speak a little about it?

WL It began after the war, maybe not in so sharp a form; but the decisive moment was in 1949. There was a Congress of Composers organized by the authorities; and the Minister of Culture spoke to us for four and a half hours, giving some ideology for Socialist Realism, and against Formalism. What those terms mean, I couldn't tell, because I don't understand them. I don't know what Formalism is. Anyway, my First Symphony was played during the Chopin Competition in 1949 at a concert for guests of the competition, the members of the jury, the participants, and for the general audience. It was the last performance of the First Symphony before it was banned. It aroused the fury of the authorities and also some members of the jury coming from the East. Ten years later, Leopold Stokowski conducted it with the Warsaw Philharmonic. During that period, after this Congress, frankly I was terribly depressed. I thought that it would last forever. So for me it meant that I would write what I wanted and never hear it—just put it in the drawer for somebody who might possibly discover it in the future. I could only earn my living by writing some functional music. By the way, I even liked it! It had nothing to do with what I wanted; but I liked writing easy pieces for schools, children's songs, or music for small radio ensembles, etc.

**RD** In your material it says that you wrote a lot for radio plays.

**WL** Yes. I couldn't say that I disliked it. I always wrote with a certain amount of pleasure. But of course, it had nothing to do with what I was aiming at. At the same time, I worked a lot on my sound language; because I decided in 1947 that what I was writing then would lead me nowhere. I decided that I must begin from scratch, from zero. I worked a lot on harmony, melody, polyphony, etc. The first examples of this work came late, after ten years of work. It was in my *Five Songs*—they were probably the first pieces in which I applied some results of my endeavors during those ten years. But at the same time I composed not only functional music, but also something that later on was taken into the repertoire by others. During the work on functional music, I developed a certain kind of style: a style consisting of folk, diatonic tunes, combined with non-tonal counterpoint and some colorful harmonies. In such a style I composed my first set of easy pieces for piano in 1945. And in 1946 the *Twenty Old Polish Christmas Carols,* of which I just recently made another version for small orchestra and choir . . . I performed it in London twice. Then one day I considered this style to be useful for composing something more serious, and that was the Concerto for Orchestra, 1950–54. But of course, this was not the music that I really wanted to compose. It was the music that I was able to compose; I was not ready yet to compose what I really wanted. Some think that it was the pressure of the government that made me compose with folktunes. No! It's absolutely not true—a sheer misunderstanding! You know, the tradition of folk-music in Polish serious music is very old. It goes back to Chopin, and was also taken over by Szymanowski, who wrote the famous ballet music *Harnasie,* in which he used some folk melodies of the Polish highlanders. He also wrote some choir songs and solo songs based on folk melodies of other regions. I began writing with folk stuff as a raw material as early as before the war.

**RD** So it was a real decision. You wanted to go through this rich body of material.

**WL** Yes! Absolutely. It interested me, but never very profoundly or very deeply. It was composing functional music. Yes, it was very useful; but it didn't interest me as profoundly as it interested Bartók, for instance. Bartók was so involved in research and also in using folk melodies in his most serious compositions. This was not my case. I used this kind of material in the Concerto for Orchestra because I was not ready yet to realize what I wanted. It had nothing to do with the regime or with pressure. It's very often misunderstood. Some people write comments in program notes that I was compelled to use folk melodies. It's not true at all. How could they make me write something that *they* wanted? With a pistol at my head? No! So it's really ridiculous to say that. Very often the writers don't really understand what happened in Poland. There were very few people who really wanted to follow those official ideologies . . . there were some . . . I mean, unimportant composers who made some compromises . . . or maybe not even compromises . . . they wanted it. But they were unimportant composers. Otherwise, there was a strong opposition to all that. And disdain for this ideology which was entirely artificial.

**RD** I have asked some composers to describe their musical genetic code. That doesn't mean just who they studied with, but rather, what went into the blood. And you say you started with the folk elements of Polish music.

**WL** No, I wouldn't say that was my starting point.

**RD** Was it one of the aspects of your genetic code?

**WL** Perhaps, but not terribly important. After all, nothing of that has remained in my present music. Because I never wanted to include folk stuff into my, as you said, blood. It was sort of an episode. It served as something that replaced what I was not able to do.

**RD** Well, that's different.

**WL** Absolutely. Very, very different. Different from Szymanowski's approach, and especially Bartók's approach. Entirely different. When I was ready to realize my first examples as a result of my work on sound language, I just abandoned folk stuff, because it didn't interest me so deeply. So that was the end. In 1955 I got rid of it. Since then I have never used it.

**RD** Can you speak about the concept of a musical genetic code?

**WL** Yes, I can, of course. You know, I think that there are two sources of tradition in 20th-century music, or the music of our time, now, in the second half of our century. I think that one source is obviously, known to everybody, the Second Viennese School, the Schoenberg doctrine and his music, pupils and followers. I had very little to do with that. There's practically no trace of twelve-tone doctrine in my music. Even if I use some rows containing twelve different notes, I think the very idea of the twelve-tone row doesn't belong to Schoenberg exclusively. It was in the air. It's quite natural that total chromaticism is something which is a question of our time. It is a natural step in the development of music in our time. But I think the other source from the past is the Debussy tradition. Debussy-early Stravinsky-Bartók-Varèse, that's a sort of line to which I clearly belong.

RD Did you know about Varèse, did you meet him?

WL Yes, I did, a little bit. I met him in 1962. Of course we spoke French. He was very skeptical about France. "Nothing happens there," he said. I think he left France very early.

RD Yes indeed. He went to Berlin and then to America. I've been a great, great fan of Varèse's music.

WL Yes, a great composer.

RD What struck you about Varèse?

WL I'm a composer of harmony. The notion of harmony plays an enormous role in my musical thinking. And Varèse was a harmonist as well. A very peculiar one. Which is not the case with the Schoenbergian school, because they based their musical thinking on something different. Harmony was a secondary element.

RD Well, you're leading into a very big topic that I want to get to, but let's come back to the Warsaw Autumn . . . certainly some of Varèse's music was performed there. In comments by Ligeti and Penderecki, I am astonished when they reveal that they heard *The Rite of Spring* for the first time in 1956 or 1957, and on the radio, yet!

WL Well, it was in the first Autumn festival program and really strangely enough, it was the first performance in Warsaw. I think it was due to the very high cost of it. Earlier we didn't have all of the instruments that are required in *The Rite of Spring;* no funds to engage such a huge orchestra.

RD But you wouldn't have heard it for the first time in 1956.

WL Oh, well, I knew it! Perfectly.

RD From before?

WL From the radio or records . . . after all, I was in a different situation from the people you mention. They are younger. Before the borders were closed in 1949, in Poland, I was already traveling. Between 1946 and 1949, I was twice to France for stays of three months each. And I was in Copenhagen and Amsterdam for the ISCM Festivals. So my contact with Western music was much earlier than those composers you mention. And that's why 1956 was not such a revelation for me. Although it had a big influence on me, because of the state of depression that I mentioned before. I was depressed, and when the borders were opened, when the music of the whole world came to Warsaw, and we became one of the centers of contemporary music, then of course, that had an influence on my psyche. I couldn't say that it influenced my work on sound language. I was not influenced by the Darmstadt avant-garde. It was terribly alien. I should say that I felt terribly lonely when I realized that everything around me was like that. Followers of Webern's music and those people like Boulez and Stockhausen were the leading figures . . . I was terribly sad. Because it was so alien to me. When I was on the jury of the ISCM Festival in 1959 in Rome, and all the scores that were sent in were of this kind, I thought to myself, "I have nothing to do with this. I'm a lone wolf!"

RD That is the area I want to open up. It could be called a kind of "Age of Disenchantment" in one way, because of the crisis that many of these composers I have spoken to have gone through. Crisis in the sense of accepting or not accepting total serialism or finding themselves being or having been led down a path which they finally had to reject or find something else.

WL Find something else! Because the Darmstadt avant-garde of the 1950s to my mind went bankrupt. There's very little music of that time that has remained. There are very few things. *Le Marteau sans maître,* perhaps, or *Gesang der Jünglinge* or *Kontakte* or *Zeitmasze.* The pieces that really remain are something important. But Ligeti, for example, didn't really belong to this movement.

RD No. He divorced himself quite early, and he speaks about it quite clearly. But there are parallels that get tossed about. For instance, these things that were probably in the air: multiple conductors, which you have used, the idea of signaling in the composition . . . signaling the entrance-exit, which you use, which Boulez used, for example, in *Improvisation sur Mallarmé II . . .*

WL But it's rather different. My use of cues is to start some ad libitum sections, and this is absent in Boulez's music. He rather has a sort of disdain for it. He doesn't want to leave anything to chance. While for instance in my music, in my ad libitum sections, I'm not interested in different versions of the same piece. It's an enrichment of the means of expression. Enrichment of texture especially, and richness of rhythm, which is not to be achieved in any other way . . . this combination of ad libitum parts together, you know perfectly well, because you have conducted my music. And you know what it consists of. The result is always something very flexible, maybe not quite clear to foresee, but there is no difference, basically, between particular performances. Because the very nature of the music is such that you couldn't notice the difference. I think the origin of my thinking with the element of chance was different. It was triggered, in a way, by a very significant moment in my life, when I just by chance caught through the radio the Concerto for Piano and Orchestra (1957–58) of John Cage.

**RD** Yes. You mentioned that Cage lit some sort of fire in your mind about chance. His music obviously is so different, but tell me about what it was from Cage that set you on fire.

**WL** To understand it fully, we must admit one thing. You know perfectly well that composers sometimes listen actively. That means what they actually hear is just a stimulus to set in motion their own imagination. They really don't hear what comes to the ear. This is just an impulse to set in motion their imagination. That was the case with Cage's Concerto. When I heard it the next time, I couldn't recognize it! Because it was my music that I really heard. But you know, Cage's music made this spark for the gunpowder in me! And when I finished the first piece written in this vein, *Venetian Games,* John Cage had asked me to send something for his book, *Notations* (1969), a page or a sketch or a whole score, I sent him *Venetian Games.* And I wrote to him, "I owe this piece to you!" And he wrote back a wonderful letter. Then he sold the manuscripts that were sent to him to a Foundation . . . I think it was for a fund for some young students in composition.

**RD** There's another name that also appears in your material, aside from Cage, and that is Xenakis, with whom I also spoke. He does claim certain priority on some technical devices: complex divisi strings and the glissandi and such things. It is mentioned that your approach had nothing to do with Xenakis . . . even sound?

**WL** Well, maybe there are some similarities, because you mention glissandi. Yes, I use glissandi. Divisi is not a discovery of Xenakis, after all. There were divisi before him! But of course, early Xenakis works especially appealed to me because I found them original. And especially with the background of the Darmstadt post-Webernism it was something that was fresh. I liked it. Now his approach is purely mathematical and terribly alien to mine, because I'm a composer of the ear. Although my scores may seem to be very carefully calculated and organized, the priority of the ear and sensitivity is present in my music. I think that's obvious to you—you have conducted my music. And it's not the case of Xenakis. I think he doesn't need such a great sensitivity of the ear for his music. I saw and I heard many times a piece that he composed for the competition for cello players. Rostropovich asked him to compose something as the required piece for the competition. And in the score it read, "Avoid a beautiful sound." That's entirely alien to me. Because I would like to restore the beauty of music which was so neglected in recent decades. There was such a lot of ugly sound, and I want to restore the beauty!

RD I asked Xenakis quite pointedly. I said, "When you don't use a system, what is there, Iannis?"

WL What did he say?

RD He said, that after you get used to the theoretical means you have developed, it becomes second nature so you don't have to calculate. Then he feels he knows what it means musically. I guess you could say that is the moment his musical intuition takes over.

WL Thanks to *that* he is such a good composer. Because without that he wouldn't be. If he composed only according to his doctrines, I don't think he would be really Iannis Xenakis. I'm not terribly interested in his doctrines, his methods of composing. I think he is a good composer in spite of them.

RD You use this word "chain" as a title for pieces. Does it have anything do with Markov Chains?

WL No. I use the word "chain" because mainly in my *Chains,* the music consists of two strands. There are two independent plots in which sections begin and end, not at the same time. That means, they overlap each other. And if you look at *Chain III,* for instance, the last piece . . . I did it for the first time with the San Francisco Symphony in December of last year . . . you can see that something is being played by a group of instruments, and in the middle of it, another group begins something entirely different. When the first group ends, the second is still going on. So it's a *chain,* a linking, I wanted to avoid this tradition that is present in almost all music of our time, the division into sections with common ends and common beginnings. I wanted to create something new in the field of sequences of musical thought. And that was original, I hope. I used this kind of thinking even in earlier pieces; but *Chains* are specific and maybe this method is used in them in a greater extent.

RD Yes, I can see flashes of the method even in the Concerto or the Third Symphony.

WL Yes, of course. But even in the Concerto, the Passacaglia is composed in chain-form; because the theme never begins together with the variations, with the exception once in the middle. But the theme repeats in the same way all the time, and the beginning is not the beginning of the variation. The variation always starts somewhere in the middle of the theme and goes beyond the end of it. It's not entirely new, but I found some new possibilities. In *Chain III* for instance, the combination of absolutely different moods in each section produces "multi-moods."

RD Was "mood" something that was given much consideration in Darmstadt?

WL No! Disdained.

RD You would agree, however, that Darmstadt caused a lot of questioning, a lot of thinking about musical problems?

**WL** Oh, absolutely. It was a great contribution to the development of music of our time; because it was a strong reaction, especially in Germany after Hitler's time, where *all* progressive music was banned.

**RD** However, you did adapt some of the chance elements which were in the air.

**WL** Yes. I *adapted* them in my own way. Because what I am interested in is not leaving important things to chance. The full control of pitch is present in my chance operations. Of course, it limits the use of chance ... but also it gives the music a certain staticism. That means that if I keep full control over the pitch organization in ad libitum sections, they become harmonically fairly static. And that's why I can't use it as only one single means for a piece of music. You have conducted my Fugue for Thirteen Solo Strings, and there is a classic example of what is ad libitum and abatutta as a role in the harmonic structure of the piece. The traditional Baroque fugue consisted of two kinds of music: static (harmonically static) and moving. The themes were static where it was in one single key, possibly modulating at the very end. And the bridge passages were modulating—so, harmonically active. I composed my fugue in such a way that the themes were aleatoric, so, fairly static harmonically. And the bridge passages are *conducted,* in order to introduce the harmonic changes. That was the analogy with Baroque fugue. But understood in my own way.

**RD** There is another piece that I conducted, and that was not in Aspen, this was the Concertgebouw—your *Novelette*—I did it right after Rostropovich did it. It was the first European performance. What about *Novelette,* in terms of what we're talking about now?

**WL** It was a piece that Rostropovich wanted to have during his first season in Washington with the National Symphony. But I was late, and finally when I gave it to him, it was no longer the first season, it was already the third season, I think. 1979. He conducted it; he conducted first a Dvořák Symphony. After the intermission he conducted my *Novelette,* and then he played my Cello Concerto with me conducting!

**RD** Amazing man!

**WL** Superhuman!

**RD** I don't think the *Novelette* took other paths than what we are talking about here.

**WL** No. I think it's very characteristic. It was composed during the period of my work on the Third Symphony. This period for the Third Symphony was pretty long—about ten years—because I discarded the whole main movement after two years' work. And I put it into the drawer and started a new one, which is now present in my score. During that period, I composed quite a few pieces like *Mi-parti,* also *Les Espaces du* *sommeil,* then Double Concerto, *Novelette,* and some smaller pieces like *Grave* for cello and piano, also the piece for Paul Sacher, *Anniversary,* and also a piece for oboe and piano, *Epitaph.* All that was composed during the period of work on the Third Symphony.

RD All right. Just to retrace once more: you mention Debussy and Stravinsky as part of your musical code. How?

WL I think that especially Debussy and on the other side Schoenberg were people who in a way put an end to the tonal system. Although Debussy was much more connected with the tonal system than Schoenberg, in his later works. But still he discovered things that gave an enormous prospectus for the future—to go beyond the tonal system and develop total chromaticism. He pointed out a way that could be followed. And I think that just the composers I mentioned—that means early Stravinsky, not the late but early Stravinsky and Bartók and Varèse followed—and then Messiaen in a way as well. So that means approaching total chromaticism from quite a different side. And I think that Debussy was the first to partly abandon the tonal system. Partly of course, because elements of the tonal system are still present in his music.

RD Another composer, older than you—Michael Tippett—speaks of it in a linear sense. Because he flatly denies that he has any harmonic system either. But he conceives of his "harmony" as linear rather than vertical. I don't know if that makes any sense to you. . . .

WL Well it does, yes, of course, but I'm a composer of harmonies, myself. That's why I'm closely connected with the Debussian tradition, and the Bartók tradition, and Varese and Messiaen as well. They created their own harmonies. I'm very sensitive to the vertical. It doesn't mean that I don't use polyphonic music, as you can see in my works. There is a lot of polyphony. But it's always a polyphony in the sense not of the Renaissance, but rather a polyphony of the Baroque, where the harmony was established in advance.

RD Would you say that of your *Musique Funèbre?* Because that's a rather slow polyphony.

WL Well, no . . . partly. But in my later works there is a lot of polyphony, which is as in the figured bass. There is still harmony present, as in Bach's music. You can trace even the most sophisticated polyphonic textures of Bach, and the figured bass is present. In my music it is the same. Of course, the harmony is different. And the principle of sequences of vertical aggregations has its own rules and principles, different from traditional ones. Obviously, I use aggregations containing more sounds than just three or four. Up to twelve! I often use a great amount of sounds together.

RD So at this particular time, you're not really going through any compositional crisis!

**WL** Oh, no. On the contrary. I think that it's only in recent, I shouldn't say years but maybe decades that I reached a certain maturity of my means of expressions. That's why I can compose a little faster now than what was the case before. Always before, composing every particular work, there was a dark wall in front of me to be pierced, and I had to find the principle for each particular section! And now it's not the case, because there are certain rules that I have discovered that enable me to do what I want.

**RD** What about the length of your pieces? You have written, I think, pieces of up to thirty-four minutes.

**WL** Thirty-five minutes is the maximum. It's a piece you conducted, the Preludes and Fugue, which is thirty-five minutes. I think that I could think about a longer piece, for example, if I were to write an opera.

**RD** I was wondering if you had anything cooking in the back of your mind for a large piece or theater work or a larger form . . .!

**WL** Yes, I would love to write an opera. Because I have finally found a subject.

**RD** Can you speak about it?

**WL** I would rather not! Because I don't know whether I will approach it. You know, at my age, it is a very difficult decision. Although, in spite of my age, I don't feel terribly old. Maybe I can think about that after completing the Piano Concerto and some other pieces which will not take so much time. The next will be a soprano piece with orchestra. I don't think that will take much time. It's a commission of the French government, and they postponed the deadline until 1989.

**RD** But you do like writing for the voice. I did the early *Five Songs* from 1956–58. Dorothy Dorow was supposed to sing them in Aspen, remember, and she got ill and finally arrived but we already had a substitute. I thought those were beautiful.

**WL** I've always liked voice and tried to write for voice more humanly—treating the voice as voice and not as an instrument with keyboard! And maybe I'll think about the opera. Because I have the text. Finally I found something for me . . . I am a very bad opera-goer. I find realistic opera ridiculous. I can't help laughing when I see love scenes with both people embracing and looking with one eye at the conductor and singing instead of saying something . . . no! It's absolutely inadmissible! I can't do something like that! It makes me laugh. The only thing that would really fit in my vision of an opera would be something absolutely unreal . . . that means a fairy tale or surrealism, absurd, a dream . . . I found something absolutely incredible; but I won't tell you a word about it because maybe I won't write it . . . it's too early!

**RD** We'll hope that it does come about. There are two things, just to wind up. One is the performance last night. I thought it was a marvelous opportunity to hear really important pieces of yours and the breadth of your expression. After all, to hear the Cello Concerto and the Third Symphony and *Mi-parti*, was quite impressive. How did you find the musicians?

WL The players of the Philadelphia Orchestra? Very good. I think they belong to the top five. They were very cooperative. Of course, there were some problems, because after all, it's not a very young orchestra. For young people it's much easier to cope with my methods . . . with this ad libitum playing. For some very experienced musicians who have played in orchestras a lot, it's a little embarrassing to suddenly approach something entirely new. But they were very nice and very willing. And they really ambitiously wanted to do their best.

RD And you must be pleased with Yo Yo Ma.

WL Oh, enormously.

RD This piece was written for Rostropovich.

WL Yes. Commissioned by the Royal Philharmonic Society.

RD But I thought Yo Yo Ma was very impressive.

WL He is a great talent.

RD Well, lastly . . . what about Poland today? For example, there is a new wave of American-Polish relations sort of unfolding. I've read a little about it in the paper. What about Poland today, artistically?

WL There was a very difficult period during the martial law, which stopped any artistic manifestations. It was for us a period of mourning. One didn't want to appear in public, to perform anything. We did not really feel prepared just to lead a normal life after such shock.

RD Psychologically very devastating . . .

WL Very devastating! And especially after the period of Solidarity which was a great revelation psychologically for the whole nation. It was like an awakening after a period of gloom. And of course this shock in 1981 put us back again into a very gloomy atmosphere.

RD Do you feel that the Church has been a salvation for the general public in times of these devastating acts?

WL Yes, certainly. It was a tremendous help psychologically.

RD And the writers . . . like for instance Michnik.

WL Oh yes! Extraordinary man! I know him. He is an extraordinary writer. I read several political essays of his.

RD In any case, do you have a kind of optimism at this point since now things are changing a bit in Poland?

WL We are waiting. We are not so terribly optimistic. Because we have already experienced some "thaws" you know? And it's always a curve, a wavy line! We'll see. Of course, one must hope, and one must preserve the goods that we have. I mean the cultural life and the cultural goods.

I think it's the best way to preserve the national identity. That's why I couldn't leave, you know, although I could live anywhere, as my works work for my living.

RD But you chose to stay with your native land.

WL Absolutely. I think if we all left, it would be a desert. We mustn't do that. It's important for our people. For the young people.

# Narrative XII

The composers in this book (thirteen Americans and thirteen Europeans) range in age from their fifties to their eighties. Their attitudes towards compositional and artistic sources vary considerably. For example, we have heard that to reject the past is easy, but to allow it is more difficult. On the contrary, we have heard that the past is not to be repeated, nor is it desirable for a creative artist to repeat himself in a compositional or artistic sense. We have also heard that composition can be based on indeterminate principles that mirror our own diverse hereditary existence.

Some of these attitudes provoke questions: how can the composer be purely an artistic "vessel" and not *possess,* but rather, simply *transmit* or *translate?* If he is not purely a "vessel" or "receiver," to what extent are artistic (or even scientific) materials or energies *borrowed, stolen,* or, perhaps more sensitively, *lent?* This involves the possessive act of transformation, of making something "one's own," as part of the creative process. And, as a consequence in both cases, how does this affect the possibility to discover (what is *there*) or to invent (what seemingly is not *there*)?

Specifically, we have heard about recent theories of composition based on a *super-formula,* a *tone-clock,* computer-supported composition based on *design* and the inclusion of *live electronics* assisted by computer technology.

In the course of these discussions, we have touched upon notions of originality, creativity, spirituality, power, utopia, ambiguity and relativity, objectivity and subjectivity, irreducibility, and musical character.

Some composers have spoken about their reactions to organized religions, society, love, war, politics, public relations, critics, and the current pop scene. Others acknowledge influences that contain jazz, African, Asian, medieval, and Renaissance aspects.

These discussions also reflect a certain historical accounting of what went on in music after the Holocaust, when the hope for humanity made all things seem possible. The "Big Bang" that occurred in the arts after World War II was a source of technical and human investigation, a time of high idealism, abundance, complexity, and many "Green Lights." I offer no conclusion, for that would imply an ending. We go on, *it* goes on, and we can only try to find where we stand, there and then.

# Index

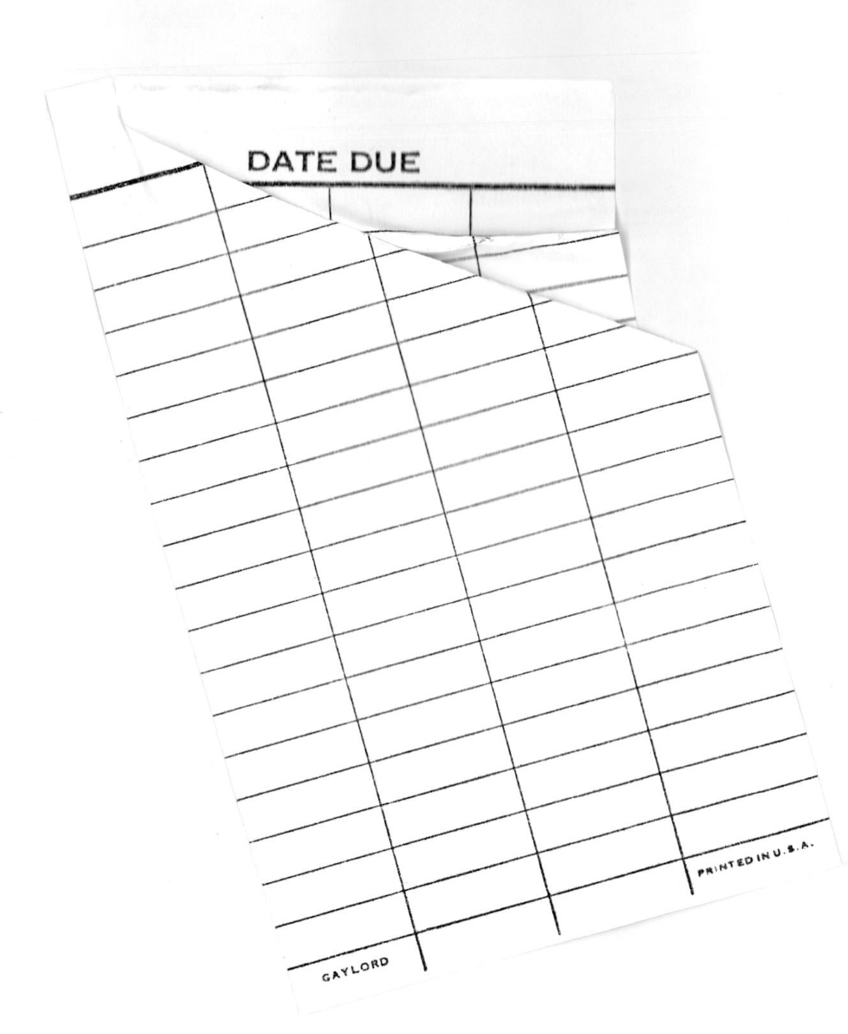
DATE DUE
GAYLORD
PRINTED IN U.S.A.